GEOPHYSICAL AND GEOCHEMICAL TECHNIQUES FOR EXPLORATION OF HYDROCARBONS AND MINERALS

Geophysical and Geochemical Techniques for Exploration of Hydrocarbons and Minerals

Edited by Marshall Sittig

NOYES DATA CORPORATION
Park Ridge, New Jersey, U.S.A.
1980

Copyright © 1980 by Marshall Sittig
No part of this book may be reproduced in any form
without permission in writing from the Publisher.
Library of Congress Catalog Card Number: 79-24469
ISBN: 0-8155-0782-8
Printed in the United States

Published in the United States of America by
Noyes Data Corporation
Noyes Building, Park Ridge, New Jersey 07656

Library of Congress Cataloging in Publication Data

Sittig, Marshall.
 Geophysical and geochemical techniques for exploration.

 (Energy technology review ; no. 52)
 Bibliography: p.
 Includes indexes.
 1. Prospecting--Geophysical methods. 2. Geochemical prospecting. I. Title. II. Series.
 TN269.S535   622'.1   79-24469
 ISBN 0-8155-0782-8

FOREWORD

The detailed, descriptive information in this book is based on 389 U.S. patents that deal with geophysical and geochemical techniques useful for the exploration of hydrocarbons and minerals. Where it was necessary to round out the complete technological picture, a few paragraphs from cited government reports have been included.

This book serves a double purpose in that it supplies detailed technical information and can be used as a guide to the U.S. patent literature in this field. By indicating all the information that is significant, and eliminating legal jargon and juristic phraseology, this book presents an advanced, industrially oriented review of modern methods of geophysical and geochemical exploration techniques.

The U.S. patent literature is the largest and most comprehensive collection of technical information in the world. There is more practical, commercial, timely process information assembled here than is available from any other source. The technical information obtained from a patent is extremely reliable and comprehensive; sufficient information must be included to avoid rejection for "insufficient disclosure." These patents include practically all of those issued on the subject in the United States during the period under review; there has been no bias in the selection of patents for inclusion.

The patent literature covers a substantial amount of information not available in the journal literature. The patent literature is a prime source of basic commercially useful information. This information is overlooked by those who rely primarily on the periodical journal literature. It is realized that there is a lag between a patent application on a new process development and the granting of a patent, but it is felt that this may roughly parallel or even anticipate the lag in putting that development into commercial practice.

Many of these patents are being utilized commercially. Whether used or not, they offer opportunities for technological transfer. Also, a major purpose of this book is to describe the number of technical possibilities available, which may open up profitable areas of research and development. The information contained in this book will allow you to establish a sound background before launching into research in this field.

Advanced composition and production methods developed by Noyes Data are employed to bring our new durably bound books to you in a minimum of time. Special techniques are used to close the gap between "manuscript" and "completed book." Industrial technology is progressing so rapidly that time-honored, conventional typesetting, binding and shipping methods are no longer suitable. We have bypassed the delays in the conventional book publishing cycle and provide the user with an effective and convenient means of reviewing up-to-date information in depth.

Foreword

The table of contents is organized in such a way as to serve as a subject index. Other indexes by company, inventor and patent number help in providing easy access to the information contained in this book. There is also a short bibliography of the U.S. Government reports used in this book.

> Some of the illustrations in this book may be less clear than could be desired; however, they are reproduced from the best material available to us.

15 Reasons Why the U.S. Patent Office Literature Is Important to You —

1. The U.S. patent literature is the largest and most comprehensive collection of technical information in the world. There is more practical commercial process information assembled here than is available from any other source.
2. The technical information obtained from the patent literature is extremely comprehensive; sufficient information must be included to avoid rejection for "insufficient disclosure."
3. The patent literature is a prime source of basic commercially utilizable information. This information is overlooked by those who rely primarily on the periodical journal literature.
4. An important feature of the patent literature is that it can serve to avoid duplication of research and development.
5. Patents, unlike periodical literature, are bound by definition to contain new information, data and ideas.
6. It can serve as a source of new ideas in a different but related field, and may be outside the patent protection offered the original invention.
7. Since claims are narrowly defined, much valuable information is included that may be outside the legal protection afforded by the claims.
8. Patents discuss the difficulties associated with previous research, development or production techniques, and offer a specific method of overcoming problems. This gives clues to current process information that has not been published in periodicals or books.
9. Can aid in process design by providing a selection of alternate techniques. A powerful research and engineering tool.
10. Obtain licenses—many U.S. chemical patents have not been developed commercially.
11. Patents provide an excellent starting point for the next investigator.
12. Frequently, innovations derived from research are first disclosed in the patent literature, prior to coverage in the periodical literature.
13. Patents offer a most valuable method of keeping abreast of latest technologies, serving an individual's own "current awareness" program.
14. Copies of U.S. patents are easily obtained from the U.S. Patent Office at 50¢ a copy.
15. It is a creative source of ideas for those with imagination.

CONTENTS AND SUBJECT INDEX

INTRODUCTION 1
   Prospecting for Oil 1
   Prospecting for Coal 2
   Prospecting for Oil Shale 9
   Prospecting for Tar Sand 10
   Prospecting for Minerals 12

GEOCHEMICAL PROSPECTING 14
   Prospecting for Minerals 14
      Bittern Salts 23
      Fluorite 24
      Gold 25
      Mercury 26
      Silver 28
      Sulfide Ores 30
      Uranium 31
   Prospecting for Hydrocarbons 33
      Determination of Hydrocarbon Presence in Soils and Rocks 33
      Analysis of Hydrocarbon Types in Soils 39
      Correlation with Subsurface Rock Types 53
      Detection of Metals and Metallic Compounds as Indicators of Hydro-
         carbon Deposits 55
      Detection of Hydrocarbon Deposits Under Water 56
      Analysis of Hydrocarbon Types in Water 62
      Detection of Fatty Acids as Hydrocarbon Indicators in Water .. 67
      Aerial Geochemical Prospecting Techniques 68

GEOBIOLOGICAL PROSPECTING 73
   Geobotanical Prospecting 73
   Geomicrobial Prospecting 73

GEOPHYSICAL EXPLORATION 84
   Spectral Sensing Methods 85
   Remote Sensing Techniques 87

Contents and Subject Index

MAGNETIC GEOPHYSICAL PROSPECTING 94

GRAVITATIONAL GEOPHYSICAL PROSPECTING 100

ELECTRICAL GEOPHYSICAL PROSPECTING 110
   Naturally Occurring Field Techniques............................111
      Self-Potential Field Techniques................................111
      Telluric Techniques..112
      Magnetotelluric Techniques....................................114
   Artificially Produced Field Techniques..........................115
      Resistivity Techniques..115
      Induced Polarization Methods..................................124
      Electromagnetic Methods 127

NUCLEAR GEOPHYSICAL PROSPECTING..........................141
   Surface Contact Prospecting 142
   Surface Aerial Prospecting..150
   Borehole Prospecting...153
   Natural Gamma Ray Logging 155
   Gamma-Gamma Logging...159
   Neutron-Gamma Logging 162
   Neutron-Neutron Logging..166
   Improved Radiation Detectors 172
   Improved Neutron Sources 174
   The Use of Radiotracers 179
   Trends in Nuclear Geophysical Prospecting......................180

SEISMIC GEOPHYSICAL PROSPECTING.............................182
   The Basic Patents 185
   Production of Seismic Waves.....................................187
      Solid Explosives 187
      Implosive Seismic Wave Generators............................194
      Gas Exploders..198
      Air Guns..202
      Seismic Vibrators...207
      Miscellaneous Land Seismic Wave Generators 212
      Miscellaneous Marine Seismic Wave Generators................220
   Seismic Data Reception and Processing..........................227
      Seismic Detectors 229
      Recorders and Displays...237
      Signal Amplification 243
      Multichannel Signal Recording 243
      Magnetic Signal Recording 246
      Three-Dimensional Representation 246
      Signal Enhancement 248
      Digital Data Processing 253
   Seismic Borehole Logging Techniques...........................254
   Offshore Seismic Prospecting 257
   Trends in Seismic Geophysical Prospecting......................262

EXPLORATORY WELL DRILLING 263
   For Mineral Deposits..263
   For Hydrocarbon Deposits 264

Determining Formation Characteristics..........................264
  Detecting Hydrocarbons..269
BIBLIOGRAPHY 288
COMPANY INDEX..290
INVENTOR INDEX 292
U.S. PATENT NUMBER INDEX 297

INTRODUCTION

Prospecting once evoked the image of the grizzled old man leading a mule which carried his equipment. Equipped with a hand axe, he looked for the telltale view of gold which might make his fortune. Since the time of the earliest settlement, the need for iron for tools and guns, lead for bullets and copper for utensils has prompted a search for the sources of these metals. The lure of gold and silver provided much of the impetus for the development of the western United States between 1850 and 1900. Later, as the country's industrial demands for metals expanded to include zinc, molybdenum, tungsten, chromium, vanadium and many others, prospectors have sought and found these metals.

As the world's energy requirements have expanded from fueling early kerosene lamps to supplying the huge heating and power requirements of homes and industry today, so have techniques for prospecting for petroleum expanded. Whereas exploratory or "wildcat" wells were once drilled more or less at random in areas where seeps of petroleum had been found, such random efforts gave way to systematic efforts using sophisticated techniques based on the latest advances in chemistry and physics.

Thus, geophysical and geochemical methods have come to the forefront in providing the metals and fuels for modern housing and industry.

Prospecting not only implies simply locating a deposit of ore or hydrocarbons, but defining it in terms of area, zones and scope.

PROSPECTING FOR OIL

In this petroleum-hungry world, prospecting for petroleum continues to occupy a key place in the futures of industry and indeed of whole nations.

As discussed by Moody (4), there are a number of steps involved in petroleum exploration. In the first place, it must be determined that the bare essentials of a geologic setting favorable for the accumulation of oil or gas are present. Then a series of steps must take place:

The leasing of the area and its logistics must be favorable;

Aerial photographs must be purchased or made;

Gravity and magnetic surveys can be made as an aid in determining the configuration of the basin;

Core drilling or seismic surveys may be made if initial work indicates the area to be favorable for the accumulation of oil or gas;

The first wildcat test well is drilled. Depending on its success, a second test well may be drilled, the area may be abandoned, or commercial production may be planned.

It is desirable to make the most of the exploration techniques which precede actual test well drilling since, as of 1979, the cost of test well drilling is $1,000 per foot or more.

Figure 1 illustrates the trend in petroleum reserves over the years in conjunction with indications of application of various types of prospecting. Thus, one can see that seepage prospecting was the earliest technique employed, along with surface geologic mapping. The progression in use of various types of magnetic, gravitational and seismic methods is also indicated.

PROSPECTING FOR COAL

The first international coal exploration symposium was held in 1976 and its proceedings have been published (5). It is interesting to note that, while coal might be considered as an ancient fuel, modern sophisticated exploration techniques have only recently received such attention. A second symposium on this topic followed in 1978 (15).

A scheme developed by *W.W. Givens; U.S. Patent 4,066,892; January 3, 1978; assigned to Mobil Oil Corporation* relates to radioactivity well logging and more particularly to a method and system for locating a coal-bearing zone in the earth and identifying its quality based primarily upon the Btu and ash contents.

The coal mining industry has primarily used coring and core assaying in locating coal-bearing zones, in defining the thickness of such zones, and in determining coal quality from Btu and ash contents.

In this process, a bore logging tool employs two neutron sources of differing energy outputs, two γ-ray detectors, and a neutron detector in the radioactivity logging of a coal-bearing formation. A first of the γ-ray detectors measures the γ-rays produced by the formation in response to the higher energy neutron source.

A second of the γ-ray detectors measures the γ-rays produced by the formation in response to the lower energy neutron source. The neutron detector measures

Figure 1: Crude Oil Production and Proved Reserves in Relation to Developments in Petroleum Exploration Technology

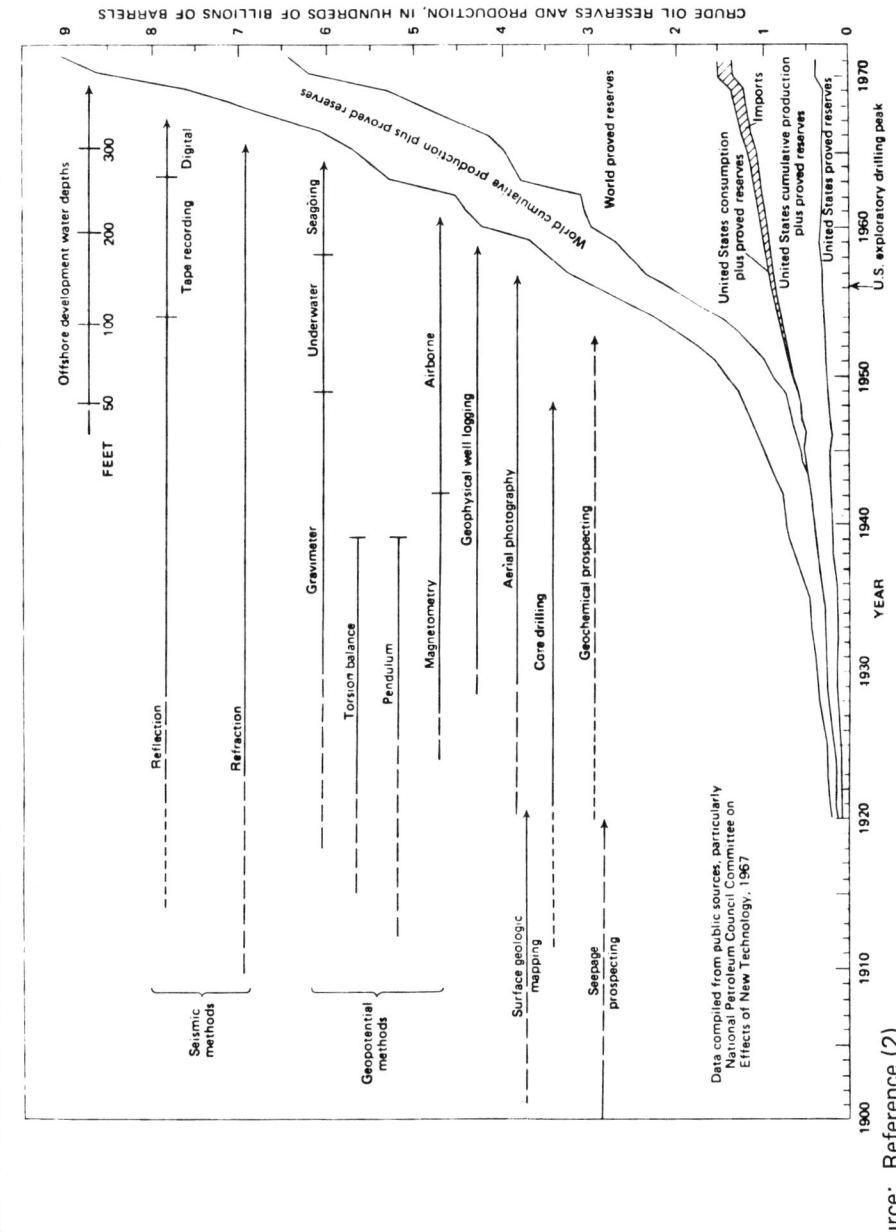

Source: Reference (2)

thermal or epithermal neutrons produced by the formation in response to the neutron sources. These measurements are selectively transferred to a recording system employing a plurality of energy level discriminators, count rate meters, and comparators. The recording system, in response to these measurements, provides output signals representative of the carbon, sulfur, ash and neutron moderating power of the coal in the formation being logged. The carbon and neutron moderating power provide a means of measuring the energy content of coal.

This technique is illustrated diagramatically in Figure 2.

Figure 2: Mobil Oil Corp. Coal Logging System

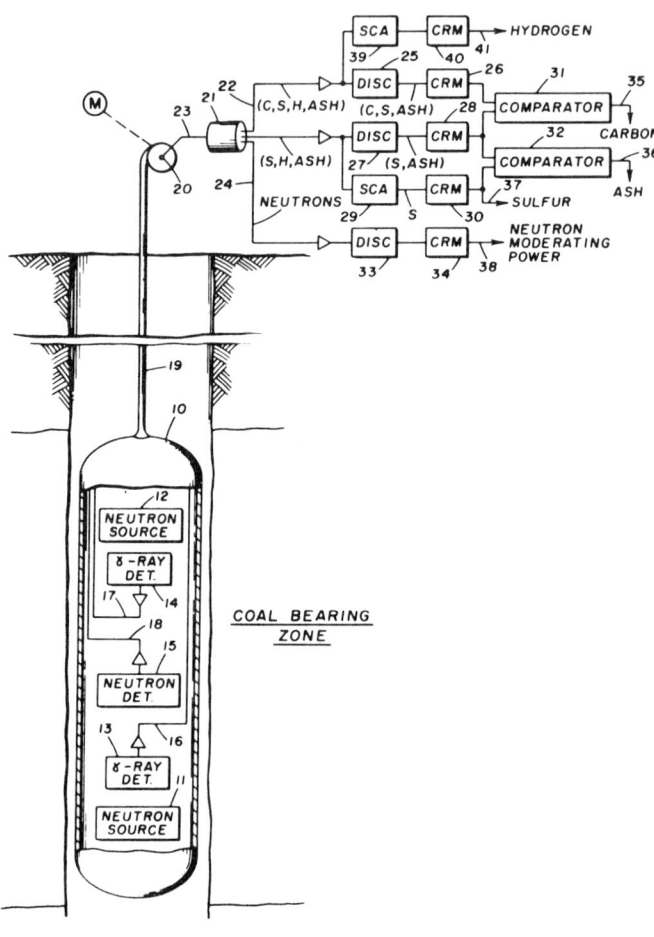

Source: U.S. Patent 4,066,892

Introduction

There is illustrated a borehole logging tool and recording system for carrying out the logging operation when the logging tool is located opposite a coal-bearing zone. The logging tool **10** includes neutron sources **11** and **12**, γ-ray detectors **13** and **14**, and neutron detector **15**. Neutron source **11** is preferably an americium-beryllium (Am^{241}-Be) source.

The average energy of neutrons from this source is about 4.5 Mev. Neutrons with energy greater than about 4.43 Mev will produce 4.43 Mev γ-rays by way of inelastic scattering in the carbon of the coal-bearing zone. The detector **13** is preferably an NaI scintillation detector which measures these 4.43 Mev γ-rays from carbon as well as captures γ-rays from the remaining major coal components of hydrogen, sulfur and ash.

Neutron source **12**, on the other hand, is preferably a californium (Cf^{252}) source. The average energy of neutrons from this source is about 2.348 Mev. Most of the neutron output of this source is of insufficient energy to produce 4.43 Mev γ-rays by way of inelastic scattering in carbon. The detector **14** is preferably an NaI scintillation detector which measures γ-rays as does detector **13**. However, since neutron source **12** is of insufficient energy to produce a significant quantity of 4.43 Mev γ-rays from carbon, the output of detector **14** is a measure primarily of the hydrogen, sulfur and ash content of coal.

The spacing between the neutron source **11** and γ-ray detector **13** is selected to yield maximum response by the γ-ray detector **13** to the 4.43 Mev γ-rays produced by the carbon content of coal. The spacing between the neutron source **12** and γ-ray detector **14** is selected such that the response of the γ-ray detector **14** to γ-rays in non-coal-bearing zones is always proportional to the γ-ray response of detector **13** in non-coal-bearing zones.

Also included in the logging tool **10** is a neutron detector **15**, preferably an He^3 proportional counter, to provide a measure of the total neutron moderating power of the coal-bearing zone. Hydrogen and carbon are both good neutron moderating materials and the carbon and non-moisture-related hydrogen in coal correlate strongly with the energy content of the coal.

Therefore, the neutron moderating power should also correlate strongly with the energy content and, therefore, provide a supplemental measure of the energy content of coal. The He^3 counter can be operated with or without a cadmium shield to provide a thermal neutron measure or preferably with a cadmium shield to provide an epithermal neutron measure.

The output signals of the detectors **13** through **15** are amplified and coupled to conductors **16** through **18**, respectively, included in the cable **19**. At the surface, these signals from conductors **16** through **18** are applied by way of the slip rings **20** and brushes **21** to conductors **22** through **24**. The signals on conductors **22** through **24** are amplified and processed by the units of **25** through **34** of the recording system to provide on lines **35** through **38** four output signals representative of the carbon, ash, sulfur and neutron moderating power, respectively, of the coal-bearing zone being logged.

The carbon signal is provided in response to the signals on conductors **22** and **23** from the γ-ray detectors **13** and **14**, respectively. As discussed above, the signal from detector **13** is a measure of the 4.43 Mev γ-rays from inelastic neutron scattering in carbon as well as capture γ-rays from the hydrogen, sulfur and ash components of the coal-bearing zone.

On the other hand, the signal from detector **14** is a measure of only the hydrogen, sulfur and ash components since the output of neutron source **12** is of insufficient energy to produce 4.43 Mev γ-rays from carbon. Therefore, the difference between these two signals is a measure of the carbon content of the coal-bearing zone. This difference is determined by operation of the discriminators **25** and **27**, the count rate meters **26** and **28**, and the comparator **31**. Both the discriminators are biased at slightly above 2.23 Mev so as to pass the amplified signals above this energy level from the detectors **13** and **14** to the count rate meters **26** and **28**.

This removes the 2.23 Mev energy level hydrogen component from the counting rates. The counting rates from these count rate meters **26** and **28** are then applied to the comparator **31** which is preferably an operational amplifier biased to produce the carbon signal in response to the counting rate differential.

The sulfur and ash signals are provided in response to only the signal on conductor **23** from the γ-ray detector **14**. In addition to being applied to discriminator **27** and count rate meter **28** as described above, the amplified signal from conductor **23** is also applied to the single channel analyzer **29** and count rate meter **30**. Single channel analyzer **29** is biased to pass to the count rate meter **30** only those amplified signals from detector **14** that are in an energy band from about 4 to 5.5 Mev. It is in this energy band that the most intense 5.43 Mev γ-rays from sulfur are produced.

Consequently, the signal on line **37** is a measure predominantly of the sulfur component of the coal-bearing zone. The counting rates from the count rate meters **28** and **30** are then applied to the comparator **32** which is preferably an operational amplifier biased to produce the ash signal in response to the counting rate differential.

The signal representing the neutron moderating power is provided in response to the signal on conductor **24** from the neutron detector **15** which responds only to neutrons. The amplified signal on conductor **24** is applied to the discriminator **33** which is biased to reject electronic noise. These signals are applied from discriminator **33** to the count rate meter **34**. The counting rate from count rate meter **34** is, therefore, a measure of the neutron moderating power content of the coal-bearing zone.

An indication of hydrogen may also be made by measuring capture γ-rays in an energy interval from about 1.95 to 2.40 Mev. This may be done by applying the signal from either γ-ray detector **13** or **14** to a single channel analyzer and count rate meter. As shown, the signal on conductor **22** from γ-ray detector **13** is applied to single channel analyzer **39** and count rate meter **40** to provide the hydrogen signal on line **41**.

In order to determine the continuity, thickness and structure of coal deposits, developments have been made in the use of mirror symmetrical Rayleigh waves, as described by *B. Stas, L. Dlouhy, L. Siska and A. Skrabis; U.S. Patent 3,858,167; December 31, 1974; assigned to Vedeckovyzkumny uhelny ustav, Czechoslovakia,* and pulsed monochromatic coherent sound waves as described in the text which follows. These wave modes are injected into the strata, detected, processed, and recorded in order to reveal the lithological characteristics of coal beds.

A technique developed by *H.-J. Hochheimer, H. Haas, H.-L. Jacob, P. Helling and B. Wülk; U.S. Patent 3,961,307; June 1, 1976; assigned to Ruhrkohle AG, Germany* is one in which pulses of monochromatic coherent sound waves are emitted at an underground test station into a coal seam from several closely juxtaposed sources to form a beam whose reflection at a discontinuity in the seam is detected by a group of receivers. The time lapse between the outgoing and incoming sound pulses is a measure of the distance of the seam boundary from the test station and, upon a controlled shifting of the phases of the outgoing and/or incoming beam components, can be displayed on an oscilloscope screen as a function of sweep angle in an azimuthal and/or an elevational plane.

Figure 3 shows a diagram of the applications of such a technique.

Figure 3: Ruhrkohle AG Technique for Exploration of Boundary of an Underground Coal Seam

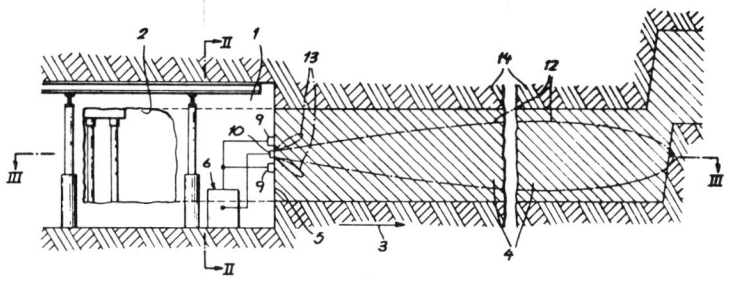

Vertical section through an underground mine drift showing a test station

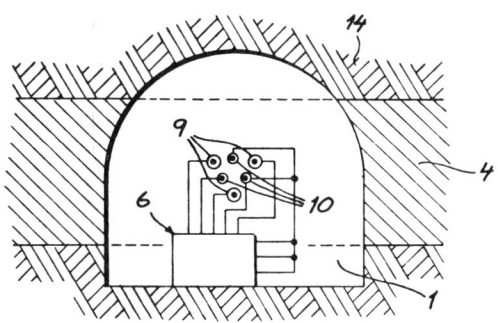

Cross-sectional view taken on line II–II

(continued)

Figure 3: (continued)

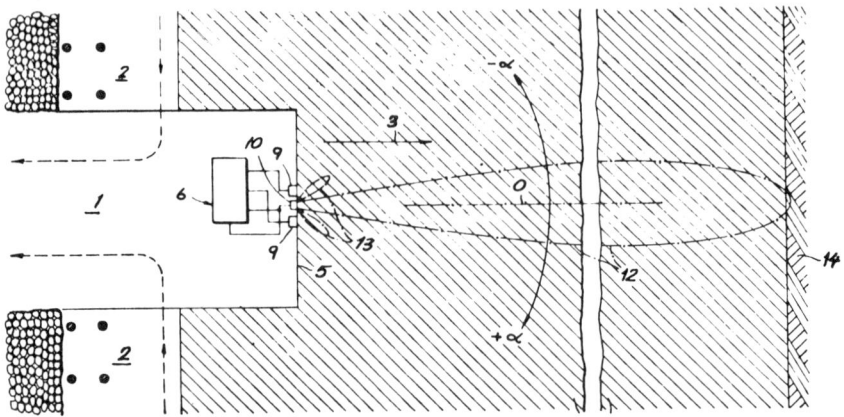

Cross-sectional view taken on line III—III

Source: U.S. Patent 3,961,307

It shows a drift **1** of an underground coal mine provided with an adit **2**. The drift **1** adjoins an underground coal seam **4** whose extent in various directions is to be explored. For this purpose there is disposed in the drift a test station, generally designated **6**, comprising electronic equipment for the operation of electroacoustic transducers, namely a first array of sound-emitting transducers **9** and a second array of sound-receiving transducers **10**. The three emitting transducers **9** are located at the corners of a vertical equilateral triangle with downwardly pointing apex, whereas the three receiving transducers **10** define the corners of a similar, but smaller triangle with upwardly pointing apex in the plane of the first triangle, i.e., at the face or breast **5** of the seam **4**.

Test station **6** comprises an electric pulse generator whose output is a succession of bursts of a carrier wave of a fixed sonic or ultrasonic frequency. This pulse-modulated carrier wave is fed in parallel via adjustable phase shifters to respective sound-emitting transducers constituting the array of transducers **9**. These transducers, therefore, generate coherent and monochromatic sound waves at the carrier frequency, the relative phasing of these sound waves being determined by the adjustment of the associated phase shifters. With proper adjustment, these waves will form a directive pattern with a major lobe **12** and some minor lobes **13**, only the former being of interest.

At the boundary of seam **4** with the surrounding soil or rock **14**, the sound waves will be reflected back toward station **6** with a directive component reaching the receiving transducers **10**. The direction of the outgoing sound waves, indicated by an arrow **3**, is here assumed to be horizontal, but this is not necessarily the case. Through adjustment of these phase-shifting means it is possible to shift this major lobe in one or two planes to obtain an indication of distance over a desired sweep range. With the aid of an oscilloscope screen, a graphic representation of distance as a function of sweep angle can be produced.

Introduction

A device developed by *D.E. Godfrey; U.S. Patent 4,143,552; March 13, 1979; assigned to General Electric Company* is an improved horizon control sensor for detecting engagement of the cutting tool of a coal mining machine with rock or other material enclosing a coal seam.

Underground coal mining is done with longwall drum shearers and continuous miners which are controlled by an operator to maintain the cutting tools in the undulating coal seam. The most critical need for increasing the productivity and safety of the underground mining of coal is an improved sensor to control the elevation of the drums on the shear loader machine of a longwall mining machine and the elevation of a continuous mining machine.

Such a device would allow mining to be done at the maximum operating speed of the machine and not be slowed down by the operators, would reduce the waste material taken from the floor and ceiling while maximizing the amount of coal taken, and is the key to automated coal mining machines which will allow the operator to run the machine while located outside the hazardous face area.

Two systems for the automatic detection of the interface between coal and surrounding rock are the sensitized pick and the nucleonic sensor. The sensitized pick is a strain gage transducer and earlier attempts to develop a working device proved feasibility, but it was not further developed. The nucleonic sensor relies on a γ-ray back scattering technique that measures the remaining coal thickness by reflecting γ-rays off the coal-rock interface. This system was developed into a marketable product, but does not properly do the job according to Godfrey.

In the device, the frequency response of the cutting tool of a coal mining machine is monitored at a resonant frequency of the cutting tool, using a vibration transducer mounted on a nonrotating support arm, to detect the difference between coal and rock or other enclosing material. The horizon control sensor maintains the mining machine within the undulating coal seam.

PROSPECTING FOR OIL SHALE

Vast untapped reserves of oil entrapped in oil shale exist in this country as well as other locations throughout the world. Increased attention is being devoted to the exploration of oil shale because of the current need for new energy sources.

The term "oil shale" as used in the industry is in fact a misnomer, because it is neither shale, nor does it contain oil. It is a sedimentary formation comprising marlstone deposits interspersed with layers containing an organic polymer called kerogen, which, upon heating, decomposes to produce carbonaceous liquid and gaseous products. It is the formation containing kerogen that is called oil shale herein, and the carbonaceous liquid product is called shale oil.

Oil shale deposits occur in generally horizontal beds, and within a given bed there are an extremely large number of generally horizontal deposition layers containing kerogen known as varves. The varves typically are nonuniformly dispersed throughout a given bed. In the higher grade oil shales, the varves are commonly cemented together into relatively thick, compact layers. The lower grade oil shales have much thinner varves spaced apart through the bed. For example, a core sample taken from a typical bed may vary from thick, solid oil shale sections to layered sections having the appearance of poker chips. In the core, small layers or nodules of other minerals and rock materials are sometimes found interspersed between the varves containing kerogen. Typical of these other minerals are nahcolite, dawsonite, other salines, dolomite, mudstone, sandstone, tuffs, analcite and bentonite.

Techniques for preparing shale oil for retorting generally comprise explosively expanding a subterranean oil shale formation to form a fragmented, permeable mass of oil shale particles. Shale oil then may be recovered from the particles by in situ retorting techniques, or by retorting in surface retorts, for example.

Prior to preparing the oil shale for retorting, the formation is explored to determine the locations within the formation containing the highest grades of oil shale. Core samples are taken from the formation and subjected to laboratory analysis to determine the kerogen content of the sample. One such analytical technique is the Fischer assay in which a sample customarily weighing 100 g and representing 1 ft of core is subjected to controlled laboratory analysis involving grinding the sample into small particles which are placed in a sealed vessel and subjected to heat at a known rate of temperature rise to measure the kerogen content of the core sample. Kerogen content is usually stated in units of gallons per ton, referring to the number of gallons of shale oil recoverable from a ton of oil shale heated in the same manner as the Fischer analysis.

Such analytical techniques are generally done in laboratories far from the drilling site. This causes a considerable delay before analytical results are available to field personnel conducting the exploration tests. Thus, immediate field decisions on the progress of the exploration program cannot be based on accurate analyses of core samples.

A technique developed by *R.T. Chew, III; U.S. Patent 4,149,805; April 17, 1979; assigned to Occidental Oil Shale, Inc.* provides a method and apparatus for rapid determination of the kerogen content of oil shale. It does not require calculations or destructive laboratory techniques characteristic of the Fischer assay and other known methods for measuring kerogen content. Further, it makes it possible to accurately analyze a core sample of oil shale in the field and still have the core sample available for other purposes, such as corroborating laboratory measurements to be conducted later at a more convenient time. Thus, field decisions on the progress of the exploration program can be made immediately, rather than waiting for several days, which is a common delay for current kerogen analysis techniques.

PROSPECTING FOR TAR SAND

In various localities throughout the world, deposits of bituminous sand are found in which the oil or tarry matter has a density greater than 1.0. The most extensive and best-known deposits of this type occur in the Province of Alberta in the Athabasca district.

These Alberta tar sands extend for many miles and occur in thicknesses varying up to more than 200 ft, and probably constitute the largest known petroleum deposit of any type. Typically, the composition of these sands contain, by weight, from about 6 to 20% of oil (or bitumen); about 1 to 11% of water; and from about 70 to 90% of inorganic solids. However, tar sand containing larger or smaller quantities of these ingredients is not unusual.

The major portion, by weight, of the inorganic solids is fine grain quartz sand having a particle size greater than about 0.1 mm and less than about 1.0 mm. The remaining inorganic solid matter has a particle size of less than 0.1 mm and is referred to as fines. The fines contain clay and silt including some very small particles of sand. The fines content typically varies from about 10 to 30% by weight of the solid inorganic content of bituminous sand.

The phrases "tar sand" or "bituminous sand" are used to refer generally to all granular solid materials soaked with liquid or semiliquid hydrocarbonaceous material although they specifically refer to a characteristic type of bituminous solid consisting of discrete particles of sand bound together by a continuous heavy hydrocarbon oil phase.

The oil obtained from these sands is a viscous, tarry material and in its crude state does not command a high price; hence, any method suitable for recovering it must involve a minimum expense to be attractive for commercial practice. Various methods are known for extracting the oil from bituminous sand. Generally the oil is separated from the tar sand by extraction such as with a hydrocarbon solvent, or the formation of an emulsion with water or combinations of these procedures with or without the use of various chemicals to facilitate recovery of the bitumen.

These techniques can be further implemented with various separation techniques such as centrifuging or settling. The two best-known methods for recovering bitumen from tar sand are known as the hot water and cold water methods.

In the former, the tar sand is jettisoned with steam and mauled with a small proportion of water at about 176°F and the pulp is then dropped into a turbid stream of circulating water and carried to a separation cell maintained at about 185°F where the oil rises to the top as a froth which is then drawn off.

The so-called cold water method does not involve heating the tar sand other than whatever heating might be required to conduct the operation at a temperature of about 73° to 81°F. This process involves milling a mixture of the tar sand, kerosene, water and soda ash and then settling at a temperature within the specified range. A mixture of oil and kerosene floats to the top of the settling zone and is removed.

In addition to the hot or cold water methods, methods are known which again depend on extracting the bitumen from the tar sand, whereby the process is practiced in situ. It is apparent, however, that the separation of the oil from the tar sand is not a very easy matter and the success of an operation for recovering such oils will depend on the ease with which the oil can be separated.

Various techniques are employed in the mining of the tar sand. For instance, mining on-the-surface or near-surface deposits of tar sand can be performed by the usual procedures to produce a raw tar sand material consisting of chunks or pieces of tar sand not exceeding about 12" in average dimensions. This may be done by open-pit mining in which the over-burden is stripped away and the tar sand is mined by means of bulldozers, clamshell shovels, and similar equipment.

Drilling and blasting may also assist in the breaking up of the tar sand into the aforementioned particles. The mined material is then transferred to processing for separation of the oil which as described above can be accomplished by various means.

Extensive tests have shown that the tar sand and particularly Athabasca tar sand vary greatly in their processability, i.e., the ease with which the bitumen is separated from the tar sand. Some sands process easily, while others tenaciously hold on to the oil. Thus, when employing the hot water process, recoveries of oil can range from about 50 to upwards of 95%.

The processability of the tar sand varies greatly even within a fairly localized area such as within an acre or two within an exploration zone. Although the quantity of oil may be the same throughout the entire zone, the ease with which the oil can be separated from the tar sand differs. Also areas which are separated by as little as 100 or 200 yards show different processability characteristics.

In the past it has been difficult to determine whether a sample of bituminous sand would process easily or not; hence, it can be seen that a simple and convenient method whereby a sample of tar sand can be taken and subjected to analysis to determine its processability is of great value.

A technique developed by *O.L. Wilson; U.S. Patent 3,273,967; September 20, 1966; assigned to Cities Service Research and Development Company* is based on the discovery that the ease with which the oil can be separated from tar sand is dependent on the amount of iron in the tar sand. There is a correlation between iron and processability of the tar sand with the higher iron content being indicative of the more easily processable tar sand. Thus, by employing the same processing method higher recoveries are obtained from sands which contain larger quantities of iron, but the same quantity of oil.

This technique can be employed for geochemical prospecting. In geochemical prospecting, tar sand samples are taken at spaced distances over the prospect area. The samples may be taken along a traverse or a plurality of traverses or otherwise over the prospect area in accordance with usual procedures in the art for determining bitumen content. The tar sand samples can then be analyzed for iron content and the iron content of the samples is correlated with the sampling locations to determine the areas having the more easily processable tar sand.

PROSPECTING FOR MINERALS

More than a billion tons of surface and underground material are mined annually in the U.S. alone to recover somewhat more than a half-billion tons of metallic ores, principally ores of iron and copper.

Greater amounts of these ores and of ores of more exotic metals must be found in the future to meet the increasing needs of industry and to replace worked-out deposits.

Because the easily found deposits have been discovered, new deposits will be more difficult to find and success will depend more and more on modern prospecting techniques.

Introduction

One good review of mineral prospecting techniques is that by Kuzvart and Bonmer (6). Another is that by Peters (7).

One example of the broadening search along with the use of sophisticated scientific techniques is found in the search for sea floor minerals.

The great demand for nickel and copper makes necessary the mining of even those deposits that are accessible only with difficulty, such as manganese lumps embedded in the sea floor at a depth of from 3,000 to 6,000 m. A decision whether these lumps are worth mining depends upon the precise knowledge of their content of usable metals. Large investments required for the necessary conveying equipment make mining appear profitable only where the fields have a size of several 10,000 km^2. No method of analysis is known, however, which would permit verification of a sufficiently dense measuring grid on such sea floor surfaces with a justifiable expenditure of time.

In order to be able to make manganese lump analyses, for example, it is the practice to bring these manganese lumps from great depths (such as 6,000 m) and lift them on board ships by means of dredges. The time required to obtain a sample amounts to several hours. Further, the sample cannot be accurately associated with the sea floor coordinates. The lumps are then analyzed on board (and also in land-based laboratories) with wet chemical processes and by x-ray fluorescence methods, for which the samples must undergo extensive preparations.

Thus, it would be very desirable to have a technique for in situ analysis of marine ore concretions and such a technique has been developed by *W. Apenberg, G. Böhme, H.-U. Fanger, B. Glaser, K. Hain, J. Hubener, W. Stegmaier, V. Prech and J. Vagner; U.S. Patent 3,942,003; March 2, 1976; assigned to Gesellschaft fur Kernforschung m.b.H., Germany* for the in situ analysis of marine ore concretions.

Ore samples are aspirated from the sea floor and introduced in a sample container aboard a carrier body travelling underwater along and in the vicinity of the sea floor. While in the sample container, the ore concretions are exposed to radioactive rays for inducing a secondary radiation emanating from the ore concretions. The secondary radiation is measured to determine the quantity of the individual elements contained in the ore concretions.

GEOCHEMICAL PROSPECTING

Geochemical prospecting relies on the detection of traces of hydrocarbons or minerals. Some of the fundamentals of exploration geochemistry have been reviewed by Levinson (8).

PROSPECTING FOR MINERALS

Geochemical prospecting is based on systematic measurement of one or more of the chemical properties of rock, soil, glacial debris, stream sediment, water, or plants. The chemical property most commonly measured is the content of a key trace element. Zones in the soils or rocks of comparatively high, or anomalous, concentrations of particular elements may guide the prospector to the elements in rocks or soils that constitute a geochemical anomaly. The actual amount of the key element in a sample may be very small and yet constitute an anomaly if the element's concentration is high relative to its concentration in the surrounding area. For example, if most samples of soil are found to contain about 0.00001% (0.1 ppm) silver, but a few contain as much as 0.0001% (1 ppm), the few "high" concentrations are geochemical anomalies. Plots of analytical results on a map may indicate zones to be explored further.

Geochemical anomalies are classified as primary or secondary. Primary anomalies result from outward dispersion of elements from mineral-forming solutions. The high concentrations of metals surround the deposit, and the dispersion of metals laterally or vertically along fractures or faults may result in a leakage halo that extends hundreds of feet away from the deposit. Halos of this type are especially useful in prospecting because they may be hundreds of times larger than the deposit they surround and hence are easier to locate (1).

Secondary anomalies result from dispersion of elements by weathering. Some primary minerals, such as cassiterite, are resistant to chemical weathering and are transported by the streams as fragmental material. Other minerals may be dissolved and the metals may be either redeposited locally or carried away in solution in ground and surface waters. Some of the metal in solution may be taken up by plants and trees and can be concentrated in the living tissue. Many studies have been made of the metal content of residual soils over sulfide deposits, and in general the distribution of anomalous amounts of metal in the soil has been found to correspond closely with the greatest concentration of metals in the underlying rock (1).

Most products of weathering in a drainage basin enter the streams and rivers that flow across it. The weathered products occur as chemicals in solution and in sediments. Either or both can be sampled and tested, and composition of the samples will reflect the chemical nature of the rocks in the drainage basin. The presence of an ore deposit in a drainage basin may be determined by sampling water and sediment from each successive tributary and by analyzing the samples for anomalous amounts of metals. This procedure narrows the search for ore deposits to the most favorable areas.

Contamination of surficial material is an ever-present hazard in geochemical surveys. The most common sources of contamination are materials derived from mine workings. Such materials may oxidize and go into solution, contaminating the soil, stream sediment, and water, thus masking natural anomalies. Similarly, smelter fumes, wind-blown flue dust, and metallic objects introduced to the natural environment by man may also contaminate the soils and rocks.

Analytical methods used in geochemical prospecting must be sensitive enough to determine minute amounts of key elements, accurate enough to show small differences in concentration, fast enough to permit large numbers of samples to be analyzed in a day, and inexpensive.

A technique developed by *J.S. Ryss, T.M. Ovchinnikova, J.G. Gavrilov, D.V. Voronin and V.M. Panteleimonov; U.S. Patent 3,758,846; September 11, 1973* is based on the selective excitation and registration of electrochemical reactions of certain minerals contained in the ore bodies. The successive excitation and registration of the electrochemical reactions are achieved by gradually varying the intensity of the current passed through an ore body, by simultaneously registering in the form of polarization curves the current intensity and the ore body potential relative to the enclosing rocks taking into account a drop of voltage in the rocks and by determining from the curves the potential values and the limit current intensity of the electrochemical reactions, on the basis of which the mineral composition, position and size of the ore body are established.

It has been found that a number of volatile components are present at an unusual concentration in the atmosphere overlying certain classes of mineral deposit. Thus, for example, mercury vapor increases from typical backgrounds of one to ten nanograms per cubic meter to concentrations varying between ten and fifty nanograms per cubic meter over gold and silver deposits as well as over many types of base metal ore body. This is attributed to the fact that mercury tends to concentrate in most metallic ores to an amount which is greater than the surrounding rocks and since mercury and some of its compounds have an appreciable vapor pressure at normal atmospheric temperatures, some of this mercury is dispersed into the overlying atmosphere above ore bodies. Similarly, there is a geochemical association of chloride, fluorine, bromine and iodine with many classes of mineral deposit, and since a number of these elements and their compounds have a significant vapor pressure, these elements and their compounds can also occur in the atmosphere over mineral deposits. In the case of the halogen elements, there is also a strong association with phosphate deposits and oil fields so that these elements can also be used to prospect for oil and phosphates. In addition to these elements there are a number of others that have much lower vapor pressures, but nevertheless can occur in minute but measureable traces in the atmosphere. These include the halide compounds of elements such as tellurium and selenium which have close affinity with certain types of copper deposit, and the halide compounds of elements such as arsenic, antimony and bismuth, all of which have sufficient vapor pressure to be present in minute traces. Given sufficient sensitivity in an analytical system, even more elements can be detected in the form of volatile inorganic compounds. In general, the halides of most metals are substantially more volatile than the metals themselves and can occur in minute quantities in the atmosphere.

In addition to the occurrence in the atmosphere of volatile inorganic compounds there are also present significant amounts of organic vapors. Organic vapors are generated by living forms of all types including plants, soil bacteria, insects and animals. Large quantities of volatile organic vapors are generated by trees in the form of natural oils known as terpenes. Some of these organic vapors oxidize in the atmosphere to form minute liquid and solid particulate matter, a fact which can cause the development of atmospheric haze over forests under some weather conditions.

A fractional component of organic vapors in the atmosphere is comprised of metallo-organic compounds. It has been shown that certain metals that are present in soils are converted into volatile organic compounds by bacteria. A well-established example is that of mercury which is converted into dimethyl-mercury by a variety of bacteria and other microorganisms in soils. Dimethyl-mercury has a lower boiling point than water and readily evaporates into the atmosphere as it is formed. It has been demonstrated that soils containing living microorganisms liberate substantially more mercury into the atmosphere than the same soils that have been sterilized.

In a technique developed by *A.R. Barringer; U.S. Patent 3,768,302; October 30, 1973; assigned to Barringer Research Limited, Canada* atmospheric particulates are collected and subsequently heated to drive off adsorbed gases and vapors, which are then analyzed. The particulates may be collected continuously from a moving aircraft, for example, and concentrated if necessary. A number of analytical techniques may be employed for the purpose of identifying the adsorbed gases and vapors.

Figure 4 shows a suitable form of apparatus for the conduct of such a process. The air enters through a sampling duct **1**, goes through an electrostatic charging grid **2** and then through a grounded collecting grid **3**. The charging grid **2** may consist of a set of grounded wires alternating with a set of wires connected to a source of negative high voltage. The grounded collecting grid **3** is made of Nichrome resistance wire which can be heated to a temperature of up to 1000°C by the application of a current through the wires. The air stream is split by a perforated baffle **4** which can be rotated into an alternative position **5** about a pivot point **6**. The perforations are arranged in the baffle such that a small percentage of the air stream passes through the perforations and the remainder is diverted into the other half of the duct. The collecting grid **3** is split into two portions which can be heated independently and are both kept at ground potential so that at all times they will collect and retain particles. A second unperforated baffle **7** is arranged to swing about pivot point **8** and can be rotated to point **9**. Two butterfly valves **10** and **11** can be operated to open and close exit tubes **13a, 13b**.

Figure 4: Barringer Research Ltd. Apparatus for Sensing Substances by Analysis of Adsorbed Matter Associated with Atmospheric Particulates

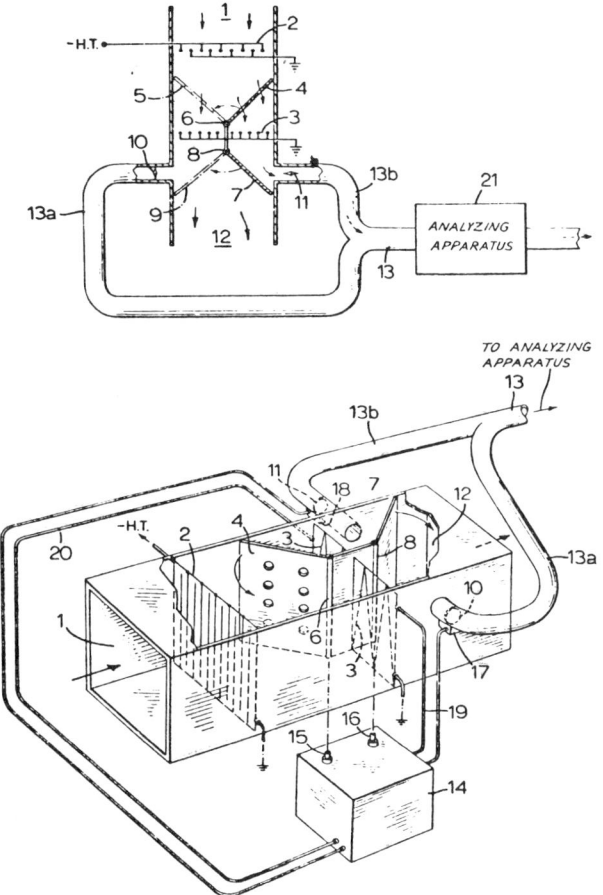

Source: U.S. Patent 3,768,302

The particulates in the air entering at **1** are charged on the grid **2** and collected on the grid **3**, the air substantially devoid of its particulates then being expelled out of the duct at **12**. With the baffles in the position shown, a small portion of the air passes over the right-hand side of the grid **3**, and a heating current is applied to this side. Adherent particulates are raised to a sufficiently high temperature to drive off adsorbed gases and volatile components which then pass out through the open valve **11** along the right-hand exit tube **13b**. Baffle **7** is closed to seal off this small flow of air and channel it out through the tube 13b. At the same time the main body of air passes over the cold left-hand portion of the collection grid 3 and particulates are collected. After a short period (e.g., 1 to 5 seconds more or less), the perforated baffle 4 is swung to the position 5, the unperforated baffle 7 is swung to position 9, the valve 10 is opened and the valve 11 is closed, and heat is applied to the left-hand side of the collection grid 3 and removed from the right-hand side. The particulates collected on the left-hand side of the collection grid 3 are now heated to a temperature sufficient to drive off adsorbed gases and volatile components and a small air stream through perforated baffle 4 carries them through the exit valve 10 and out to the tube 13a. Also, the heating current in the right-hand side of the collection grid 3 is switched off and particulates are collected on this portion of the grid 3 from the main body of the air stream. The cycle is then repeated.

Appropriate control means is provided to control the functions described. The details of these control means are well within the skill of those skilled in the art, and the control means is therefore shown in block form. Typically, the control means includes a control box 14 having shafts 15, 16 connected to the baffles 4, 7 at pivot points 6, 8 to swing the baffles in unison from one side of the device to the other, to close off first one grid-containing chamber and then the other. Electric valve actuators 17, 18 are connected to the butterfly valves 10, 11 and are connected by electrical leads (not shown) to the control box 14 and are controlled thereby so that the appropriate exit tube 13a or 13b will be connected to the device. Control leads 19, 20 also extend from the respective halves of the grid 3 to the control box so that each half grid will be heated when the chamber in which it is located is closed.

Although only two chambers, each containing half of the collection grid 3, have been shown, additional collection chambers can be used if desired. Alternative means may also be used to collect and heat the particulates, but desirably a continuous record should be provided as the air over the terrain under investigation is traversed.

In this fashion, a continuous air stream flows through the exit tube 13 containing a concentration of the volatile components that were originally present in adsorbed form in the incoming air stream. The rate of flow through the perforated baffle 4 can be arranged to be very small such as 1% of the total incoming air or less. This provides a high degree of concentration. Furthermore, in a simple modification of the system, the perforations in the baffle 4 can be entirely removed and the closed chamber so formed when the baffles are in position can be flushed out with argon from a gas cylinder if so desired. This can provide an inert carrier gas through the heated particulates instead of using oxygen. This can be particularly convenient for some types of emission spectrographic analysis of the vapors. Since the baked particulates accumulate the collection grid 3 must occasionally be cleaned or replaced to maintain efficient operation.

A variety of analytical techniques can be used for analyzing the gases and vapors emerging from the exit tube 13. For example, the gases and vapors can be scrubbed with a water spray and the liquid solutions so obtained can be passed over a specific ion electrode. Electrodes are available which can have high sensitivity for the halogen elements such as fluorine and can provide a continuous electrical reading of fluorine concentration. In the case of measuring for mercury vapor, the vapors can be passed through an optical cell and the absorption of a 2537 A beam of light can be measured. This absorption is related to the concentration of mercury vapor present. In a more sophisticated analytical arrangement, an argon carrier gas can be used as described above and the gas stream can be passed through a microwave cavity in order to generate a microwave plasma. The emission light from this plasma can be passed into a spectrometer and measurements made of specific emission lines corresponding to elements such as mercury, fluorine, iodine, bromine, chlorine, tellurium, arsenic, antimony, bismuth, etc. Since various kinds of analytical apparatus may be used, the analytical apparatus connected to exit tube 13 is indicated in block form at 21.

A technique developed by *J.R. Rhodes, R.D. Sieberg, M.C. Taylor, J.C. Westkaemper and R.C. Young; U.S. Patent 4,137,751; February 6, 1979; assigned to Columbia Scientific Industries, Inc.* is an improved technique for collection of airborne particles in an airborne prospecting method. A high volume of air is sampled and processed through a cyclone separator to achieve a high concentration of particle collection on a collecting surface. The collected particles may be analyzed as to mineral element type and correlated with the geographical position of collection to permit a deduction as to the type of minerals and soil in the terrain below.

In conventional geochemical prospecting, samples of rock, soil, vegetation, stream sediments or water are collected and such samples are analyzed for predetermined elements for the purpose of revealing anomalous geochemical distributions of such elements, related to mineralization or the existence of hydrocarbon deposits. Commonly, the samples are taken in soil at depths of between about 10 cm to 1 m. When samples are taken nearer to the surface it is usual to discard the top 1 to 2 cm layer of the soil on the theory that the very surface may be contaminated to some extent, due, for example, to the presence of animals or deposition of windswept material. In addition, the collection, storage and analysis of a large number of samples is very time-consuming and expensive, so that at present it is practical to take samples only at fairly wide-spaced intervals. As a result, it is often difficult to assess the significance of some apparent geochemical anomalies.

In a technique developed by *A.R. Barringer; U.S. Patent 4,056,969; November 8, 1977; assigned to Barringer Research Limited, Canada* contrary to conventional practice, it is particulates which are contained in the very surface, or surficial layer of the earth, or of vegetation, or water that are collected and analyzed. Samples of the surficial layer are taken rapidly, in quick succession, and at relatively low cost. More particularly, particulate or finely divided material comprising the surficial layer of soil, vegetation or water, such as mineral grains, clay minerals, saline evaporative residues, plant fragments, microorganisms and the like are sampled, for example, by applying suction to a tube positioned near to the surface to be sampled, and preferably depositing the particulate or finely divided material which is drawn up by suction through the tube, onto a suitable collection surface such as a tape that is adapted to be moved step-wise or else continuously past a collection station, whereby the collected material may be stored for subsequent analysis.

Alternatively, or in addition, at least some of the more easily volatilized elements and compounds such as mercury, iodine and hydrocarbons may be analyzed immediately after the particulates are collected. Preferably, prior to the deposition of the particles on the tape, the material is graded in size and only the large particles are collected, the remainder being discarded as the smaller particles are more likely than the larger particles to have migrated a considerable distance due to wind.

The geochemical basis of the technique can be illustrated by an example of a surface dust traverse carried out across a known mineral deposit. Samples of approximately the top millimeter of dust were collected at 105-meter intervals across the well-known copper-nickel-sulfide deposit in western Australia known as Poseidon, Mt. Poseidon, Mt. Windarra ore body. This deposit, which contains more than 4.2 million tons of nickel does not outcrop at the surface at the line of traverse. The collected samples were sieved through a nylon screen to exclude all coarse, sand-sized material and ensure that the sample was confined to dust material only. Copper and nickel were extracted with perchloric acid and analyzed by standard atomic absorption techniques.

The analyses in this case in relation to the mineral deposit are shown in Figure 5. It will be noted that a strong anomaly in both nickel and copper is obtained over the deposit. Similar results have been obtained in tests over a number of other deposits. Thus, it is apparent that the technique has considerable potential in the field of exploration for metalliferous deposits.

In addition to the foregoing, surface particulate materials exhibit an adsorptive capacity for gases and volatile materials emanating from the ground such as mercury vapor, iodine vapor, hydrogen sulfide, sulfur dioxide, methane, ethane, propane and volatile organic materials of biological origin. These gases are often associated with mineral, geothermal or oil accumulations and therefore their surface emanations tend to be an indicator of subsurface economic deposits.

Figure 6 shows, from top down, a land vehicle for sample collection, the sample tube tip, the impingement device and the resultant sample tape. The vehicle **10** is of rugged four-wheel type suitable for operation over rough terrain. A vacuum pump sampling assembly **11** is supported on shock mounts in the back of the vehicle **10** and is connected to a strong flexible plastic tube **12** which is able to withstand to a reasonable extent the extensive wear encountered in trailing along the ground behind the vehicle. The tube **12** terminates in a removable meshed cup **13** as shown. This meshed cup serves the function of sieving out very coarse particulates and allowing only relatively fine particulates sucked off the surface of the ground to be passed up the pipe **12**, i.e., material below about 200 μ in size.

The sampling assembly **11** consists of a vacuum pump **14** which is connected to an inertial impaction device **15** which in turn is connected to the tube **12** by means of a pipe **16**. The impaction device **15** is similar to that shown in U.S. Patent 3,868,222. By positioning jet **15a** of the impaction device **15** a short distance from the outer surface of collection tape **17**, for example, 2 to 3 cm, the larger particles will tend to travel in a straight line to impact on the tape **17**, but the smaller, lighter particles will be swept away with the air and will not reach the tape **17**. The separation between the jet **15a** and the tape **17** can be adjusted until most particulate material of a size below about 50 microns is discarded. Air in the pipe **16** carrying particulates is directed through the jet **15a** against the surface of the tape **17**, the outer surface of which is preferably coated with a suitable adhesive material such as silicone adhesive.

Figure 5: Plots Showing Surface Particulate Analyses Versus Metalliferous Deposit Location

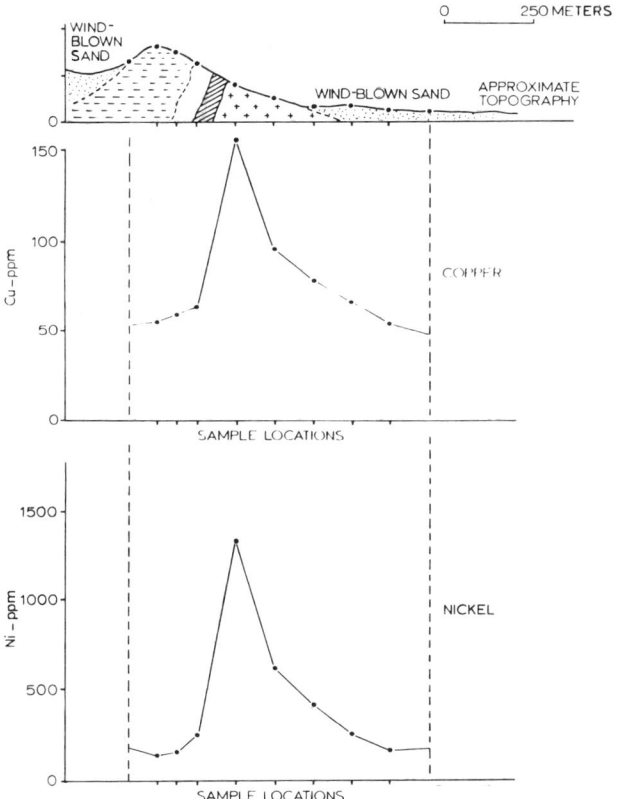

Source: U.S. Patent 4,056,969

The tape 17 is supported by a block 18 to which is fixed a numbering device which prints a mark and a number on the inner or back side of the tape each time the tape is incrementally advanced. A supply of tape is carried on a reel 19, which feeds the tape past the inertial impaction device 15 onto a pickup reel 20. The adhesive surface of the tape is covered with a cover tape 21 from reel 22. Tape 21 is made of a suitable plastic material which will not adhere strongly to the adhesive surface of the tape 17, such as Teflon. The cover tape 21 protects the particulate samples prior to analysis. The samples are collected as circular spots 23 on the surface of the tape 17 and the tape 17 may be advanced incrementally at equal time periods such as every 10 seconds or equal distance intervals of traverse as determined by distance measuring equipment on the vehicle 10.

Several alternative methods of analysis of the particulates on the tape can be employed. For example, the spots 23 on the tape 17 may be bombarded with

22 Geophysical and Geochemical Techniques for Exploration

x-rays and secondary x-ray fluoresence may be detected with a lithium drifted germanium detector cooled to liquid nitrogen temperatures. Pulse height analysis of the pulses produced by the detector enables identification of the elements present in the spots **23** to be carried out. This method has the advantage of being nondestructive so the tape can be reanalyzed for additional elements at any time. The determintion of concentration of elements is obtained from the ratio of the counts obtained for a given energy representing an element, against the Compton Scatter from the particles.

Figure 6: Barringer Research Ltd. Apparatus for Analysis of Surface Particulates

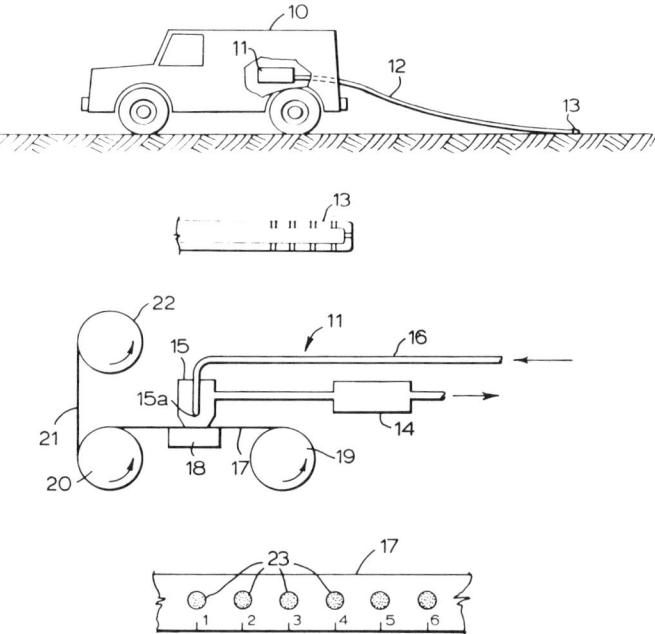

Source: U.S. Patent 4,056,969

An alternative method of analysis is the use of leaching methods in which the spots **23** are treated with a suitable leaching liquid such as 0.5 N HCl. After leaching for a standard period of time such as 2 minutes, the sample drop is removed and analyzed by atomic absorption or by atomic emission methods. In the case of atomic emission techniques, the sample drop may be nebulized into a radio frequency plasma and the optical emission from the plasma may be passed to a spectrometer with multiple exit slits and photomultipliers. Each exit slit is positioned appropriately for a given element, and by such means it is possible to analyze simultaneously for large numbers of elements. The observation of element ratios allows for the detection of subsurface mineral deposits. Thus, in detecting a copper deposit, copper is proportioned against a suitable reference element such as titanium or zirconium.

An apparatus developed by *A.R. Barringer; U.S. Patent 3,868,222; February 25, 1975; assigned to Barringer Research Limited, Canada* is a high-resolution geochemical prospecting system in which particles in large volumes of air are received, highly concentrated and then analyses are made of particles collected during periods of not more than 15 seconds each.

Bittern Salts

The search for bittern salts by geochemistry has heretofore been conducted by seeking evidence of high bromine content in the solid sodium chloride always associated with such deposits. One of the disadvantages of this process is that the salt section must be drilled or cored to obtain samples for analysis. Often wells drilled for hydrocarbons are bottomed above, or just when they enter salt sections and thus there is a dearth of the necessary geological information.

A technique developed by *G.C. Egleson; U.S. Patent 3,653,837; April 4, 1972; assigned to The Dow Chemical Company* provides a method of geochemical prospecting wherein information concerning the location of bittern salt deposits is derived by collecting samples of subterranean aqueous fluids from prospect areas and determining the concentration of dissolved ions which are affected by the processes of seawater evaporation and brine diagenesis. The presence of these dissolved ions in specific amounts or in specific ratios has been discovered to indicate that these subterranean waters have been in contact with, or are near, a bittern salt deposit since the reaction between bittern salts and its associated formation waters results in unique concentrations and ratios of certain ions common to both.

In other words, the waters contain certain concentrations and ratios of ions that they could not contain had not bittern salts been precipitated from them. Specifically, increasing concentrations of bromine or rubidium, or decreasing ratios of bromine to potassium, (magnesium plus calcium) to rubidium or (magnesium plus calcium) to bromine indicate the direction, in both plan and profile, toward deposits.

Thus, by measuring the concentration of ions associated with the process of seawater evaporation and brine diagenesis in fluid samples or cores containing such fluids it is possible to determine whether a particular underlying salt bed may contain bittern salts.

In practicing this technique, any suitable method of analysis may be employed, for example, atomic absorption. The necessary samples may be obtained by swabbing gas and oil well boreholes, from drill-stem test fluids, as core fluids or any other convenient source of subterranean fluids. The samples to be utilized for a regression in the plan view of the formation will be obtained from a given stratigraphic zone. Samples utilized as profile indicators will be generally taken from a given localized area but from any stratigraphic level. A stratigraphic zone is one consisting of a persistent stratum of some one kind of rock in which there is some porosity.

Generally, in the process, the formation fluid samples are analyzed for the abovementioned ion values. Having determined the specific concentrations of bromine and rubidium and the indicated ratios, the results of such analyses are plotted. This plot may take the form of an actual graphic representation of the data or any mathematical simulation of the same. When plotted, either in plan or in profile, or in both, the regression analysis indicative of the location of bittern deposits will show a pattern of generally increasing concentrations of bromine and/or rubidium or decreases in the specific ratios of Br/K, Mg'/Rb or Mg'Br (wherein Mg' equals the sum of magnesium and calcium ion values expressed as magnesium). A ratio of Br/K less than 0.17, Mg'/Rb less than 3,000, or Mg'/Br less than 14 each indicates underlying layers of bittern deposits. Each such ratio therefore serves also to indicate vertical direction to deposits.

In plan view, two or more data points wherein the specific ratios are more than these values but wherein one or more of the ratios are close to these values indicate horizontal directions to the deposits. Although the number of data points utilized is not critical so long as at least the two necessary to a directional vector are utilized in plan, preferred practice is to seek a similar or coincident regression of two or more of the indicator parameters. Reliability is significantly improved, if the indicator parameter is based on at least one fluid sample for every 100 square miles included within the plan. As previously indicated, the samples utilized for developing a plan view should be taken from a given stratigraphic layer. In this manner a relatively highly definitive plan view of bittern locations is developed.

It is, of course, true that there will be variations in the concentrations of the ions associated with the processes of seawater evaporation and brine diagenesis in the formation waters from different localities due to the history of the waters and formations surrounding the waters, even in the absence of bittern salt deposits; however, larger variations will be found due to the presence of buried bittern deposits and as a rule the small natural variations are negligible as compared to the larger variations resulting from the presence of the deposits. Thus, in the process, both the total amounts of ions as well as the trend, i.e., directional increases in the amount of ions in the water or the trend of certain ratios, are indicative of the presence of bittern salts.

Fluorite

Considerable difficulty has been experienced in attempting to detect the presence of fluorite by macroscopic inspection of drill cores where the ore body contains fluorite in fine veinlets. The detection of fluorite by macroscopic inspection is made difficult in specimens wherein the fluorite is present in small quantities because of the presence of other minerals of a similar chemical nature. A calculation of calcium fluoride based on total fluorine is very suspect because of the presence of other fluorine-bearing minerals such as topaz or idocrase which may be present.

Because of the increase in value of fluorite it has become practical to mine reserves in remote areas even though the fluorite is present in relatively small quantities. One problem in mining fluorite-including minerals in remote areas is the difficulty of conducting a careful macroscopic examination. While it has been common practice to effect a color sorting by staining minerals for a beneficiation purpose or process, it has not previously been suggested to use a fluorite staining process to identify the fluorite in fine form in a gangue matrix.

A technique developed by *W.P. Reid; U.S. Patent 3,957,438; May 18, 1976; assigned to Lost River Mining Corporation Limited, Canada* provides a method of detecting fluorite in an ore body comprising the steps of activating and dissociating the fluorite by the application of a cation exchange resin reagent, substituting the sodium ions of the activated and solubilized fluorite with a rare earth mineral and reacting the substituted rare earth mineral with an alizarin complexone to color-stain the fluorite whereby the presence of the fluorite may be readily detected.

Gold

A method developed by *H.W. Lakin and H.M. Nakagawa; U.S. Patent 3,397,040; August 13, 1968; assigned to the U.S. Secretary of the Interior* provides a geochemical exploration method for detecting and determining the amount of gold present in geologic materials such as soil, rocks and minerals.

Present gold determination methods employed during prospecting and studies of the chemical composition of the earth's crust either involve expensive equipment or are slow and tedious. Furthermore, although some of these methods such as the classical fire assay process are very accurate, they lack sensitivity towards the presence of minute amounts of gold in geologic materials, and lack the necessary mobility for a successful gold determination field process that can be employed in the area where the geologic sample is taken.

The object of the method is to provide an inexpensive method sensitive to the presence of minute quantities of gold in the geologic sample being tested, which method is highly suitable for field applications.

The process is as follows. A geologic sample is initially ground to a powder, for example, about 80 to about 200 Tyler mesh size, so that during later acid treatment most of the gold present in the sample will come into contact with the acid solution. After roasting the ground sample to burn off organic materials and to volatilize certain elements such as mercury and selenium, it is contacted with a concentrated hydrobromic acid-bromine solution which dissolves elemental gold and gold compounds present in the sample. Sodium bromate powder can be preliminarily mixed with the roasted sample (prior to contact with the acid) to provide the bromine for the oxidation of the sample and dissolution of gold.

No attempt is made to dissolve the entire sample, only to bring the acid and bromine in contact with most of the gold present in the sample by the preliminary grinding of the sample to a powder. Aqua regia could be used as the dissolution agent in which case it would be then necessary to effect a transfer of the gold in the nitric acid-hydrochloric acid to a hydrobromic acid solution.

Pregnant acid solution resulting from the hydrobromic acid dissolution step is diluted with water and then mixed with ethyl ether which is capable of extracting the gold present therein. Dilution of the aqueous solution prior to extraction prevents other elements such as iron from being extracted. About a 1.5 N to about 8.0 N hydrobromic acid solution concentration yields a maximum gold extraction by the ether solution. Since gold tends to be adsorbed on particles of insoluble residue remaining after the dissolution step, the insoluble residue is not removed from the acid solution during the ether-contacting step. As a result, when the ether removes gold from the aqueous acid solution, adsorbed gold goes into solution and likewise becomes extracted by the ether. The extracted ether phase is separated from the aqueous acid raffinate phase and remaining residue.

Ethyl ether extract phase is then extracted with about a 1.2 N to about 1.8 N hydrobromic acid which extracts iron, silver, mercury and palladium which may be present in the pregnant ethyl ether solution, but does not remove gold therefrom. The ethyl ether raffinate phase formed is separated from the acid extract phase. Afterwards, the ether solution is completely evaporated; the solid residue is dissolved in dilute ammonium acetate or nitrilotriacetic acid; and 4,4'-bis(dimethylamino)thiobenzophenone [thio-Michler's ketone (TMK)] in an organic carrier solvent such as isoamyl alcohol is added thereto to yield a solution with a red color whose intensity is proportional to the amount of gold present. If a spectrophotometer is available, the amount of gold present can be determined by a light absorbance test. If a test out in the field is desired, standard solutions each containing known quantities of gold may be prepared and compared red-color-intensity-wise with the test solution.

Mercury

Mercury is found in combination with other valuable elements in many minerals. Furthermore, apparently as a result of the mercury's relatively high volatility, some of the mercury migrates from and forms halolike configurations around such mercury-containing minerals in the depths of the earth. Since the discovery that the presence of mercury values in geologic materials such as soil and rock is a clue to the presence of concealed deposits of ores containing other valuable elements such as gold, silver, lead and zinc, there has been considerable interest in the use of mercury as a pathfinder for sources deep in the earth containing these valuable elements.

To measure the minute amounts of mercury occurring in the halos around such deposits, very sensitive spectrographic and vapor absorption procedures have been developed. However, both of these procedures require elaborate and expensive instruments, which have tended to reduce the potential usefulness of mercury surveys in geochemical prospecting.

A technique developed by *F.N. Ward; U.S. Patent 3,434,800; March 25, 1969; assigned to the U.S. Secretary of the Interior* is a simple process for economically determining the presence of as little as 30 ppb of mercury, in elemental or compound form, in a 1-gram sample of geologic material such as soil or rock.

Further objects and advantages will be had from the following description of the process. The procedure is based on the catalytic effect of mercury on the reaction of ferrocyanide with nitrosobenzene to produce a violet colored compound the intensity of which is related to the mercury concentration present. A spectrophotometer or visual comparison of the test solution with standard solutions of known mercury content can then be used to determine the presence and amount of mercury present.

Initially a small particulate sample of geologic material such as rock or soil is heated to vaporize substantially all the mercury therein and thereby effect the separation of mercury from elements which would otherwise interfere with the subsequent treatment. A temperature ranging from about 650° to about 750°C is adequate for these purposes. Evolved mercury vapors are then trapped in a dilute bromine solution. Although the exact nature of the reaction is not known, it appears that the mercury may be trapped as a mercuric bromide complex. After all the vapors evolved are trapped, the solution is heated to drive off excess bromine and a pH of about 3.2 to about 3.5 is established. Ferrocyanide and nitrosobenzene are then added to the solution and the mixture is heated to cause a reaction between these two compounds, catalyzed by the presence of mercuric ions. As stated previously, a spectrophotometer or standard mercury solution can then be used to determine the presence and amount of mercury in the purple solution since the optical density of the test solution is proportional to the mercury concentration present therein.

Any bromine solution which supplies bromine as an anion and will not interfere with the subsequent ferrocyanide-nitrosobenzene color-forming reaction can be employed. Sodium bromide and potassium bromide solutions are suitable for the purposes of this process. A dilute bromine solution efficiently traps the mercury vapors.

The proper solution pH (about 3.2 to about 3.5) necessary for the color-forming reaction can be attained by adding a base such as sodium hydroxide to the mercury-containing bromine solution. However, the addition of a buffer solution to the bromine solution prior to the mercury vapor trapping step more readily obtains the proper pH. Such a buffer solution could also be added after the trapping step, but adding it beforehand is preferable. Any buffer solution which will not interfere with the reaction or cause precipitation of mercury can be employed. Sodium and potassium acetate buffer solutions are suitable.

A specific procedure for the entire process is as follows:

(1) Place 0.1 to 0.5 g of –40 Tyler mesh soil or rock sample into an 18 x 75 mm test tube; fit on a delivery tube and insert the end of the tube containing the sample into a tube furnace maintained at a temperature of about 650°C.
(2) Heat the sample for 2 minutes and collect evolved vapors in 4 ml of bromine-buffer reagent contained in a 16 x 150 mm test tube.
(3) Remove the sample tube from hot furnace; disconnect the delivery tube, and rinse the delivery tube with 1 ml of bromine-buffer reagent. Add rinse to the collecting reagent.
(4) Place a magnetic stirring bar in the test tube containing collected vapors; insert the tube in an aluminum heating block fitted over a magnetic stirrer and heat the test tube at 95°C until bromine distillation is complete and solution colorless. Heat an additional 2 minutes. Cool the solution to about 25°C and remove the magnet.
(5) Add 1 ml of potassium ferrocyanide and 2 ml of nitrosobenzene solutions to the contents of the test tube and mix; place the test tube in a water bath at 50°C for 45 minutes.
(6) Remove the tube from the water bath, cool rapidly, and compare with standard solutions by viewing axially. Alternatively, use a spectrophotometer to measure the absorbance at 528 mμ and ascertain the mercury content by reference to a previously established standard curve.

Mercury, occurring as mercury ore alone or in association with gold, silver, copper, zinc, and other metallic ores, gives rise to free elemental mercury through geochemical and biochemical actions. The free mercury has a significant vapor pressure which gives rise to a mobile gaseous phase which diffuses through the earth's structure into the atmosphere, thereby serving as an indicator of the presence of mercury ore. Since mercury is commonly found in association with gold, silver, copper, zinc, and other metallic ores, the mercury vapor phase also serves as an indicator of the presence of these ores. Conventional prospecting techniques include traversing the earth's structure with a magnetometer which measures variations in the local magnetic field, or the use of induced polarization surveys which reflect the distribution of sulfide ores as an indicator of other ore presence. Shortcomings of these techniques include short operational range, interference due to variations in subsurface structure, and blocking by overburden in the earth's surface. Another shortcoming of such techniques resides in the necessity for being physically present on the earth's surface, or airborne in the atmosphere, above that portion of the earth's surface at which the ore is located.

There has been no prior art attention given to the use of mobile vapor phase tracking or the combination of such tracking with combining meteorological considerations. For example, no earlier investigators have proposed horizontal tracking of density flow of vapor phase emanations downwind of an ore deposit, then marking of the ore deposit as density flow and diffusion vertically of the vapor phase through the earth's structure coincide.

A technique developed by *G.H. Milly; U.S. Patent 3,730,683; May 1, 1973; assigned to Geomet Mining and Exploration Company* involves prospecting for mineral deposits of the type having a gaseous vapor phase which diffuses through the earth's structure into the atmosphere by initially sensing the gaseous phase in the atmosphere; discriminating between density flow of gaseous phase arising from ore bodies and that from ambient background, horizontally tracking the density flow towards the ore deposits and marking the ore deposits as density flow and diffusion of gaseous phase through the earth's structure from the ore deposit coincide.

Refinements of the technique include initially conducting a line survey downwind of the area being prospected; outlining katabatic flow as a function of topography; estimating bulk atmospheric drift in heavily wooded areas on the basis of above-canopy wind direction; dust and vapor sampling during tracking, and capping of the earth's structure adjacent the diffusion of gaseous phase through the earth's structure.

Silver

A technique developed by *H.M. Nakagawa and H.W. Lakin; U.S. Patent 3,395,987; August 6, 1968; assigned to the U.S. Secretary of the Interior* relates to the field determination of the presence and amount of silver in geologic materials such as soils and rocks.

Spectrographic analysis has been used for silver determinations, but high instrumentation costs as well as the limited sensitivity of spectrography are distinct disadvantages. Present chemical-reaction-type field methods possess inadequate sensitivity and selectivity.

The object of the technique is to provide a rapid, sensitive, selective field method useful in geochemical exploration for the determination of silver in geologic materials.

It has now been discovered that the presence and amount of silver in geologic samples may be readily and accurately obtained by (1) digesting the sample in nitric acid; (2) extracting silver from the resultant acid solution with tri-isooctyl thiophosphate (TOTP) in organic carrier solvent; (3) stripping the resultant organic extract phase with dilute HCl; and (4) determining the amount of silver in the HCl by its catalytic action on the persulfate oxidation of manganous ion to purple permanganate.

A more detailed understanding of the technique is to be had in the following description. A finely ground sample of the geologic material is digested in nitric acid to dissolve out any silver compounds and elemental silver therein. If it is suspected that the sample contains silver chloride in excess of 20 ppm (silver chloride being relatively insoluble in nitric acid), the sample should be fused with ammonium persulfate prior to acid digestion to convert the silver present as chloride to a more soluble form.

After all the silver compounds and elemental silver are digested in acid, silver is then extracted from the acid solution with triisooctyl thiophosphate in a carrier solvent such as benzene, carbon tetrachloride, or toluene. TOTP is added to the carrier solvent in amounts ranging from 15 to 50% by volume of the organic extractant solution. Preferably 30% by volume TOTP is employed. 0.5 to 1 part by volume of organic solution per part of acid solution is suitable for the purposes of this process.

After separating the aqueous raffinate phase from the organic extract phase, silver is extracted from the organic phase with approximately an equal volume of dilute HCl. An optimum concentration of about 0.3 N HCl is employed because the stripping ability of the acid at this concentration is the highest that can be obtained without having any adverse effects on the later catalytic reaction.

The organic raffinate phase resulting from the HCl extraction operation is discarded and potassium persulfate is added to the dilute HCl extract phase, to oxidize the silver in the extract phase, which in turn, catalytically aids in the oxidation of manganous ion to permanganate. The rate at which manganese is oxidized (the intensity of the purple permanganate color imparted to the solution after a set period of time) is therefore representative of the amount of silver present. To determine the amount of silver present in the sample, the purple color of the test sample is compared, intensity-wise, to a series of standard permanganate solutions each prepared with varying known amounts of silver.

Since silver ions tend to be adsorbed on glass, the analysis must be carried out with dispatch before the silver in the dilute HCl solution has a chance to be adsorbed by the glass container (e.g., test tube) in which the solution is held. Test tubes should be carefully cleaned between analyses with dilute cyanide solution to remove any silver that may have been adsorbed on the previous tests.

Some substances may interfere with each phase of the procedure but the overall method is virtually specific for silver. For example, TOTP may extract other elements such as palladium from the raw sample. However, palladium has no catalytic effect during the manganous oxidation step. Incomplete separation of the organic and aqueous phases during the acid stripping step offers the most serious interferences to the test. This interference is caused by TOTP that may remain in the HCl extract phase and which forms a yellow wax during persulfate treatment.

This condition is remedied by a preliminary heating of the HCl extract phase with potassium persulfate after taking an aliquot for the catalytic procedure.

Sulfide Ores

Many commonly used metals, such as copper, zinc, lead, silver, cadmium, mercury, and the like, frequently occur as sulfide ores in subterranean deposits. Numerous methods have been used in attempts to locate such mineral deposits. Most such methods require extensive drilling, coring, and the like to determine the presence of subterranean mineral ore deposits. Other methods for locating subterranean mineral deposits include seismic prospecting, drilling, coring, and the like.

Clearly, such methods leave much to be desired, since they are cumbersome and involve expensive testing techniques. Accordingly, a continuing search has been directed to the development of a method which is relatively inexpensive to use and which reliably indicates the presence of subterranean mineral ore deposits.

A process developed by *J.W. Wimberley; U.S. Patent 4,067,693; January 10, 1978; assigned to Continental Oil Company* is one whereby subterranean sulfide mineral ore deposits are readily located by a method consisting of (1) analyzing the soil at a plurality of locations in an area of interest to determine a ratio between mercury present in the soil as mercury sulfide and the total mercury present in the soil at such locations, and (2) comparing the ratios to determine the location of sulfide mineral ore deposits in the area of interest.

The method is useful in locating sulfide mineral ore deposits generally. Commonly occurring sulfide mineral ore deposits are the sulfide ores of metals selected from the group consisting of copper, zinc, lead, silver, cadmium, mercury, mixtures thereof, and the like. Such metals are commonly found as mixtures in varying proportions in naturally occurring sulfide mineral ore deposits.

In the practice of the method, soil samples are collected at a plurality of locations distributed over an area of interest. The samples should be representative of the soil at the location. In some instances, it may be desirable to take the soil samples at some distance beneath the soil surface, although it is believed that unless the surface has been contaminated with mercury and the like, it will be unnecessary to take the samples at any substantial distance beneath the surface. In particular, the samples are desirably taken at depths ranging from 2 to 13 inches, although deeper samples could, of course, be used if desired. It has been observed that with samples taken at varying depths at closely adjacent locations, while the total mercury in the samples may vary, the ratio of the mercury sulfide to the total mercury present in the soil in each instance was remarkably constant.

The mercury sulfides are present as HgS, Hg_2S, and mixtures thereof. Desirable results have been obtained wherein the ratio of HgS to the total mercury present in the soil was used. Of course, the ratio may be expressed in varying ways, such as the ratio of HgS to the total mercury present in the soil, the ratio of HgS to the other forms of mercury present in the soil, and the like. It will be recognized that the ratio of mercury sulfide to total mercury is considered in all such ratios, however expressed. The use of different ratios is possible since the primary use of the ratio is for comparative purposes only, as will be shown more fully hereinafter.

The determination of the mercury sulfide and the total mercury present in the soil may be by any accepted laboratory technique. Desirable results have been obtained wherein the soil samples were heated progressively, with the evolved mercury being absorbed in an apparatus similar to that shown in U.S. Patent 3,693,323. The collected mercury is then analyzed. It is known that the various forms of mercury, such as mercury sulfide, mercury oxide, mercury halide, and the like, thermally decompose at different temperatures. It is thus possible to determine the particular form of mercury which has decomposed to yield the mercury collected in the mercury trap at a given temperature.

Such calibrations are well known to those skilled in the art and need not be discussed extensively. It is pointed out, however, that such calibrations are readily accomplished by merely heating a soil sample to a temperature such that no further mercury is evolved, thereafter cooling the soil sample and adding a particular form of mercury, such as $HgCl_2$ and thereafter heating the sample again and observing the temperature at which the mercury is evolved. It is clear that a calibration for commonly occurring mercury salts can be prepared by the technique discussed as well as by other techniques known to those skilled in the art. It is necessary that the decomposition temperatures for the mercury sulfide be determined with the other mercury being allowed to thermally decompose and evolve without particular concern as to the form of the mercury in the soil. The ratio of the mercury sulfide to total mercury is then readily calculated.

The higher ratios of mercury sulfide to the total mercury indicate the presence of subterranean sulfide mineral ore deposits. It is postulated that over the millions of years, the sulfide mineral ore deposits release sulfur slowly, thus providing reactive sulfides which move upwardly through the soil and react with the mercury in the soil, thus increasing the ratio of mercury sulfide to the total mercury present in the soil in areas over sulfide mineral ore deposits. Obviously, in some instances, the higher ratios may occur in only a portion of the area overlying sulfide mineral ore deposits, since the reactive sulfides would tend to move upwardly through the more porous earth formations and the like.

It is recognized that in many instances, the total amount of mercury contained in the soil samples will be quite low, and it is pointed out that the total amount of mercury is of no particular concern. The information of interest in the analysis of the soil is the ratio of the mercury sulfide to total mercury present in the soil. It is desirable that efforts be made to avoid the collection of nonrepresentative samples in areas which have been sprayed for crop treatment and the like by taking samples from a suitable depth in the soil.

Uranium

A great deal of attention has been given to the detection of uranium deposits by both geophysical and geochemical techniques as described, for example, by Campbell (9).

A process developed by *P.R. Gray; U.S. Patent 4,066,891; January 3, 1978; assigned to Phillips Petroleum Company* is one in which valuable information on the presence or absence of uranium-enriched ores, geothermal reservoirs, or natural gas sources is obtained by analyzing earth samples for their relative ^{210}Po content.

The results of this analysis constitute a valuable tool which, together with other geophysical information, enables a prospector to locate, e.g., uranium-enriched ores, or geothermal reservoirs or natural gas sources. It is particularly advantageous to convert the results obtained by sample analysis into a ^{210}Po profile. The location of increased ^{210}Po concentration is related to the underground location of other uranium-enriched ore, or a geothermal reservoir or a natural gas source.

The physical mechanism on which this process is based is the radioactive decay of uranium (^{238}U). This decay progresses through various elements finally ending with the stable lead (^{206}Pb). Among the various elements of this decay series, only radon (^{222}Rn) is a gas. The ^{222}Rn diffuses through the formation to the surface. For this reason ^{222}Rn has been used for environmental analysis, e.g., in uranium mines. ^{222}Rn decays further and in its decay series one stage is ^{210}Po. This ^{210}Po is a solid material under normal temperature and pressure conditions and does not to any substantial extent leave the formation once it is formed. The ^{210}Po is an alpha emitter and has a half-life of about 138 days. Since ^{238}U has an extremely long half-life time as compared to all the following decay elements thereof, a uranium ore can be envisaged as a producer of ^{222}Rn gas for practical purposes at a constant rate.

^{222}Rn in turn decays with a half-life of about 3.8 days into polonium 218 (^{218}Po), a solid material under normal temperature and pressure conditions. This element ^{218}Po in turn decays and becomes via a series of short-lived intermediates, lead 210 (^{210}Pb) with a half-life of 22 years, which decays via bismuth 210 (^{210}Bi) (5 days half-life) to ^{210}Po. The half-life time of ^{210}Po of 138 days is on the one hand sufficiently long that earth samples can be taken and efficiently analyzed and at the same time the alpha activity is sufficiently high since the half-life is not too long so that meaningful and accurate results can be obtained.

Although ^{222}Rn has a rather short half-life of 3.8 days, it diffuses to the surface to some extent. Thus, earth samples at the earth surface can be analyzed for ^{210}Po to obtain valuable information on the eventual presence of the specific subterranean formations mentioned above. Preferably, however, the earth samples are not taken from the surface but are taken from a layer which is about 2 inches to about 5 feet below the earth surface. Thereby interferences with randon diffusion through the atmosphere and correspondingly depositions of ^{210}Po from this interfering radon gas are effectively avoided. At the same time meaningful information on the nature and composition of the earth far below the sampling location is obtained in a fairly inexpensive and reliable manner.

The process is mainly applicable to locate uranium-enriched ores. Uranium is an element that is very widely spread across the earth. Whether mining of uranium-containing ores is economical depends primarily on the uranium concentration. These uranium concentrations in various parts of the United States range from about 0.1 to 0.4% U_3O_8 in the uranium-containing ore. At a uranium price of \$20/lb of U_3O_8 and under normal mining conditions, the uranium concentration would have to be above 0.15% U_3O_8 uranium-containing ore to render mining of such ores economical. Generally, uranium ores mined today have concentrations in the range of about 0.1 to 0.30% U_3O_8 in the uranium-containing ore.

PROSPECTING FOR HYDROCARBONS

Determination of Hydrocarbon Presence in Soils and Rocks

In the location of subterranean petroleum reservoirs, use has been made of geochemical methods. These methods are based on the theory that hydrocarbons migrate upwardly from the subterranean reservoir and the presence, in the surface or near-surface zones of the earth or in the earth strata intermediate to the surface or near-surface zones of the earth, of hydrocarbons or organic derivatives thereof is indicative of the presence of a subterranean petroleum reservoir. Thus, as a prospecting method or a well logging method, samples of earth material have been collected and analyzed for hydrocarbons or organic derivatives thereof and the concentrations of hydrocarbons or organic derivatives thereof have been correlated with the locations from which the samples were taken to obtain information as to the probable location of the petroleum reservoir.

A technique developed by *H. Hoover, Jr.; U.S. Patent 2,305,384; December 15, 1942; assigned to Consolidated Engineering Corporation* is one for detecting and measuring significant constituents which have migrated to the surface of the earth from valuable subterranean deposits. It involves locating the top of the subweathered layer which exists beneath the soil, collecting portions of earth from the subweathered layer at systematically distributed points throughout a region under investigation and analyzing the samples for constituents significant of subterranean deposits.

This technique enables one to obtain accurate results free from any of the errors or objections inherent in previously used methods due to irregularities in surface soil characteristics.

In drilling wells in search of oil and gas, samples of the subsurface formations are commonly obtained, either as cuttings produced by the bit, or as cores especially taken by core bits, or both. Commonly also, such rock samples are inspected and examined for their content of gaseous, liquid and solid hydrocarbons as well as for other constituents of diagnostic importance. Examination for liquid and solid hydrocarbons usually takes the form of inspection under ultraviolet light, solvent extraction, distillation or the like, and examination for gaseous hydrocarbons usually takes the form of disintegration of the rock sample often in the presence of water, so that the pore spaces are opened up and the gaseous hydrocarbons thereby liberated, whereupon they can be withdrawn or otherwise collected for testing, generally with a so-called hot wire gas detector.

A difficulty with systems of determining gaseous hydrocarbons in rock samples currently in use is that the heavier gaseous hydrocarbons tend to become lost by sorption and condensation in the apparatus employed, and these heavier hydrocarbons are the very ones, as is well known, that are particularly diagnostic for oil-bearing strata.

A technique developed by *P.J. Moore; U.S. Patent 2,871,105; January 27, 1959; assigned to National Lead Company* overcomes these problems by carrying out disintegration of the rock sample by mechanical action under a suitable liquid such as water and in a container enclosing within a single space the rock sample, the mechanical disintegration means, the water, an air space above the water, and the gas detection means in the air space.

A method developed by *E. McDermott; U.S. Patent 3,120,428; February 4, 1964;* is a method of geochemical survey in which the measurements of surface anomalies of a geochemical survey are corrected for the inherent variations of entrapping and holding capacities of the various soils present in the analyzed samples, as well as the permeability of the ground which controls the leaching effect due to rainfall and surface waters. More specifically, the method involves:

(1) testing a plurality of soil samples obtained from spaced portions of an area under geochemical exploration to determine detector concentration values therefor;
(2) identifying for each of the soil samples a characteristic soil property indicative of its retentivity of detector concentration values;
(3) recording a value for each of the soil samples indicative of its identified characteristic soil property for retentivity and its detector concentration value;
(4) statistically analyzing the recorded values for the soil samples to establish a mean background value;
(5) eliminating those of the soil samples which are characteristically indicative of negative qualifications as anomalies by having recorded values substantially the same as the mean background value; and
(6) locating the portions of the area corresponding to the remaining soil samples in which the presence of detector concentration values having their background values eliminated are characteristically indicative as anomalies.

A method developed by *C.S. Hall, Jr., W.B. Huckabay and M.C. Kelsey; U.S. Patent 3,180,983; April 27, 1965; assigned to Rayflex Exploration Co.* involves:

(1) drilling a plurality of test holes at each of a plurality of stations located in spaced relation over the area being investigated;
(2) withdrawing a sample of soil gas from each hole;
(3) measuring and recording the percentage of methane in each soil gas sample;
(4) determining the mean of all the methane values;
(5) plotting the locations of the station on a map of the area being investigated;
(6) plotting on each station location on the map the percentage of samples at the respective station having a methane value exceeding the mean by a predetermined minimum; and
(7) contouring the values obtained in step (6), whereby the area closed by the highest values indicates the area overlying a petroleum deposit.

A method developed by *E.E. Bray and J.R. Zimmerman; U.S. Patent 3,446,597; May 27, 1969; assigned to Mobil Oil Corporation* is a geochemical exploration method which involves the identification of source rocks of gaseous petroleum hydrocarbons.

In attempting to ascertain whether structures contain petroleum hydrocarbons, geologists have in recent years turned to exploration techniques based upon the so-called "source rock" concept. Under this concept, it is assumed that petroleum hydrocarbons are formed during long periods of burial in sediments of high organic matter content, the petroleum hydrocarbons being derived from matter of biologic origin which was deposited with these sediments. The exact mechanism by which oil and/or gas are formed within such sediments, i.e., source rocks, is not known with certainty. After the hydrocarbon-forming mechanism has taken place, it is thought that the petroleum hydrocarbons in the source rock then migrated to more permeable reservoir rocks where they accumulated in the concentrated deposits found today.

In Bray's process, a sample obtained from a formation in the earth's crust is subjected to thermal action in order to determine if organic matter originally present in the formation has, during the course of geologic history, undergone degenerative reactions yielding gaseous hydrocarbon products. By the analysis of certain gases produced during this treatment of the sample, it can be determined whether the formation from which the sample is taken is in a relatively early stage of maturation in which certain hydrocarbon gases have not been generated, or in a relatively late state of maturation where there is a strong possibility that such hydrocarbon gases have been generated during the course of its geologic history.

The production during this thermal treatment of gases such as nitrogen and carbon dioxide, and in particular nitrogen, is associated with low activation energies. On the other hand, the production of hydrocarbon gases, and in particular methane, is associated with higher activation energies. Thus, the production of relatively large amounts of nitrogen and carbon dioxide during the early stages of the thermal treatment is indicative of an early state of maturation. On the other hand, relatively low production of nitrogen and carbon dioxide during the early stages of thermal treatment is an indication of a late stage of maturation. The rock formation thus is indicated as a possible source rock, particularly where the production of significant quantities of methane is observed during the later stages of the thermal treatment.

A technique developed by *R.R. Thompson; U.S. Patent 3,539,299; November 10, 1970; assigned to Pan American Petroleum Corporation* is one in which earth samples are caused to release their content of hydrocarbon gases for analysis by treatment with a hot ethylenediaminetetraacetic acid solution preferably at a pH of about 7 or higher. Under these conditions, carbonate minerals are decomposed without creating the large volume of gaseous CO_2 that complicates the analysis when conventional strong-acid treatment is used.

A technique developed by *H.L. Wise; U.S. Patent 3,685,345; August 22, 1972; assigned to Shell Oil Company* is one in which the sampling of the concentration of a selected fluid in a subterranean earth formation is improved by circulating a fluid such as air into repetitive contacts with the earth formation and measuring the concentration of the ethane in the circulating air when the concentration is at a level at which its rate of change is low relative to that exhibited in the initial stage of the fluid circulation.

The boreholes used can be relatively deep or shallow and can be freshly drilled or preexisting. Freshly drilled shallow boreholes are preferred. Particularly suitable boreholes are drilled with an augering-type device which requires no circulation of drilling fluid.

A technique developed by *L. Horvitz; U.S. Patent 3,700,407; October 24, 1972* is a method of geochemical prospecting by the taking of samples of the earth formation at spaced locations, in an area to be investigated, which samples are analyzed to determine the concentration therein of significant hydrocarbon gas leakage products from subterranean petroliferous deposits. The results of such analyses are correlated with sample location to provide information concerning the presence of such deposits in the area.

Valuable indications of the possible presence of natural gas, petroleum or radioactive mineral deposits may be obtained by the detection of subsurface anomalies of gas concentration, particularly of helium.

It is known that trace quantities of helium (approximately 5 ppm) are fairly uniformly distributed throughout the troposphere and throughout the earth's crust. Higher concentrations of helium are, however, often associated with hydrocarbon accumulations such as natural gas and petroleum as well as with radioactive mineral deposits such as those of uranium and thorium.

Although the origin of the normal or background composition of helium in the earth's crust may not be completely known it is known that helium is produced as a result of the process of radioactive decomposition which takes place constantly throughout the earth's crust. No doubt this is one source of the helium which is found uniformly distributed throughout the troposphere and throughout the earth's crust.

Helium is often found in abundance in hydrocarbon and bitumen deposits. The presence of anomalously high concentrations of helium associated with natural gas and petroleum deposits may be explained by the probability that those physical features which entrap such fluid substances as natural gas and petroleum also tend to entrap quantities of naturally occurring gases such as helium. These traps may be formed in many ways; commonly they are caused by the occurrence of a dense impermeable zone which forms a barrier preventing the fluids from flowing away through more permeable layers.

A technique developed by *L.A. Pogorski; U.S. Patent 3,835,710; September 17, 1974* is a method for prospecting for hydrocarbon, bitumen and radioactive mineral deposits by collecting relatively small subsurface soil gas samples and analyzing same to determine areas of anomalously high helium content therein. The samples are collected by using a long slender shaft capable of being driven into the ground which has a capillary bore running substantially its entire length.

The inlet to the lower end of the capillary bore is controlled by a cap which may be opened or closed by turning the shaft when the device is in the ground and the upper end of the bore is sealed by a septum comprising a flexible sleeve fitting over the end of the shaft and a flexible plug closing the sleeve. The sample is collected in a container having a sharpened hollow tube as an inlet port. The sharpened tube may be inserted through the flexible plug to communicate with the capillary bore of the shaft for withdrawing a gas sample from the subsurface soil. The container is the type which may be deformed into a collapsed configuration and will return to its expanded configuration to draw in a gas sample.

A technique developed by *L.A. Pogorski; U.S. Patent 3,862,576; January 28, 1975* is a method of geochemical prospecting by locating areas of anomalously high concentrations of elements indicative of subsurface deposits such as hydrocarbons or radioactive minerals. The presence of indicative elements such as helium is detected by placing a gas reservoir in communication with the subsurface environment and allowing the elements to transfer under partial pressure from the subsurface environment to the gas pocket and then removing a sample of the gas pocket for analysis to determine the presence of the indicative element.

A method developed by *K.S. Schorno; U.S. Patent 3,847,549; November 12, 1974; assigned to Phillips Petroleum Company* is a method of geochemical exploration wherein the rock sample from a subterranean formation is treated under conditions such as to artificially accelerate the aging of the sample. Following the aging period the sample is analyzed to compare the percent extractable oil, the ratio of extractable oil to total oil, and the n-alkane distribution of the original and aged rock to determine the maturity of the rock and its capability for generating oil.

A technique developed by *J. Espitalie and B. Durand; U.S. Patent 3,953,171; April 27, 1976; assigned to Institut Francais du Petrole, des Carburants et Lubrifiants, France* comprises the steps of heating a sample of sediments, determining a group of two parameters S_1 and S_2 which represent, respectively, the amounts of at least one hydrocarbon compound and of at least one oxygen-containing compound formed by heating the organic material contained in the sediment, and evaluating from the determination of this group of parameters the capacity of the sediments to constitute a good mother rock (i.e., source) for hydrocarbons.

A technique developed by *M.E. Smith; U.S. Patent 3,975,157; August 17, 1976; assigned to Phillips Petroleum Company* is a geochemical exploration technique wherein petroleum source rocks are characterized on the basis of an Isoprenoid Generation Index (IGI) which is the ratio of the relative concentrations of the 20-carbon isoprenoid and the sum of the 18- and 16-carbon isoprenoids. Organic extracts of rocks in which diagenesis has not proceeded far enough to have produced migratable petroleum have an IGI of 3 or greater, whereas crude oils and extracts of source rocks have lower values. The IGI is used to calculate other, origin-diagnostic parameters useful in correlating oils and extracts of marine origin.

An apparatus developed by *H.E. Aine; U.S. Patent 4,051,372; September 27, 1977* is particularly useful for petroleum exploration by detecting minute quantities, i.e., ppm or ppb or more, of hydrocarbon gases, such as ethane or propane or both, diffusing into the atmosphere through the overburden from underground petroleum deposits. The gas detector system is carried over the region to be explored as by aircraft, land vehicle or ship and the concentration of the gaseous component or ratios thereof are detected as a function of position over the area to be explored, whereby plumes of such gases are detected to permit pinpointing of the underground petroleum deposits.

In this apparatus, the optoacoustic detector cell of an infrared laser optoacoustic spectrometer is interfaced to a source of gas under analysis, such as the atmosphere, by means of a relatively large area membrane separator. Gas to be analyzed is passed through the membrane separator, compressed by a compressor and then fed at a pressure of between 10 and 1,000 torrs into the optoacoustic detector cell of the infrared spectrometer for analysis therein. A method of tracing is disclosed wherein deuterium is incorporated into a system to be traced. The deuterium diffuses out of the body in extremely minute quantities which are then inducted into the optoacoustic detector cell of an infrared laser optoacoustic spectrometer via a palladium membrane separator and reactor, whereby an extremely sensitive detection or tracing system is obtained.

A technique developed by *F.F. Holub and H.W. Alter; U.S. Patent 4,065,972; January 3, 1978; assigned to Terradex Corporation* is one for collecting gas samples associated with underground minerals. A plurality of small gas sample containers are planted in an inverted position in shallow holes in the surface of the earth in a predetermined pattern, and upwardly migrating gases associated with the underground minerals being sought are collected in the containers for a predetermined time period, to obtain a time-integrated representation of the amount of one or more gases of interest migrating to the surface over that time period.

Two types of containers are alternately employed: A total gas sample container in which all upwardly migrating gases are admitted into a sealed compartment at a preselected flow rate; and a specific gas sampler container in which strips of detector material which are sensitive to predetermined specific gaseous substances are mounted within the inverted container and exposed to the upwardly migrating gases over the predetermined time interval. The containers are removed from the shallow holes after the lapse of the predetermined time interval, and the detector material alone, or the containers with the detector material still mounted, are subjected to qualitative and quantitative analysis. The gas collection period is maintained substantially constant for all containers.

This information is then interpreted to identify potentially valuable deposits of petroleum, gas, or other substances of interest.

One of the most important geochemical parameters to evaluate in oil exploration certainly is the amount of available organic matter, including both soluble and insoluble organic components. The organic carbon content of a kerogen rock, which represents the insoluble organic matter closely associated to this rock, should advantageously be determined and made visible with a diagraph on the very site of an oil drilling.

Up to now, the organic carbon content has always been determined according to a very time-consuming method which requires previous physicochemical treatment of the sediments, and especially an extraction by a solvent and an acid treatment to eliminate the carbonates. The sample is then submitted to a pyrolysis in the presence of a catalyst and the amount of CO_2 which is formed from the organic matter is determined. It is obvious that with such a method, the carbonates present in the rock must be completely eliminated, because they also emit some CO_2 which is not representative of the organic carbon content of the rock. The time which is needed for these various operations is at least one day, and the advantage of a method allowing determination of the organic carbon content in a short period of time is easily understood.

An apparatus developed by *P.A. Leplat-Gryspeerdt; U.S. Patent 4,106,908; August 15, 1978; assigned to Labofina SA, Belgium* allows such determination from small samples of the raw material, e.g., in the range of several milligrams. It further provides such a process which permits determination in a short period of time, e.g., within about 10 minutes.

A method developed by *J. Espitalie, J.-L. Laporte, M. Madec and F. Marquis; U.S. Patent 4,153,415; May 8, 1979; assigned to Institut Francais du Petrole, France* comprises heating a rock sample successively to a first temperature comprised in the range from between 200° to 400°C, then to a second temperature preferably from 550° to 600°C, measuring the amount of native hydrocarbons vaporized at the first temperature and the amount of hydrocarbon compounds produced by pyrolysis of the organic material of the sample at the second temperature. Oil-related characteristics of the sediments are derived from the two measured values.

Analysis of Hydrocarbon Types in Soils

It has been recognized that hydrocarbons may be present in the earth as the result of conditions other than migration from a subterranean petroleum reservoir. For example, it is recognized that methane, a hydrocarbon constituent of petroleum can be present in the earth as a result of vegetative decomposition and that heavier hydrocarbons, which could be constituents of petroleum, can be present in the earth as a result of bacterial action. Thus, a geochemical prospecting method, to be positive with respect to the probable location of a subterranean petroleum reservoir, should be capable of identifying hydrocarbons or organic derivatives thereof in earth samples as having their origin in a subterranean petroleum reservoir. It would appear then that it would be a simple matter to obtain a positive indication of the presence of a subterranean petroleum reservoir by merely analyzing earth samples for hydrocarbons or organic derivatives thereof that could be present in the earth samples only because of migration of hydrocarbons from a subterranean petroleum reservoir.

However, the quantity of hydrocarbons or organic derivatives thereof present in earth samples, even when taken from localities overlying a subterranean petroleum reservoir, is so small as to make it difficult to identify even the simple hydrocarbons therein much less to identify the more complex hydrocarbons or their organic derivatives. But of greater difficulty is the fact that no means have been heretofore known to ascertain whether any hydrocarbons or organic derivatives thereof in earth samples have had their origin in a subterranean petroleum reservoir.

A technique developed by *E.E. Bray; U.S. Patent 2,742,575; April 17, 1956; assigned to Socony Mobil Oil Company, Inc.* for overcoming this problem is based on the discovery that earth samples taken from localities in proximity to a subterranean petroleum reservoir will contain constituents capable of selectively absorbing infrared radiation having wavelengths of 12.35±0.10 and 13.40±0.10 microns whereas earth samples taken from localities where there is no possibility of the proximity of a subterranean petroleum reservoir will contain costituents capable of selectively absorbing infrared radiation having these wavelengths. Thus, the detection in earth samples of a constituent or constituents capable of selectively absorbing infrared radiation having wavelengths of 12.35±0.10 microns and 13.40±0.10 microns identifies the constituent or constituents as having an origin in petroleum oil.

It should be indicated that not all constituents of crude petroleum oil or organic derivatives thereof are capable of selectively absorbing infrared radiation having wavelengths of 12.35±0.10 microns and 13.40±0.10 microns. For example, the paraffin constituents of crude petroleum oil or their organic derivatives do not exhibit this property. However, while the property of selectively absorbing infrared radiation having the wavelengths of 12.35±0.10 microns and 13.40±0.10 microns is not shared by all constituents, or their organic derivatives, of crude petroleum oil, crude petroleum oil contains a constituent or constituents exhibiting this property and the constituent or constituents exhibiting this property migrate through the soil from a subterranean petroleum reservoir as postulated for any constituent of the crude petroleum oil.

In the practice of this technique, earth samples are collected from the prospect area. Where exploration is conducted by prospecting over a surface area of the earth, the samples are preferably collected from points set out in the area according to a definite plan or pattern. For example, the sampling points may be spaced at regular intervals, about one-fifth of a mile along a traverse, or along a series of traverses, or along a grid pattern. However, any suitable interval spacing may be employed.

Sampling over a surface area is generally conducted by taking the earth samples either from the surface zones or the near surface zones. Preferably, the samples are taken at a distance of at least one foot from the surface in order to avoid as much as possible any contamination or excessive content of organic debris. Where exploration is conducted in connection with the drilling of a well, the earth samples, comprising cores or cuttings, may be obtained from successive depths in the well as desired. Generally, sampling may be carried out either over the surface of the earth or from successive depths in a well in accordance with any of the patterns and any of the procedures heretofore employed in the art.

After collection, the earth samples are dried and crushed and a portion of each, which may be of the order of 150 g, is extracted with a suitable solvent. Any solvent capable of desorbing and dissolving hydrocarbons from the earth samples may be employed for extraction. Suitable solvents include carbon tetrachloride, chloroform, carbon disulfide, acetone, methanol, ethanol, propanol, diethyl ether, benzene, etc. Mixtures of solvents may also be employed, such as an equimolar mixture of an alcohol and carbon tetrachloride or a mixture comprising about ten volumes of benzene and one volume of methanol. The extraction of the soil samples may be carried out in any known manner, as, for example, by employing a Soxhlet extractor or a ball mill extractor.

The solutions of extract thus obtained may be filtered, if required, and without further treatment, subjected to analysis by passing the infrared radiation therethrough and obtaining the transmittance curve or otherwise measuring the absorption of the infrared radiation, assuming that the solvent in which the extract is dissolved does not absorb the infrared radiation so as to mask any absorption bands over the range of wavelengths measured.

Preferably, however, the solutions of extract are weathered in order to obtain extracts which are uniform with respect to the removal therefrom of any volatile constituents extracted from the earth samples. Weathering may be accomplished by filtering, if required, or, in lieu of filtering, decanting from any earth sample residue, heating to a temperature sufficiently high to vaporize the solvent and any volatile constituents of the earth sample, and blowing a stream of air or other suitable gas thereover during the heating period. A suitable temperature for weathering is 40°C, and weathering is continued at the temperature employed until the extract comes to constant weight. The extract, after removal of the solvent, is in the form of a waxy semisolid.

Following weathering, the extract per se may be subjected to transmission of infrared radiation or the extract may be dissolved in a suitable solvent and the infrared radiation transmitted through the solution of extract. For transmission and measurement of the infrared radiation, any suitable spectrophotometer apparatus capable of measuring intensity of infrared radiation as a function of wavelength over the desired range of wavelengths may be employed.

A technique developed by *P.B. Weisz; U.S. Patent 2,755,388; July 17, 1956; assigned to Socony Mobil Oil Company, Inc.* involves the detection of hydrocarbons in a soil gas sample which comprises subjecting a mixture of argon and a soil gas sample to electrical excitation to give rise to metastable states of argon, discontinuing excitation of the argon, and observing the rate of disappearance of the metastable states of argon in the presence of the soil gas sample in comparison with the rate of disappearance of the metastable states of argon under conditions similar except for the absence of the soil gas sample.

The following is the manner by which the presence of hydrocarbons of higher molecular weight than methane may be detected. When hydrocarbons of C_2 or higher are found in the presence of the metastable argon states, those metastable states will find an outlet for their potential energy in ionizing the hydrocarbon molecules, leading to rapid deexcitation of the metastable states. Methane, carbon monoxide, carbon dioxide, hydrogen, water, nitrogen, oxygen, i.e., other atomic species present or likely to be found in soil gas have ionization potentials much higher and could not so act.

A technique developed by *E.E. Bray; U.S. Patent 2,773,991; December 11, 1956; assigned to Socony Mobil Oil Company, Inc.* is one in which earth samples are collected from a prospect area, each of the earth samples is analyzed for the quantity therein of at least a selected portion of the organic matter, and at least a representative sample of the organic matter contained in the earth samples having anomalous amounts of organic matter and at least a representative sample of the organic matter contained in the earth samples not having anomalous amounts of organic matter are analyzed for the proportion of carbon therein having an atomic weight of 14 to carbon having an atomic weight of 12.

Since the half-life of carbon having an atomic weight of 14 is about 5,568 years, the amount of carbon having an atomic weight of 14 in any matter having its ultimate origin in plant life will in, e.g., 35,000 years, decrease to approximately one-eighteenth of its original value and amounts of this order will begin to approach undetectability. Accordingly, since petroleum is considered to have a plant and animal origin and an age of hundreds of thousands of years, the amount of carbon therein having an atomic weight of 14 will, for all practical purposes, be zero.

On the other hand, since the time elapsing between the utilization of the carbon dioxide by plant life in the formation of their body structure by the photosynthesis reaction and the complete conversion of the carbon in the body structure of the plant life to carbon dioxide by bacterial decomposition and otherwise is variously estimated to be on the order of 500 to 1,000 years, the organic matter in the earth's crust derived directly from vegetative matter will contain measurable amounts of carbon having an atomic weight of 14. Thus, where at least a portion of the organic matter in an earth sample has an origin in a subterranean petroleum reservoir, the proportion in the organic matter of the amount of carbon having an atomic weight of 14 to the amount of carbon having an atomic weight of 12 will be less than where a lesser amount or none of the organic matter has an origin in a subterranean petroleum reservoir. Therefore, by analyzing at least representative samples of the organic matter for the amount of carbon having an atomic weight of 14, it can be determined whether those earth samples containing anomalous amounts of a selected portion or all of the organic matter also contain a lesser proportion of carbon having an atomic weight of 14 to carbon having an atomic weight of 12 than those earth samples not containing anomalous amounts of the same selected portion or all of the organic matter.

Where earth samples contain an anomalous amount of a selected portion or all of the organic matter, i.e., a larger amount of a selected portion or all of the organic matter than the background amounts, and also contain an anomalous proportion of carbon having an atomic weight of 14 to carbon having an atomic weight of 12, i.e., a lesser proportion of carbon having an atomic weight of 14 to carbon having an atomic weight of 12 than the background proportion, they contain organic matter derived from a subterranean petroleum reservoir and the presence of a subterranean petroleum reservoir is indicated.

On the other hand, where earth samples contain an anomalous amount of a selected portion or all of the organic matter but do not contain a lesser proportion of carbon having an atomic weight of 14 to carbon having an atomic weight of 12 than the background proportion, they do not contain organic matter derived from a subterranean petroleum reservoir, and the presence of a subterranean petroleum reservoir is not indicated.

A technique developed by *D.C. Bond; U.S. Patent 3,033,287; May 8, 1962; assigned to The Pure Oil Company* involves recovering gas samples from cores or other earth samples or from bodies of water under which petroleum reservoirs are suspected to lie and analyzing for the presence of hydrocarbon gases such as methane, and in particular, for the ratio of methane containing the ^{12}C isotope to methane containing the ^{13}C isotope.

Figure 7 illustrates the application of this technique in geologic surveys including stratigraphic correlations and geochemical prospecting. Here is shown a longitudinal cross section of a geological structure favorable to the accumulation of petroleum, illustrated in a simple asymmetric anticlinal fold. A plurality of boreholes **10** which penetrate the various strata are drilled to a depth of 10,000 feet. For each well drilled, an isotopic log is prepared employing a $^{12}C/^{13}C$ ratio for the methane constituent of the gas recovered from the drilling mud as the correlation indicator. The various formations penetrated are characterized by the isotopic ratios noted in the well logs obtained for each well.

It is seen that for the persistent strata shown, each has a characteristic isotopic ratio which serves as horizon marker for correlation purposes, thereby facilitating the developing of structural and stratigraphic relationships for the area under consideration. Although small variations in the magnitude of 0.1 unit are obtained, the use of sensitive apparatus such as that described above permits distinctions to be made for the various strata traversed. From wells surveyed in this manner, the declination and course of formations can be determined and valuable maps and records thereby derived. Also illustrated is the more definitive change in isotopic ratio which is encountered at the diffusion front of methane which is migrating to the surface.

Figure 7: Diagram Showing Application of Pure Oil Co. Carbon Isotopic Analysis to Petroleum Detection

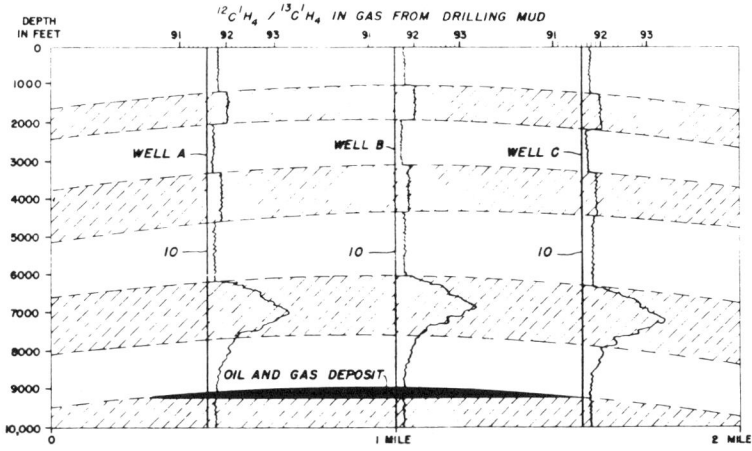

Source: U.S. Patent 3,033,287

The example illustrates the use of $^{12}C^1H_4/^{13}C^1H_4$ ratio in gas from drilling mud as an aid in correlation of strata. In the interval from 4,000 ft to 7,000 ft, there are mixed sands and shales which are difficult to correlate by the usual methods. From the distinctive shape of the ratio log (ratio vs depth) for the three wells, it is possible to correlate strata and reveal the presence of an anticline that contains a petroleum deposit. In the example shown, the diffusion front has not reached the surface, so this case cannot be used to demonstrate the application of the ratio method in exploration by analysis of surface samples of soil-gas. The enrichment with respect to $^{12}C^1H_4$ at a certain depth (e.g., 5,000 ft) illustrates the use of the ratio method in detecting a deeper petroleum deposit.

A method developed by *H.A. Slack; U.S. Patent 3,033,654; May 8, 1962; assigned to The Pure Oil Company* is a geochemical prospecting method for the exploration of oil and/or gas accumulations in porous beds overlaid by an impervious stratum or stratum less permeable than the formation in which the petroleum hydrocarbons are contained. This method employs isotopic anomalies as indexing criteria for determining the accumulation of petroleum hydrocarbons in structural stratigraphic traps which are below shaly or other impermeable or slightly permeable deposits which will prevent the substantially vertical passage of fluid emanations from the subterranean traps.

Figure 8 illustrates the use of this method. The figure shows a cross-sectional view of a geological structure containing an accumulation of oil in a rock reservoir 10 which consists of sandstone. Overlying reservoir 10 are several permeable limestone and siliceous formations 13 which are capped by impermeable shaly formation 14. The fluid emanations 15 from reservoir 10 diffuse upwardly into formation 13. Because their upward travel is blocked they are laterally deflected through formation 13 in which the abovedescribed fractionation continues to be effected as the fluid moves radially.

To carry out the process, samples of gas are collected from the selected rock stratum as each of a plurality of exploratory, or "wild-cat," wells are drilled into the formation subcontiguous with the impermeable stratum. The selected indexing criterion, e.g., methane, contained in each of these gas samples is analyzed. If methane is selected, the ratio of heavy methane ($^{13}C^1H_4$) to ordinary methane ($^{12}C^1H_4$) is determined. A significant lateral variation in this ratio indicates that the gas has been fractionated by diffusion from a parent accumulation. Determination of the gradient of the surface formed by contouring these ratios indicates the direction from which they have diffused, and hence the direction of the accumulation.

Figure 8: Schematic of Pure Oil Co. Method for Hydrocarbon Deposit Location

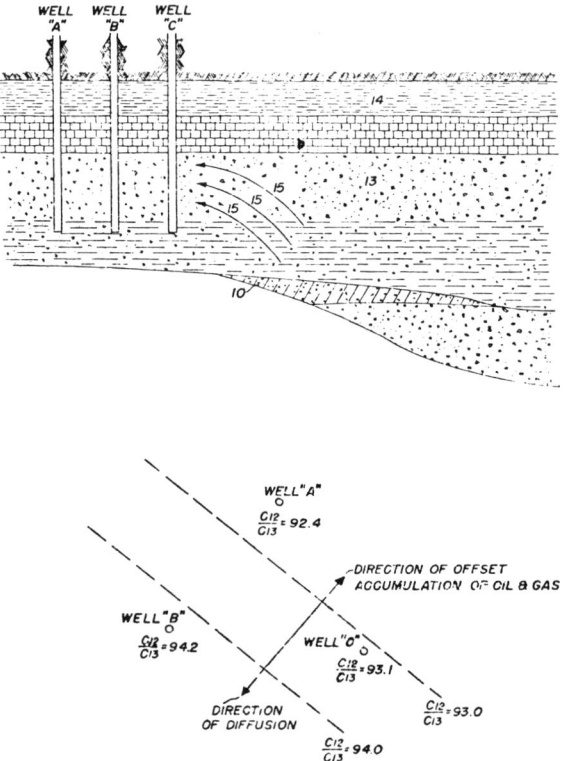

Source: U.S. Patent 3,033,654

A method developed by *E.W. Biederman, Jr. and B. Heinze; U.S. Patent 3,149,068; September 15, 1964; assigned to Cities Service Research and Development Company* is one in which the presence of petroleum constituents in earth samples may be determined with a high degree of accuracy by first applying chromatographic separation techniques to separate from the earth sample those constituents which are not mobile in polar chromatographic carrier solvents but which are mobile in nonpolar chromatographic carrier solvents. The fluorescence of these constituents is then checked to determine the presence of fluorescent petroleum constituents in the earth sample.

This process is capable of distinguishing between those fluorescent constituents of earth samples which indicate the presence of petroleum and those fluorescent constituents which may or may not be present in petroleum but which do not necessarily indicate the presence of petroleum.

A method developed by *W.S. Fallgatter and B. Heinze; U.S. Patent 3,254,959; June 7, 1966; assigned to Cities Service Oil Company* is a paper strip chromatogram technique in which the evaporation of a greater amount of solvent extract on the chromatographic paper enables more hydrocarbons to be placed on the paper than is feasible employing the two-dimensional technique. Preferably, a strip of chromatographic paper is immersed in a bottle containing the soil sample and solvent, or a measured volume of solvent extract, so that the solvent is carried by capillarity up the strip and past the mouth of the bottle. The solvent evaporates as it is carried past the mouth of the bottle leaving a concentrated band of soil extract on the paper.

Alternatively, the band may be placed on the paper by slowly dropping the solvent extract on the paper at a rate which allows the solvent to evaporate before spreading over too much area. The evaporation of the solvent as it passes the mouth of a bottle is preferred since this method gives unattended, uniform evaporation and smooth deposition of extract. The band of soil extract may be spread out by means of various chromatographic carrier solvents so as to separate those constituents of the earth sample which are not mobile in polar chromatographic carrier solvents but which are mobile in nonpolar chromatographic carrier solvents. The fluorescence of these constituents is then checked to determine the presence of fluorescent petroleum constituents in the earth sample.

The greater sensitivity of this technique enables analysis to be made of surface soil samples which contain only trace amounts of extractables for which significant results are difficult to obtain using the two-dimensional chromatographic technique which was described above in U.S. Patent 3,149,068.

A technique developed by *B. Heinze; U.S. Patent 3,300,641; January 24, 1967; assigned to Cities Service Oil Company* is one whereby the presence of an underground petroleum reservoir is determined by anomalous variations in the ratio of fluorescence of nonpolar soluble constituents to that of polar soluble constituents.

It has been found that the distinction between reservoir or crudelike hydrocarbons and organic constituents extracted from nonreservoir shales, is maximized by using a weak nonpolar solvent such as heptane or hexane and a polar solvent such as ethylene glycol.

This technique has been used to investigate the San Andres formation as it occurs in the Northern Permian basin, in the State of Texas. Earth samples were extracted with toluene as the primary solvent. Approximately 0.0001 ml of rock extract was added to a vial containing 1.75 ml ethylene glycol, 1.75 ml heptane, 0.25 ml isopropanol and 0.25 ml toluene. The more polar compounds were retained in the glycol phase that sank to the bottom of the vial, and the reservoir hydrocarbons were retained in the heptane phase which floated on top. The total fluorescence of each phase was then measured under high-intensity ultraviolet light (3660 A).

It was observed that the ratio of the fluorescence of the heptane fraction to the fluorescence of the glycol fraction increased as the known or measurable amount of oil contained in the rock samples increased. It was observed that only about 4 to 5% of the total fluorescence was retained in the ethylene glycol fraction when reservoir rock was analyzed. As the ratio of the fluorescence of the nonpolar to the polar fraction increases, the area is considered more favorable for prospecting. The rapidity with which a large number of earth samples may be accurately analyzed at the exploration site, in accordance with this method, represents a significant advance in the ability of the art to quickly and efficiently evaluate a given exploration area.

Prior art methods of geochemical prospecting for hydrocarbon may involve obtaining soil gas samples from the soil over the area of interest and analyzing these samples for hydrocarbon content. Often a plot of the results of such a survey indicates that an oil-bearing formation lies under at least a portion of the area. The question then is one of how deep below the earth's surface is the oil-bearing formation. An answer to this question is desirable in order to decide how deep to drill in order to recover the oil. In many areas it will be known that several formations underlying the area are not potentially oil-bearing formations. The question is usually which of these is the one which is actually oil-bearing.

A technique developed by *R.R. Thompson; U.S. Patent 3,302,706; February 7, 1967; assigned to Pan American Petroleum Corporation* is simply to determine the ratio of two particular hydrocarbons in soil gases obtained over fields known to be producing from the various potential oil-bearing formations underlying the survey area. The ratio of the same two hydrocarbons is then measured in soil gases obtained over the survey area. A well is then drilled to a formation producing the soil gas hydrocarbon ratio like that over the survey area.

A technique developed by *J.B. Davis; U.S. Patent 3,307,912; March 7, 1967; assigned to Mobil Oil Corporation* is one in which a gas probe first is inserted into an undisturbed portion of the ground. Gas from the adjacent soil is withdrawn through the probe and passed to a chromatographic column which is calibrated with respect to the emergence times of the hydrocarbons of interest. At least a portion of the effluent from the chromatographic column then is burned in a hydrogen flame to produce ionized gases and a signal is generated which is representative of the amount of ionized gases produced by this reaction. This signal, which is indicative of the concentration of hydrocarbons in the soil gas, is recorded as a function of time.

A suitable form of apparatus for the conduct of such a technique is shown in Figure 9.

Figure 9: Mobil Oil Corporation Soil Gas Analysis Apparatus

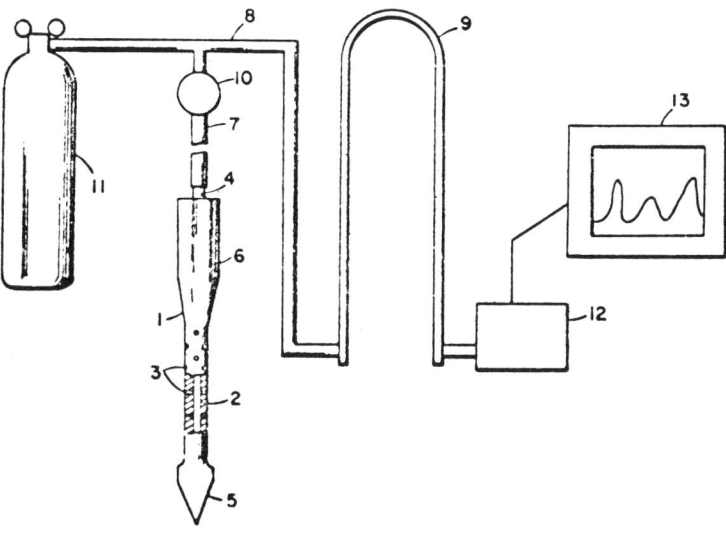

Source: U.S. Patent 3,307,912

With reference to the drawing, there is provided a gas-conducting probe **1** having an interior passageway **2** which includes a plurality of transverse ports or intakes **3** and an outlet **4**. The probe is provided with a conical cutting end **5** which at its maximum diameter is larger than the diameter of the intermediate portion of the probe in which the ports **3** are located. The opposite end portion **6** of the probe is flared outwardly to a diameter slightly greater than the diameter of the cutting end. When the probe is driven into the ground, this portion will provide a seal with the wall of the hole formed by the cutting end, thus providing a chamber in which suction may be applied in order to withdraw gas from the soil.

A fluid conduit formed by lines **7** and **8** connects the outlet end of passageway **2** to a chromatographic column **9**. At least a portion of line **7** is flexible or extensible in order that the probe may be projected with respect to the remaining elements of the system. A pump **10** is provided in line **7** in order to withdraw gases from the soil and a carrier gas source **11** is also connected to the conduit extending from the probe to the chromatographic column. The carrier gas source may be a conventional bottled gas container having suitable pressure-indicating and flow-regulating means.

Exemplary of a suitable chromatographic column is a molecular sieve column formed by a 6' ¼" diameter stainless steel column packed with cylindrical zeolite pellets. The pellets are $\frac{1}{16}$" in diameter and ¼" long (No. 10X, Linde Company). The column is calibrated with respect to the emergence times of the various hydrocarbon components of the soil gas as explained more fully below.

An ionization detector **12** which will detect small conconcentrations of methane as well as the heavier petroleum gases is connected to the discharge end of the chromatographic column. The detector is preferably of the hydrogen flame type in which an ionizing reaction is produced by a small hydrogen flame maintained within an ionization chamber. The ionization chamber is provided with electrodes which are exposed to the flame. The difference in potential across these electrodes will remain constant until such time as a hydrocarbon component of the soil gas is injected into the detector.

The hydrocarbon is burned as it is exposed to the flame and the products of combustion are ionized, thus resulting in an increase in the difference of potential across the electrodes. This signal, which is representative of the extent of ionization produced by the ionizing reaction and therefore the concentration of the hydrocarbon component, is recorded by a recorder **13** as a function of time. A suitable hydrogen flame ionization detector and recorder may be obtained as a unit (Aerograph Hy-Fi Model A-600-B, Wilkens Instrument & Research, Inc.).

Other ionization detectors such as the argon type may be used. However, a flame detector of the type described is preferred since it is highly sensitive to petroleum gases. For example, it will detect on the order of 5 to 10 ppb of such gases in 1 ml of air, and yet is insensitive to the presence of air and water vapor as well as most other inorganic compounds. This is particularly advantageous since the soil gas samples will usually contain air and water vapor, the detection of which is unnecessary and burdensome in geochemical prospecting operations. Thus, no precautions need be taken with regard to air and water vapor when using the flame ionization detector.

A method developed by *C.D. McAuliffe; U.S. Patent 3,345,137; October 3, 1967; assigned to Chevron Research Company* for geochemical prospecting as a method of finding petroleum comprises the steps including sampling and sample preservation, separating the hydrocarbons from the sample, analyzing the hydrocarbons for the presence of selected hydrocarbons containing from 2 through 7 carbon atoms, and correlating the results of the analysis with the surface location from which each sample was collected to provide an indication of the presence of a petroleum deposit.

Figure 10 shows an apparatus with which the hydrocarbons contained in a water sample may be separated from the sample and quantitatively analyzed. As is shown a flask **20** provided with a heating mantle **22** is connected by tubing **26** to a gas cylinder **23**. The system is flushed with gas, which might be, for example, nitrogen, from cylinder **23** to displace air which might contain hydrocarbon constituents. The sample is then placed in flask **20**, preferably by nitrogen gas displacement from the collection bottle. The temperature in flask **20** is raised to the boiling point of the liquid contained therein. A carrier gas, e.g., nitrogen, from cylinder **23** is flowed through a silica-gel, activated carbon filter **21** into flask **20** by means of conduit **26** through regulator **25** and bubbled through the sample. The sample is therefore subjected to evaporative conditions in the presence of the carrier gas. The filter **21** is used to remove possible traces of hydrocarbons from the nitrogen.

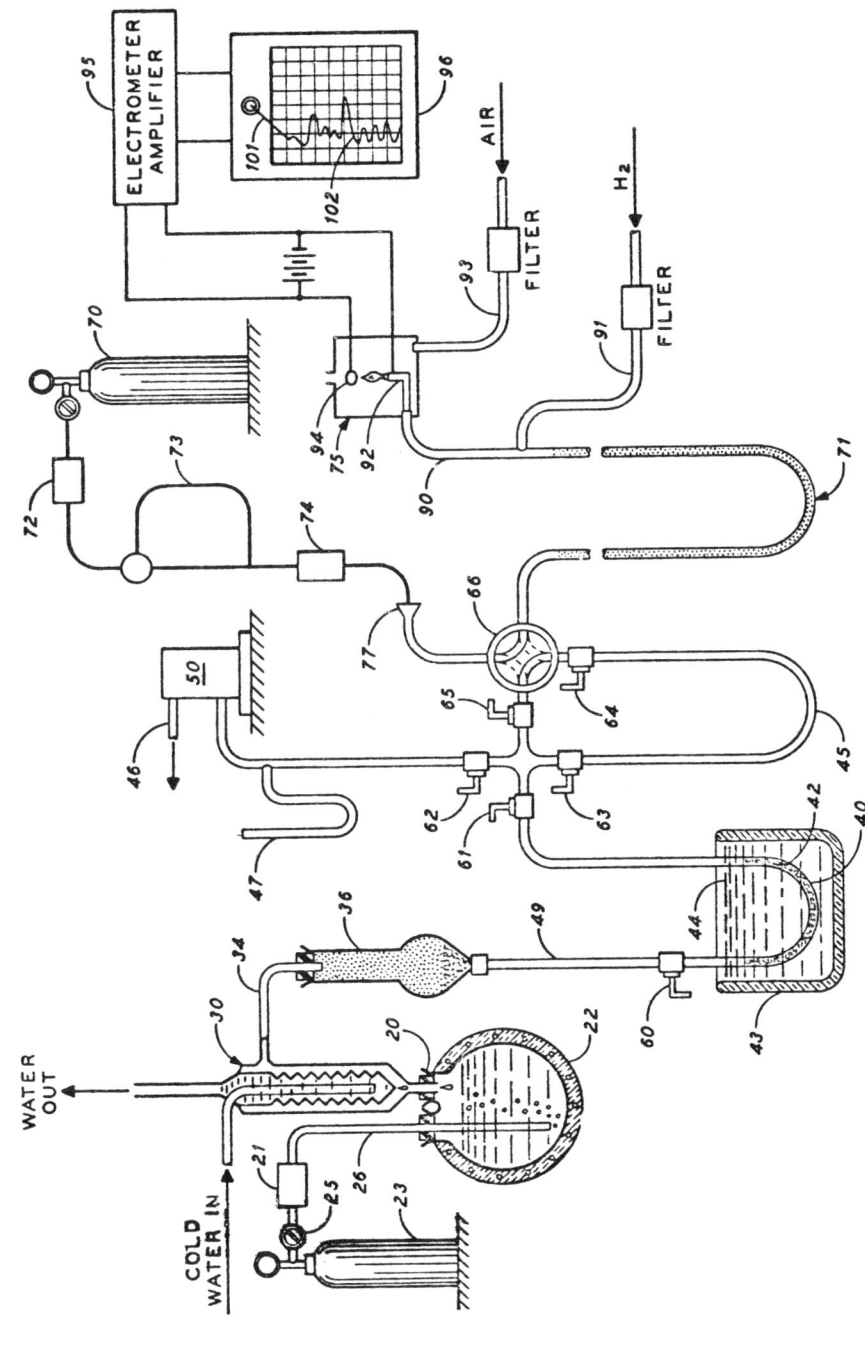

Figure 10: Chevron Research Company Apparatus for Hydrocarbon Analysis in Water Sample

Source: U.S. Patent 3,345,137

Other gases, e.g., helium or hydrogen, which may be used in place of nitrogen in the process should also be flowed through appropriate filters similar to filter **21** to remove trace hydrocarbons. The sample is subjected to evaporative conditions in the presence of carrier gas and, therefore, C_1 through C_7 hydrocarbons contained in the sample are volatilized and carried out of the sample in the effluent to a reflux condenser **30**. Water vapor is condensed in condenser **30** and falls back into flask **20**. The effluent which then flows out of condenser **30** through conduit **34** is made up primarily of carrier gas, gaseous hydrocarbons, and traces of water vapor, CO_2, and other volatile gases, e.g., oxygen, which may be present in the sample. The effluent enters drying tube **36** which is filled with a material such as ascarite which removes water vapor and CO_2 from the effluent.

Prior to analysis for individual hydrocarbons, the hydrocarbons are separated from the carrier gas. The preferred means of accomplishing the separation is described below. From drying tube **36** the effluent which now contains primarily carrier gas, oxygen, and hydrocarbons is flowed through conduit **49** to U-tube **40** which may be glass and which is packed with glass wool **42**. Conduit **49** has a restricted opening which regulates the gas flow through the system to 100 to 200 ml/min.

A preferred method of separating the hydrocarbons from the carrier gas is cooling the hydrocarbon-carrier gas effluent to a temperature at which the hydrocarbons become liquid or solid while the carrier gas and oxygen remain as gasses and can be drawn from the relatively immobile hydrocarbons. One means of accomplishing this is to submerge U-tube **40** in a container **43** of liquid nitrogen **44**. The temperature in U-tube **40** is desirably below $-190°C$. The temperature in container **43** is $-196°C$. Since the gaseous hydrocarbons of interest freeze or liquefy at temperatures above $-196°C$, they are solidified or liquefied in U-tube **40**, and the carrier gas, in this case nitrogen, passes through U-tube **40** to be exhausted from the system at vent **46**.

A vacuum is exerted on the system by means of a vacuum pump **50** to prevent oxygen gas from liquefying in U-tube **40**. The capillary **49** is of such size that the gas flow is 100 to 200 ml/min, and the vacuum measured at manometer **47** located between tube **49** and the vacuum pump **50** is 27 to 29" of mercury, a pressure adequate to prevent oxygen from liquefying in U-tube **40**. Under these conditions most of the methane (C_1) is lost. Ethane (C_2) and higher hydrocarbons, however, are completely held in U-tube **40**.

Carrier gas is flowed through the sample in flask **20** until substantially all the hydrocarbons having carbon numbers from 1 through 7, are removed from the liquid. Generally, substantially all these hydrocarbons will be removed from the sample in flask **20** in about 10 to 15 min after the sample reaches its boiling temperature.

After substantially all the hydrocarbons have been liquefied or solidified in U-tube **40** as described above, valve **60** is closed to stop the flow of carrier gas through U-tube **40**. The vacuum on U-tube **40** is increased to less than 1 mm of mercury with valves **61, 62**, and **63** open and valves **64** and **65** closed. Valve **66**, which is a two-way valve and represented schematically, is adjusted to a position to close the loop formed by the tubing connecting valves **63, 64** and **65**.

When two-way valve 66 is in this position, gas from cylinder 70 passes directly to chromatographic column 71 through pressure regulator 72, flow regulator 73, flow meter 74, tubing 77, and valve 66. By maintaining U-tube 40 under vacuum, the methane content of the hydrocarbons frozen therein is almost completely lost. However, ethane and higher hydrocarbons are retained in U-tube 40. Valve 62 is now closed. The liquid nitrogen container 43 is removed from U-tube 40 and placed in similar position to submerge U-tube 45. Valves 61 and 63 are open, and the hydrocarbons of interest gasify as the temperature in U-tube 40 is increased by placing boiling water around U-tube 40.

The hydrocarbons are moved from U-tube 40 to U-tube 45 where they are again solidified or liquefied by the low temperature therein. After a determinable time, approximately 10 min, valves 61 and 63 are closed, and the hydrocarbons are trapped in U-tube 45 ready to be flowed into the detecting apparatus for analysis of the individual hydrocarbon fractions.

To flow the hydrocarbons from U-tube 45 into the detection apparatus, valves 63, 64 and 65 are opened and valve 66 turned to permit the gas chromatograph carrier gas contained in cylinder 70 to flow through U-tube 45. The hydrocarbons contained in U-tube 45 are flash volatilized by removing the liquid nitrogen from around U-tube 45 and replacing it immediately with boiling water. The volatilized hydrocarbons are swept by the carrier gas into the chromatographic column 71. The gas chromatograph carrier gas from cylinder 70 may be, e.g., helium.

The detection apparatus for determining the concentration of the individual hydrocarbons is preferably a gas chromatograph. The gas chromatograph consists of 2 interrelated but essentially independent systems, a carrier-gas and sample system, and an electrical sensing system. It is convenient to first describe each of these systems independently and then describe the consequences of flowing the hydrocarbons from the sample into the chromatograph assembly.

The gas system begins with a cylinder of compressed gas 70. This gas is usually helium, nitrogen or air, but may be any gas appropriately chosen to be used with the detector system. The pressure in cylinder 70 is reduced to an appropriate working pressure by regulator 72. The flow rate of the carrier gas is then controlled by the flow regulator 73, and the rate of flow measured by flow meter 74. The constant flow of carrier gas leaving regulator 73 passes sequentially through the following parts of the chromatograph: flow meter 74, two-way valve 66, directly to column 71 or (depending upon the adjustment of valve 66) through valves 65 and 63, U-tube 45, and valve 64 to chromatographic column 71. The abovementioned valves and associated tubing are maintained at 100°C when the hydrocarbons are being flowed therethrough to prevent condensation of hydrocarbons.

The sensing elements may take many forms. In gas chromatographs utilizing helium or nitrogen as a carrier gas, the sensing elements are characteristically thermal-conductivity detectors. When the carrier gas is air, the sensing elements may be a catalytic combustion detector. A hydrogen flame ionization detector is illustrated and will be described. The hydrogen flame ionization principle provides a very sensitive method for measuring hydrocarbons. The detector does not measure inorganic compounds and fixed gases.

The hydrogen flame detector, represented generally by **75**, ionizes the sample by combustion in flame. The helium effluent carrying the individual hydrocarbons which have been separated in chromatographic column **71** is mixed with hydrogen gas from conduit **91** in conduit **90** and flamed in burner **92**. The mixture is burned in air entering from conduit **93**, and the resulting ions are collected by a small electrode **94** above the flame. Detection is based on the measurement of the ion current and is proportional to the number of carbon atoms present in the flame. The ion current is amplified by amplifier **95**, and the amplifier output drives a strip chart recording potentiometer **96**.

In operation the vaporized hydrocarbons are carried from U-tube **45** in the helium stream of the gas chromatograph into column **71**. In column **71** the various components of this hydrocarbon sample interact with the stationary liquid constituting the column packing of column **71**. The separation of the hydrocarbon in the sample is brought about by interaction with the stationary liquid packing in column **71** according to the principles well known in gas chromatography art. A preferred liquid stationary phase for column **71** is a silicone gum-rubber (SE-30, General Electric). The effluent from column **71** then consists of carrier gas intermittently contaminated with a single separated hydrocarbon of the sample which was placed in the apparatus.

As each of these hydrocarbons pass through detector **75**, it brings about a production of ions in the oxygen-hydrogen flame of the detector, resulting in an ion current described above. The ion current from the detector is fed into amplifier **95** and strip chart potentiometer **96** where a pen **101** traces a curve **102**. The amplitudes of individual excursions in the curve correspond to the amount of various hydrocarbons present in the original sample. If the chart of potentiometer **96** is moving in synchronism with the flow of carrier gas in the apparatus, then the position of each peak on the chart can be correlated with the specific hydrocarbon causing the peak according to the principles well known in the art. The concentration of the individual hydrocarbons is determined by measuring the area under the curve by any of several well-known methods.

Although the abovedescribed method of hydrocarbon analysis is preferred, other analytical methods can be used. For example, the separated hydrocarbons held in U-tube **45** may be introduced into an analytical mass spectrometer where the individual hydrocarbons can thereby be determined. The mass spectrometer technique permits an alternate method of analysis of C_2 to C_7 hydrocarbons to the chromatographic separation and hydrogen flame ionization detection.

A technique developed by *E.E. Bray; U.S. Patent 3,496,350; February 17, 1970; assigned to Mobil Oil Corporation* is a geochemical exploration method in which a sample obtained from an organic matter-containing formation in the earth's crust is analyzed for the relative concentration of hydrogen atoms on aromatic nuclei having one or more hydrogen atoms but not more than two adjacent hydrogen atoms attached thereto and the relative concentration of hydrogen atoms on aromatic nuclei having four or five adjacent hydrogen atoms attached thereto. The ratio of the relative concentrations of these two classes of hydrogen atoms is indicative of the likelihood of the formation being a petroleum source rock.

There are three types of carbon in organic matter. Total carbon includes all of the organic, i.e., noncarbonate, carbon in a sample. Live carbon is that portion of the total carbon which, on pyrolysis at 500°C in the laboratory or during burial to greater depths and temperatures in the subsurface, yields appreciable quantities of hydrocarbons and other volatile organic matter. Dead carbon is the remaining portion of total carbon, i.e., that which, on heating, yields essentially no hydrocarbons or other volatile organic matter.

In earth formations, mixtures of both dead and live carbon commonly occur. It is the live carbon of fine-grained rocks which represents the source of the carbon of petroleum molecules. After exposure to laboratory temperatures of 500°C or to subsurface temperatures of 250° to 300°C, the live carbon is destroyed thermally, and the remaining carbon is dead carbon.

A technique developed by *R.L. Heacock and A. Hood; U.S. Patent 3,508,877; April 28, 1970; assigned to Shell Oil Company* is one for measuring the live carbon content of an organic sample by heating the sample to pyrolyzing temperature so that vapors are given off. The vapors are condensed, the fluorescence thereof is measured and the live carbon content of the sample material is determined by correlating the measured fluorescence with the fluorescence from a material of known carbon content.

A scheme developed by *J.B. Allison, K.K. Bissada and C.M. Peterson; U.S. Patent 3,834,122; September 10, 1974; assigned to Texaco, Inc.* is one for thermal desorption of gas, including light hydrocarbons, from an earthen sample. Carrier gas mingles with the desorbed gas to carry it to a dehydrating unit. After dehydration, the gas mixture is provided to a hydrocarbon trap which substantially separates the light hydrocarbons from the gas mixture. The hydrocarbon trap is then removed for analysis of the trapped light hydrocarbons.

Correlation with Subsurface Rock Types

It is generally believed that oil and gas are associated much more often with marine or saltwater sediments than with nonmarine or freshwater sediments. Accordingly, a convenient method of classifying samples of sedimentary formations as to whether they are of marine or nonmarine origin is of distinct value in evaluating the likelihood of occurrence of petroleum deposits. Such classification is often done by study of the fossil materials contained in the recovered samples, but sometimes significant fossils are missing from the samples.

It has been observed that changes in the salinity of seawater are most marked close to shore or in shallow waters. Where streams and rivers enter the sea from the adjacent land areas, the salinity varies from nearly zero to the relatively high values characteristic of the open ocean. Large variations of water salinity are often found around reefs and sandbars that are close to shore. These bars and reefs may impede the mixing of saltwater and entering freshwater; or they may create lagoons and like areas of relatively quiet seawater, from which evaporation takes place and increases the salinity over that of the normal seawater.

After porous seashore sands, sandbars, limestone reefs, and the like have been buried by impermeable sediments, their importance as traps for the accumulation of valuable deposits of oil and gas can hardly be overestimated. Nevertheless, as these porous bodies are seldom associated with marked structural deformations of the surrounding strata, they are exceedingly difficult to locate by such surface geophysical methods as seismic geophysical surveying.

A method developed by *A.F. Frederickson, J. Hower, Jr. and R.C. Reynolds, Jr.; U.S. Patent 3,022,140; February 20, 1962; assigned to Pan American Petroleum Corporation* is a method of prospecting which involves determining the salinity of the water in which sedimentary formations have been laid down. This may be more briefly termed depositional water salinity, or simply sedimentation salinity. There is a direct relationship between depositional water salinity and the concentration of trace-element boron in the clay mineral illite.

More specifically the method comprises the steps of systematically collecting illite-containing sedimentary earth samples from spaced points of a prospecting area, quantitatively analyzing each of the samples for the ratio of boron to illite in the clay-mineral fraction thereof, and correlating the measured values of the ratio with the corresponding sample locations in the area, whereby the variations in depositional water salinity in the area, proportional to the variations in the ratio, may be ascertained.

A scheme developed by *G.M. Friedman; U.S. Patent 3,343,917; September 26, 1967; assigned to Pan American Petroleum Corporation* involves determining paleoenvironmental information for use in prospecting for subterranean deposits of petroleum. Calcareous sedimentary rock samples in the area to be explored are collected. The bulk rock samples are examined or analyzed to determine carbonate mineralogy. The rock is next microscopically examined for identification of fossil fragments to determine the existence of those composed of low-magnesium calcite and known to have been originally deposited as such low-magnesium calcite which is a stable-phase carbonate.

It is next determined what trace elements, i.e., iron, manganese and barium, are present in such low-magnesium calcite. The quantitative evaluation of such trace elements indicates the nature of the environment in which the original fossil was laid down. This is useful for further exploration work by geologists in their various studies in seeking indications of the presence of petroleum deposits.

A method developed by *R.L. Borst; U.S. Patent 3,596,089; July 27, 1971; assigned to Phillips Petroleum Company* is one in which subterranean sandy and silty sedimentary formations are examined to determine that mineral crystals, known to grow most favorably in alkaline, saline conditions that typify a marine depositional environment, are authigenic, i.e., formed in place, in the sedimentary formation. This indication of the nature of the depositional environment is useful for further exploration by geologists in seeking the presence of petroleum deposits.

A technique developed by *H.L. Overton; U.S. Patent 3,711,765; January 16, 1973* is a method of locating anomalous zones of chemical activity in a well bore by measuring one or more cationic potentials and the redox (reduction-oxidation) potential of shale cuttings obtained from the well bore at different elevations during the drilling thereof, and graphically representing the different values to obtain comparisons which are indicative of the location of petroleum formations in the well bore.

It is generally known that highly reducing conditions in an organic rich sediment are quite favorable to the genesis of large amounts of crude oil. It is not unreasonable to postulate that the rate at which reducing conditions are established is one of the determining factors as to whether the organic matter deposited in a sediment has on the one hand yielded petroleum gas and/or oil in sufficient quantity to migrate and accumulate in a reservoir, or on the other hand, moved toward the family of nonmigratable bitumens such as coals and oil shales. Under ordinary circumstances this reducing condition remains a characteristic of the source rock and indeed of oil produced from it.

Direct determination of the reducing character of a rock in a bore hole is complicated by a number of factors, e.g., presence of the agents H_2S, HS^-, and $S^=$. These agents tend to poison metallic electrodes which may be sought to be employed. Further, for complex organic systems, there appears to be no simple relation between redox potential and pH. Further, in fine-grained rocks appreciable streaming potentials may be developed.

A process developed by *J.G. Erdman; U.S. Patent 3,719,453; March 6, 1973; assigned to Phillips Petroleum Company* provides a method for determining whether a formation may be a source or petroleum-containing formation by introducing a salt such as a transition metal salt, e.g., vanadium salt, in solution in the +5 oxidation state into the formation and then later, withdrawing a sample of the liquid to determine whether it has been reduced to the +4 oxidation state in the case of vanadium. In another variation, such a determination is made by an electron spin resonance method.

A method developed by *R.R. Thompson, R.W. Duschatko and A.J. Nash; U.S. Patent 3,801,281; April 2, 1974; assigned to Amoco Production Company* is a geochemical prospecting method in which samples are analyzed for a critical carbonate mineral which contains most of the hydrocarbons in the sample. The mineral is almost always dolomite, calcium-magnesium carbonate, but may be other carbonate minerals, such as iron carbonate, or even calcium carbonate.

The critical carbonate should be separated from any other carbonates which are present in significant amounts. Gases are then released from the critical carbonate. Released gases are analyzed for hydrocarbon content. Preferably, gases are released by acid treatment. These gases are then analyzed for carbon dioxide content as a measure of the carbonate material. The ratio of hydrocarbons per unit of critical mineral is then plotted to form a geochemical prospecting map.

Detection of Metals and Metallic Compounds as Indicators of Hydrocarbon Deposits

It has been found that microorganisms present in the earth's surface over an oil deposit tend to gather over the center of the deposit, where the escape of gases and liquids is strongest. In addition, it is noted that the zone of oxidizing conditions in the earth's surface will also tend to deepen over the center of an oil deposit possibly due to such microorganism concentration.

Since the microorganisms live on and destroy or alter the hydrocarbon constituents, and since the stronger oxidizing conditions will tend to oxidize or alter the hydrocarbon constituents, the methods utilizing the detection of hydrocarbon constituents, and the methods utilizing the detection of trace element reduction due to the escape of reducing hydrocarbon gases will detect anomalies only around the fringes of the oil deposit while detecting little or no anomaly towards the center unless the sampling zone is extended downwardly beneath the zone of oxidation. However, other constituents such as organo trace constituents will not be so affected and may be detected continuously across the entire extent of the oil deposit.

Among such detectable trace constituents are organo-metallic compounds including for example the volatile metallic porphyrins which are carried upwardly with the liquid and gaseous oil escaping from the oil deposit, and on reaching oxidation zone at the surface, or near surface, become oxidized and can be detected as organo-metallic compounds present in the soil. Such organometallic compounds and their derivatives and oxidation products can also be detected in other media, such as stream, lake or marine sediments, surface and groundwater, and in rock formations both on the surface and in subsurface areas.

Thus, a technique developed by *D.R. Clews; U.S. Patent 3,285,698; Nov. 15, 1966; assigned to Barringer Research Ltd., Canada* for detecting the presence of oil, gas and bituminous deposits by sampling from the surface of the earth consists of obtaining a series of earth surface samples from spaced locations; treating the samples to separate therefrom any organometallic compounds of the deposits; determining the proportions of the compounds present in the sample; and plotting the proportions in relation to their locations.

A technique developed by *G.K. Billings; U.S. Patent 3,428,431; February 18, 1969; assigned to Sinclair Research, Inc.* provides a method of exploration for petroleum deposits which comprises collecting samples of formation waters from wells and determining the concentration of at least one transition metal in the waters as a component indicative of the presence of petroleum deposits.

Detection of Hydrocarbon Deposits Under Water

Various methods are practiced in the exploration for petroleum deposits in areas covered by water. Seismic methods have been used for locating and outlining subsurface structures which exhibit characteristics similar to previously discovered petroleum-containing structures. Additionally, measurements of gravimetric, magnetic, and radioactive properties have been made which may be indicative of petroleum deposits.

The methods hereinabove discussed are all indirect indicators of petroleum deposits. A more direct method of determining the location of a petroleum deposit involves the detection and/or measurement of elements or compounds resulting from one or more constituents of a petroleum deposit seeping to the surface.

An apparatus developed by *C.M. Mason; U.S. Patent 3,681,028; August 1, 1972; assigned to Sun Oil Company* is a hydrocarbon detection system for use in an exploration survey over water-covered areas. Samples of water are taken, in which hydrocarbon gases may appear in a dissolved state. Removal of dissolved gas from the sample is accomplished by admitting the sample to a chamber, passing it through a fog nozzle, and applying a vacuum to remove the gas from the sample, thus creating fog, liquid, and gas zones in the chamber. Gas is removed from the top and liquid is discharged from the bottom of the chamber by pumps. Valves are located in the liquid entrance line and the degasified liquid exit line, and are controlled by liquid level responsive devices. The removed gas is subjected to analysis.

A survey or map of the amount with location of a mobile reservoir fluid that might be contained within a body of water is not significant unless the bottom sediments have allowed the reservoir fluid to flow through them and enter the water. In a region containing subterranean structures having geophysical properties that may be indicative of a subterranean reservoir, the overlying water may be substantially devoid of any mobile reservoir fluid. Such a situation is common when the bottom sediments are soft and relatively impermeable, such as the clays or muds that cover much of the Gulf of Mexico.

An absence of hydrocarbons in the water above a subterranean reservoir structure that might contain hydrocarbons provides substantially no definite information. There may be no hydrocarbons present in the reservoir structure, or none that escape from the reservoir, or none that provide detectable amounts of seepage through the relatively impermeable and hydrocarbon-adsorptive layer of mud that underlies the water.

A process developed by *E.E. Daigle, R.R. Luke, J.B. Turner and H.L. Wise; U.S. Patent 3,714,811; February 6, 1973; assigned to Shell Oil Company* overcomes such problems by jetting water into soft bottom sediments, forming a slurry of the sediments in water, extracting gas from the slurry and measuring the concentration with location of a mobile reservoir fluid that was entrapped within the bottom sediments.

The construction and arrangement of the apparatus for the conduct of such a process is shown in Figure 11.

A vessel **1** is shown on a body of water **2**, above a layer of unconsolidated sedimentary earth formation or mud **3**. A mud-sampling device **4** forms a slurry of sediment in water that is pumped to a surface location through conduit **5**. The mud-sampling device contains a pumping means **6** having an intake port **7**, and a pump-discharge conduit **8**. The pump-discharge conduit is connected to jet nozzles **9** which are arranged to jet water into the mud to form a stream of slurry which flows into conduit **5** as shown by the arrows. The mud-sampling device is mounted on a base structure **10**.

As the mud-sampling device advances to increasing depths within the mud, a measuring means, such as a rack and pinion arrangement **11**, responds to the extent of such advance and telemeters an electrical signal relating to the depth of penetration by means of electrical conduit **14**.

In a near-surface location, the slurry of sedimentary solids in water is pumped through gas extractor **12** and discharged, as indicated at **12a**. The extracted gas is passed through a gas analyzer **13**, and discharged, as indicated at **13a**. An electrical signal from the gas analyzer is conveyed to recorder **15** by electrical conduit **16**. The recorder may indicate, e.g., measurements made at each of a series of depths within the sediments **3** of the concentration of ethane and the concentration to total hydrocarbons. Such measurements can be utilized to plot the concentration with depth of such components of the sedimentary earth formation.

Figure 11: Shell Oil Company Device for Marine Mud Hydrocarbon Surveying

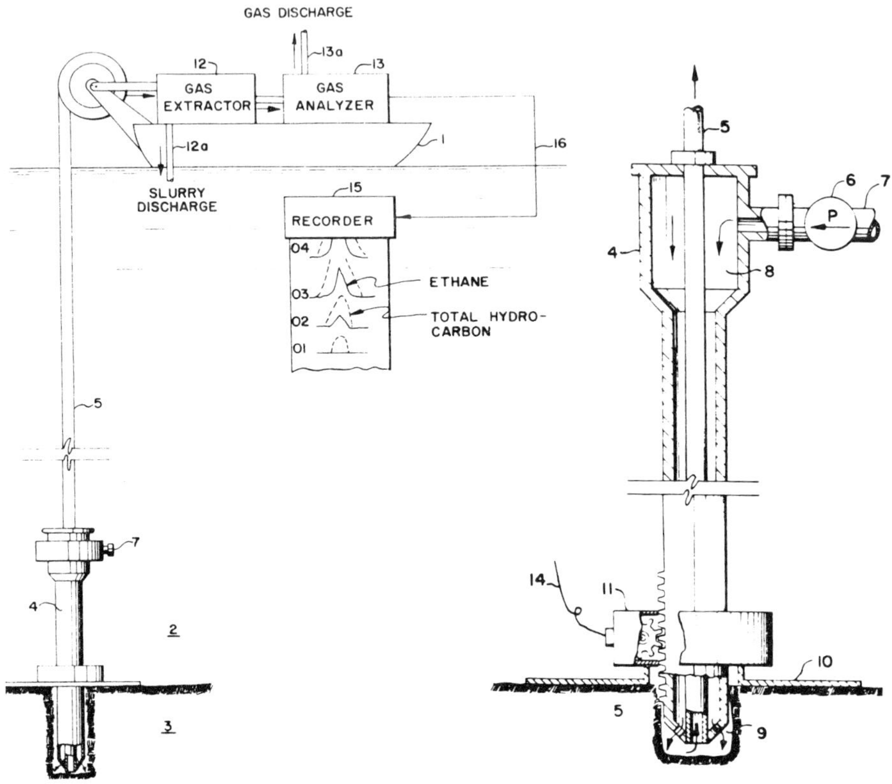

Source: U.S. Patent 3,714,811

A technique developed by *L. Horvitz; U.S. Patent 3,722,271; March 27, 1973;* is a method of geochemical prospecting in submerged areas comprising the taking of samples of the water at or close to the bottom at spaced locations in such an area. Samples are then analyzed to determine the concentration therein of leakage products indicative of the presence of subterraneous petroliferous deposits and the results of such analyses are correlated with sample location to provide information concerning such deposits.

The method also includes the taking of samples of the earth formation beneath the bottom at such sample locations from which the character of the formation and the concentration therein of products indicative of the presence of subterranean petroliferous deposits may be obtained.

A technique developed by *H.W. Alter; U.S. Patent 3,987,677; October 26, 1976; assigned to Terradex Corporation* is one in which a plurality of gas sample collecting containers are distributed in a predetermined pattern from a vehicle. The containers are cup-shaped but weighted so that the open end sinks

Geochemical Prospecting

first and engages the subaqueous earth layer. Gas migrating upwardly through the earth is caught by the inverted cup. The gas sample may be either collected in a separate container or exposed to a detector material located above the expected water level. The sample collecting containers may be retrieved by means of marker floats, magnetic grappling devices or self-actuated flotation devices.

Figure 12 shows the mode of placement of such devices as well as details of construction of such devices.

Figure 12: Terradex Corporation Apparatus for Underground Deposit Detection in Water-Covered Areas

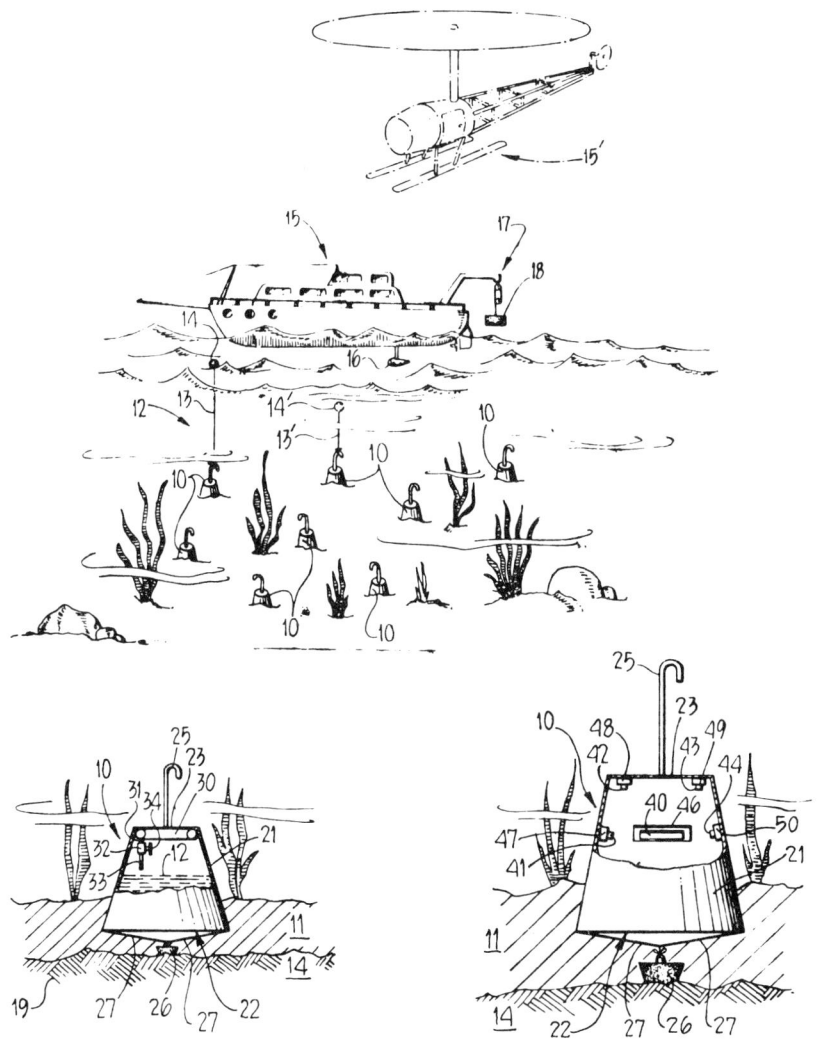

Source: U.S. Patent 3,987,677

As shown in Figure 12, a plurality of small gas sample containers **10** are situated in the sedimentary layer **11** underlying a body of water **12**. For illustrative purposes the physical size of the gas sample containers is greatly exaggerated. Attached to each of the containers by means of a flexible line **13** is a small float **14** such as a fishing bobber, which serves to demarcate the location of the associated container. The float may be attached to the associated gas sample container by a flexible line having sufficient length to permit the float to rest on the surface of the body of water, or a shorter line **13'** which causes the bobber **14'** to be held below the surface of the body of water by the superior weight of the associated container.

The detail at the lower left is a sectional view showing a first embodiment of an underwater sampling container configured as a total gas sample container in situ. The total gas sample container has a generally frustoconical shape with a sidewall portion **21** terminating in an open mouth **22**, and a closed end portion **23**. Attached to the closed end portion and extending externally thereof is a hooklike member **25** provided for handling purposes to be described and preferably fabricated from a magnetizable material. Attached to the open mouth end and depending downwardly therefrom is a weight **26** secured to the container at or adjacent the open mouth by means of a flexible attachment means, e.g., flexible lines **27**. The weight may be tapered as shown, or may be configured as a sphere, spheroid or other shapes. For small containers, common fishing sinkers may be used.

Secured to the interior of the total gas sample container and adjacent to the closed end portion is a partially evacuated toroidal tank **30** having an inlet **31** coupled to the outlet of a conventional metering valve **32** having an inlet **33** in communication with the interior volume of the container **10**, and a manually operable adjustment knob **34** which enables the metering valve to be opened and closed.

As shown, the mouth **22** of the container is embedded in a soft sedimentary layer to define an enclosed volume for containing any gases which migrate upwardly through the sedimentary layer from the earth's crust **19**. Trapped within the internal volume of container **10** is a quantity of water **12**. The upper level of water lies below the mouth of inlet **33** so that only gases are admitted into the evacuated interior of toroidal tank when the metering valve is open and the container is lodged in situ.

The physical dimensions of container **10** required to ensure that inlet **33** always lies above the level of the water is dependent upon the depth of sedimentary layer below the surface of the body of water. For example, if the sedimentary layer is at a depth of 30 ft below the surface, and the container has a height of 2 ft (measured between open mouth **22** and closed portion **23**) defining an internal volume of 2 ft^3 the mouth of inlet **33** should be arranged no more than 1 ft below closed end portion **23** to avoid contact with the water trapped in the interior of the container. In practice, the container is typically designed for use at a maximum depth and is preferably marked accordingly.

The detail at the lower right of the figure illustrates an underwater sampling container configured as a specific gas sample container. As shown, a plurality of specific gas detectors **40** through **44** are secured to the inner wall surfaces of cup **10** at a location above the level of the water at the working depth by means of suitable mounting members **46** through **50**. The choice of individual specific gas detectors **40** through **44** depends upon the specific gases to be detected over the collection period.

Containers **10** are deployed in the water from either a surface vessel **15** or a suitable aircraft, e.g., a helicopter **15'**. To deploy from a surface vessel, a quantity of containers, configured as either a total gas sample container or a specific gas sample container, or a mix of both types, is placed on board the vessel and transported to the survey site. As the vessel proceeds along the surface of the body of water, individual containers are placed overboard, preferably at regular intervals. As each container is released into the water, it descends under its own weight to the sedimentary layer **11**, where weight **26** and flexible cord **27** maintain the container in the inverted attitude illustrated so that the air trapped therein cannot be displaced by the water but merely compressed.

When the container reaches the sedimentary layer, it is partially embedded therein due to the momentum provided by the combination of its own mass and the mass of the weight. When planted in situ, a quantity of water is trapped within the internal volume of the container, as shown.

It should be noted that the combined mass of container and weight must exceed the weight of the water displaced by these elements and the buoyant force provided by bobber **14** in order to ensure that the container descends to the water-earth interface. In addition, the weight must be attached to the container in such a manner that the assembly is hydrodynamically stable so that the container maintains the inverted attitude during the downward descent. This condition may be assured by securing the weight directly to the rim of the open mouth **22** or adjacent thereto by a flexible support, such as cords.

Other equivalent attachment arrangements are a flexible net attached to the open mouth at the rim, elastic bands arranged in a substantially identical manner to lines **27**, springs secured to the rim of the open mouth at one end and to the weight at the other end.

Once all containers required for a given survey have been planted, their respective locations are determined by observing the bobbers from the vessel. If desired, the bobbers and container locations may be determined by activating a vessel-mounted conventional sonar transponder **16** as the vessel traverses the survey site. In deep-water locations where the use of bobbers is impractical, containers may be located using conventional sonar techniques. Alternatively the individual containers may be provided with a transponding device which is activated by a signal from the vessel.

After the predetermined time interval has elapsed, the containers are retrieved by traversing the survey area and withdrawing each container manually by bobber and line. For submerged bobbers, and in deep water locations not employing bobbers, the containers may be retrieved by operating a winch and boom assembly **17** to lower a conventional mechanical or magnetic grappling device **18** into the water. As the grappling device traverses the survey site, the individual containers are secured thereto by hooked members **25** and are raised to the surface of the water by operating the winch and boom assembly.

Analysis of Hydrocarbon Types in Water

A scheme developed by *N.D. Coggeshall and W.E. Hanson; U.S. Patent 2,767,320; October 16, 1956; assigned to Gulf Research & Development Company* involves geochemical prospecting for hidden hydrocarbon deposits by analyzing subsurface brines or other waters for the presence of dissolved hydrocarbon components, in particular aromatic hydrocarbons and specifically benzene.

One method of geochemical prospecting has been conducted by making systematic collection of soil samples over an area to be mapped. The soil samples have been tested for a variety of components indicative of the presence of hydrocarbon deposits. Methods of extracting the components from the soil samples have included the use of organic solvents or simply the application of heat to remove hydrocarbon gases trapped in the soil samples.

Among the components for which these samples have been tested are the lighter paraffinic hydrocarbons which are presumed to have migrated from deep subsurface hydrocarbon deposits. One of the difficulties encountered with this method of prospecting is that generally there is a wide variation of the sorptive and retentive capacities of the soil samples taken over a plurality of locations. This is due primarily to the differences in chemical and physical properties of the soil itself.

A pattern of such concentration values may thus not give a true indication of the proximity of hydrocarbon deposits. Another difficulty encountered is that some of the lower molecular weight paraffinic hydrocarbons may arise from the recent decay of organic or vegetable matter instead of from petroleum deposits.

In other known methods of geochemical prospecting, techniques have been used for detecting extremely small concentrations of crude hydrocarbons dispersed in drilling fluids. These methods have involved subjecting the fluid or cutting samples to ultraviolet light to produce fluorescence by means of which it is possible to detect extremely small concentrations of crude oil which would ordinarily be invisible to the unaided eye. One of the problems which has been encountered in this type of prospecting for crude oil deposits is that positive indications are obtained when the drilling fluid is contaminated with refined petroleum products, such as pipe-thread grease or lubricating oils.

These problems are overcome by a method of geochemical prospecting wherein information concerning the proximity of hydrocarbon deposits or source beds is derived by collecting aqueous subterranean samples from prospect areas and wildcat wells and extracting the hydrocarbon components dissolved in these samples with a suitable extracting agent and determining the concentration of these hydrocarbons by the use of electromagnetic radiation. The presence of these dissolved hydrocarbons is an indication that these subterranean waters have been in contact with or are near a petroliferous deposit.

Aromatic hydrocarbons are universal components of crude petroleum and furthermore this class of hydrocarbons is more soluble in water or brine than any other type of hydrocarbon which occurs in crude petroleum. Benzene and a considerable number of benzene homologues have been recognized in crude petroleum. The best established are benzene, toluene, the three xylenes, ethylbenzene, naphthalene and the two methylnaphthalenes.

Geochemical Prospecting

The actual percentages of these components depend on the geographical area where the petroleum is found, but they are always present. The aqueous solubility of these aromatic hydrocarbons found in petroleum decreases rapidly with an increase in molecular weight. For this reason, the lower molecular weight members of this series, such as benzene, are favored as criteria for the proximity of petroleum deposits. This is based on the observation that brines or other waters when brought into contact with petroleum will preferentially dissolve out a portion of the aromatics and particularly benzene and retain them in solution, although the waters may have been later physically separated from the crude petroleum.

A method developed by *R.L. Slobod, H.F. Dunlap and T.F. Moore; U.S. Patent 2,918,579; December 22, 1959; assigned to The Atlantic Refining Company* employs the dissolved hydrocarbon gas content of water as an indicator of the presence of hydrocarbon-bearing, subsurface formations by obtaining a sample of the water to be analyzed, passing such sample through a separator adapted to break the gas out of the solution and separate it from the water, and then analyzing the separated gas for a preselected component by nondispersive infrared apparatus.

An apparatus developed by *J.S. Bradley; U.S. Patent 3,455,144; July 15, 1969; assigned to Pan American Petroleum Corporation* is an improved marine seep equipment for detecting the presence of gas, especially methane, in seawater. A sample of seawater is continuously passed through a gas breakout portion of the system. This gas breakout portion includes an aspirator for creating a high vacuum in the flowing sample of seawater to cause dissolved gases in the seawater to break out in the form of small bubbles.

The water having the small bubbles is fed to a separator where the gas rises to the top and the degassed water flows out the bottom. The degassed water flows into a standpipe which has an outlet higher than the separator. The separator has a gas sample outlet which goes to a flame ionization detector. A bypass is connected from the gas sample outlet to a gooseneck stinger in the standpipe. The stinger gooseneck level in the standpipe controls the water level in the separator. The level may be adjusted by moving the gooseneck up or down.

J.B. Davis and H.F. Yarbrough; U.S. Patent 3,457,044; July 22, 1969; assigned to Mobil Oil Corporation have developed a geochemical exploration technique which involves the analysis of formation water for one or more saturated hydrocarbons having at least ten carbon atoms and preferably, ten to thirty carbon atoms. The presence of such hydrocarbons in the formation water is indicative of the propinquity of subterranean petroleum deposits.

A method developed by *G.W. Schmidt; U.S. Patent 3,524,346; August 18, 1970; assigned to Pan American Petroleum Corporation* is one in which the concentration of at least one aromatic hydrocarbon is determined in a sample of formation water from a reservoir formation in the earth. The measured value is compared with the so-called target value for this hydrocarbon. If they are nearly alike, the point from which the sample was obtained is close to a reservoir of crude oil. Greater differences between these two values represent greater distance to the crude oil accumulation.

The target value is determined by contacting the type of crude oil to be expected in the reservoir from which the water sample came with water solutions of salt and measuring the concentration of the aromatic hydrocarbon in the solutions. This permits determining the variation in concentration of this aromatic hydrocarbon as a function of the salinity of the dissolving water. The salinity of the sample of the water from the well is determined.

It is then possible to determine the target value, i.e., the value of concentration of this aromatic hydrocarbon, which would exist at the point of contact between crude oil and formation water of that salinity in the subsurface reservoir.

A device developed by *C.B. Craig; U.S. Patent 3,561,546; February 9, 1971; assigned to Leo Horvitz* provides an apparatus for underwater geochemical prospecting by taking samples of the bottom formation and water at or immediately above the bottom. The method comprises taking samples simultaneously of the water at the bottom and of the earth formation immediately below at spaced-apart locations for analysis to determine the concentration of significant hydrocarbon leakage products from subterranean petroleum deposits to be used in exploring for such deposits.

The sample-taking apparatus comprises a tubular body whose lower end is open and provided with means for penetrating the bottom formation and retaining a sample of the same in the body. The sample taker includes a piston movable upwardly from a lower position closing the lower end portion of the body to an upper position above when the body reaches a predetermined position at or close to the bottom during its downward travel to draw in a sample of water at or immediately above the bottom.

The apparatus is adapted to be suspended by an operating cable and means is provided for adjusting the piston-actuating means to allow predetermined setting of the apparatus to allow free fall of the body from a desired point of its downward travel before the body penetrates the bottom formation.

J.S. Bradley, W.H. Luehrmann and G.D. Roe; U.S. Patent 3,571,591; March 23, 1971; assigned to The Atlantic Richfield Company offer a method for determining the origin of hydrocarbon seeps in water-covered areas in order to establish whether the seeps are derived from depth and thereby indicative of subsurface deposits of petroleum or whether they merely represent marsh gas.

The seep gases are sampled directly and also by degassing water samples taken from sediments in the immediate vicinity of the seeps. Alpha-activity measurements are made on the seep gases and their origin is determined by comparison with alpha-activity ratios obtained using known standards.

Figure 13 shows an area of application of such a technique. Assume that gas show **10** in body of water **11** is a hydrocarbon gas of unknown origin. Gas evolved from the show may have its origin at depth or be formed by decomposition of organic matter in sediment layer **12**. A possible source of gas derived from depth is represented by petroleum deposit **13** in subsurface formation **14**. Fissure **15** represents a path which gas from petroleum deposit **13** may follow in reaching the surface. The apparent origin of the show is indicated by point **16** on the surface of sediment layer **12**.

Figure 13: Earth Section Showing Application of Atlantic Richfield Company Hydrocarbon Analysis Technique

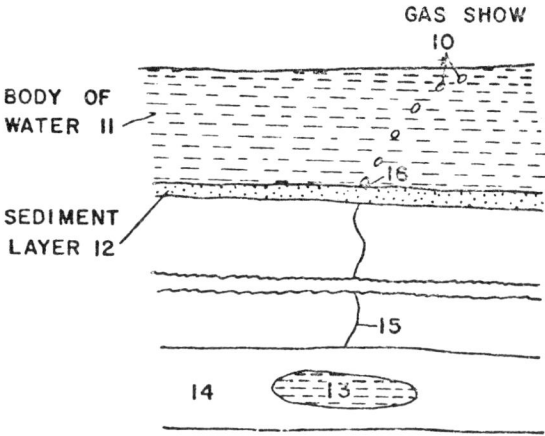

Source: U.S. Patent 3,571,591

Sediment layer **12** may contain material such as gravel, shingle, sand, silt, ash, clay, mud, etc. The mud (or ooze) may contain accumulations of carapaces, shells, skeletons, and decaying vegetable matter. Also, layer **12**, being permeable, is generally saturated with water from body of water **11**.

Wherever its actual origin, a portion of the gas from show **10** will dissolve in the body of water and the remainder will stay in the vapor or gaseous phase, rise to the surface, and disperse into the atmosphere. It follows that gas from the show or, to be more accurate, each component thereof, should tend toward an equilibrium relation between its gaseous phase and the surrounding aqueous phase.

As a practical matter equilibrium between gas originating in the sediment layer and the surrounding water can only be established in the sediment layer, i.e., below the interface between body of water **11** and the sediments. Thus, only within the confines of layer **12** can one assume that the surrounding water is saturated with the gas comprising the show.

Marsh gas generated in the sediments will show certain equilibrium relations with the surrounding water, i.e., the interstitial water in the sediments, and that gas generated at depth usually will not establish equilibrium relations. It follows that one should be able to ascertain the source of a gas of unknown origin by measuring the relative concentration of a preselected component thereof between the gas phase and the sediment water and correlating this information with the relative concentrations established for gas shows where the evolved gas is known to be gas from depth, on the one hand, and marsh gas, on the other.

One way whereby gas from depth can be distinguished from marsh gas involves the use of a radioactive substance as the preselected component. The concentration of the preselected component in different phases can then readily be measured by conventional radiological techniques.

It is not necessary that this radioactive component be a hydrocarbon gas as long as it is generally associated with both gas from depth and marsh gas. It is necessary, however, that the preselected component have at least limited solubility in water. The ideal choice for such a component is radon.

Radon possesses a particularly stable electronic configuration which gives it the chemical inertness characteristic of noble-gas elements. Therefore, the only chemical form known is the free element. Radon is appreciably soluble in water and in organic liquids. Radon is found in natural sources because of its continuous replenishment by the radioactive decay of longer-lived precursors, such as uranium. Radon has wide distribution and may be detected in the soil, surface and subsurface waters, hydrocarbon deposits, seep gas, marsh gas, the atmosphere near the ground, etc.

All isotopes of radon are radioactive with short half-lives. Radon 222, the most stable isotope of radon, decays by the emission of energetic alpha particles at such a rate that one-half of any given quantity of radon disintegrates in 3.82 days. Any surface exposed to radon 222 becomes coated with an active deposit which consists of a group of short-lived daughter products, including radium. The radiations of this deposit include energetic alpha, beta, and gamma rays. Hence, though radon is generally determined from alpha-activity measurements, beta or gamma determinations may be used as quantitative measures where desired.

One first determines the radon concentration of gas evolved from a preselected gas show and, in conjunction therewith, determines the radon concentration of water in the immediate vicinity to the apparent origin of the show in the sediments. By comparing the ratio of the radon concentrations thus obtained with standard radon ratios similarly obtained representing evolved gas known to be gas from depth and marsh gas, respectively, it is possible to predict the origin of the show, i.e., distinguish between gas from depth and marsh gas.

A process developed by *B.O. Prescott, G. Rittenhouse, A. Walters and H.L. Wise; U.S. Patent 3,957,439; May 18, 1976; assigned to Shell Oil Company* involves surveying the hydrocarbon content of a groundwater. Successive portions of the water are flowed from a selected subsurface location within the groundwater habitat to a nearby surface location, at rates such that each portion arrives without significant chemical change. The arriving portions are flash-distilled so that gas inclusive of substantially all materials more volatile than water are separated substantially as soon as the water reaches the surface location.

Measurements are made of the concentration of at least one hydrocarbon in successively accumulated slugs of the so-separated gas, substantially as soon as the slugs accumulate. In indicating a hydrocarbon concentration in the groundwater in the selected subsurface location, the measurement value used is one that remained substantially constant throughout a plurality of such measurements.

Figure 14 is a schematic of such a process. It shows a groundwater aquifer **1** overlying a petroleum deposit **2**. Water from the aquifer is produced by a conventional pumping arrangement from a water well **3**. The water is preferably produced at a rate exceeding that needed for analysis and the excess is metered off with a conventional metering valve **4**. In the gas extracting and analyzing system a gas pump **19** draws the water into a gas extraction chamber **18** from which the excess water is removed by a water pump **23**. The extracted gas is freed of water in a gravity separator **29** and subjected to hydrocarbon analysis.

In a preferred embodiment hydrogen flame chromatographs are employed for determinations of respectively total hydrocarbons, hydrocarbons of from 1 to 5 carbon atoms, and hydrocarbons of from 5 to 10 carbon atoms. A separate on-site analysis is also made of carbon dioxide.

Figure 14: Shell Oil Company Technique for Surveying Constant Value Concentrations of Hydrocarbons in Groundwater

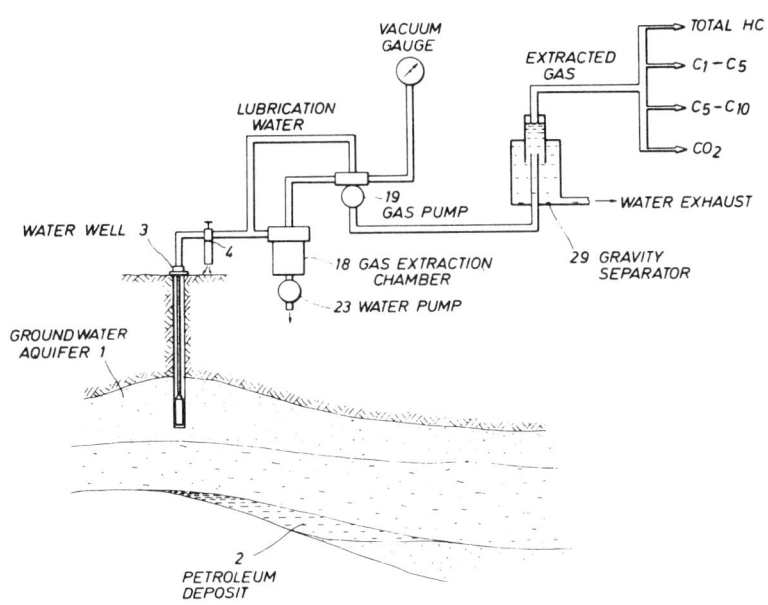

Source: U.S. Patent 3,957,439

Detection of Fatty Acids as Hydrocarbon Indicators in Water

A method developed by *J.E. Cooper and K.A. Kvenvolden; U.S. Patent 3,305,317; February 21, 1967; assigned to Mobil Oil Corporation* is one in which a sample of water is removed from a subterranean formation and analyzed to determine the ratio of the quantity therein between selected fatty acids, which ratio is indicative of the presence in the subterranean formation of a petroleum reservoir, and thereafter, where the ratio indicates that the formation contains a petroleum reservoir, carrying out drilling operations in the formation to locate the reservoir in order to produce the petroleum contained in it.

It has been ascertained that fatty acids are present in recent sediments, ancient sediments, and in waters from formations containing petroleum. It has been ascertained further that in these fatty acids, those having an odd number of carbon atoms are present along with those having an even number of carbon atoms. The fatty acids containing an even number of carbon atoms predominated in amount over the fatty acids containing an odd number of carbon atoms in each case studied.

In this connection, it was recognized that the relative abundance of the fatty acids containing an odd number of carbon atoms increased from the recent sediments to the ancient sediments to the formation waters. However, the source of the fatty acids containing the odd number of carbon atoms was not known.

This process is based upon the discovery that the ratio of fatty acids containing an even number of carbon atoms to those containing an odd number of carbon atoms in formation waters is not a random distribution of these acids. Rather, this distribution is indicative of the past history of the formation with respect to the presence or absence of petroleum. Where this ratio is not more than 1.6, it can be considered that the formation at some location therein contains petroleum.

A technique developed by *K.A. Kvenvolden; U.S. Patent 3,480,396; Nov. 25, 1969; assigned to Mobil Oil Corporation* is a geochemical exploration technique wherein petroleum source rocks are characterized on the basis of an acid distribution ratio, i.e., the ratio of relative concentrations of even carbon number fatty acids to odd carbon number fatty acids found in rock samples.

Preferably, the acid distribution ratio is determined with respect to fatty acids having more than sixteen carbon atoms, and more desirably, more than eighteen carbon atoms. Where a formation is found to exhibit an acid distribution ratio of 1.3 or less, it is indicated as a possible petroleum source rock and a well may be drilled into a reservoir rock formation in fluid communication with the source rock formation.

Aerial Geochemical Prospecting Techniques

A technique developed by *J.S. Bradley, W.H. Luehrmann, C.A. Youngman and J.R. Gunter; U.S. Patent 3,143,648; August 4, 1964; assigned to The Atlantic Refining Company* provides an airborne method for hydrocarbon seeps detection which is suited to high-speed, large-area reconnaissance missions.

The relentless search for oil deposits during the past decades has covered most of the civilized and accessible areas of the world's surface. The search has been extended into the vast reaches of the Arctic and the tropics as well as inaccessible deserts and mountainous terrain. As this search has extended into the more remote and inaccessible regions of the world, the need for an exploration tool suitable for high-speed, large-area reconnaissance has become more pressing.

Known methods and systems, while in most cases adequate for their original purposes, do not meet the need for a high-speed, highly mobile reconnaissance tool for pinpointing areas of interest in vast unexplored regions. Conventional seismic reflection and refraction prospecting, electrical prospecting, magnetic prospecting, gravimetric prospecting, radioactive prospecting, geochemical prospecting, etc., are extremely limited in their applications since they must traverse the earth's surface and, therefore, are restricted in their speed and maneuverability. Waterborne devices conducting seismic operations, seeps detection and allied explorations are similarly restricted to traversing water-covered areas. Their speed and maneuverability are likewise limited.

Although airborne, radioactive, gravimetric, and magnetic surveys have been conducted, they do not provide enough information for a satisfactory large-scale reconnaissance tool. This is true since the information they yield is limited to delineating very large areas of possible interest. These areas must be confirmed and further localized by large-scale geological or geophysical operations.

In other words, if the most satisfactory airborne survey system is utilized, it can delineate an area of interest on the order of 10,000 to 100,000 square miles. Ground-based geological or geophysical operations must then be used to check likely areas in the 10,000 to 100,000 square mile area. This is in itself a tremendous undertaking and cannot be carried out in most areas where a large-scale reconnaissance tool is designed to operate.

This technique is based on the unexpected discovery that gas plumes from naturally occurring seeps do not dissipate beyond recognition and that they can be successfully detected in the atmosphere at relatively great distances from the seeps.

This process involves the steps of conducting an airborne traverse over an area of interest by traversing a predetermined course at a predetermined altitude; isolating a portion of the atmosphere; analyzing the isolated portion for at least one predetermined component; and recording the presence of the component with relation to the geographical location of analysis.

An aerial technique which does not require the use of aircraft has been described by *G.H. Milly; U.S. Patent 3,734,489; May 22, 1973; assigned to Geomet Mining and Exploration Company.* It involves prospecting for hydrocarbons such as oil and gas on the basis of the mobile **gaseous** phase of hydrocarbons which diffuses through the earth's structure and becomes windborne.

More specifically, it involves initially sensing the gaseous phase in the atmosphere, discriminating between density flow of gaseous phase arising from oil or gas deposits and that from ambient background, horizontally tracking the density flow towards the oil or gas deposits and marking the oil or gas deposits as density flow and diffusion of gaseous phase through the earth's structure from the oil or gas deposit coincide.

Refinements include initially conducting a line survey downwind of the area being prospected; outlining katabatic flow as a function of topography; estimating bulk atmospheric drift in heavily wooded areas on the basis of above-canopy wind direction; vapor and dust sampling during tracking and capping of the earth's structure adjacent the diffusion of gaseous phase through the earth's structure.

A method developed by *A.R. Barringer; U.S. Patent 3,759,617; September 18, 1973; assigned to Barringer Research Ltd., Canada* is a method for performing a rapid geochemical survey over an area of the earth consisting of collecting and concentrating atmospheric particulates, preferably transferring the particulates into a stream of an inert carrier gas, heating the particulates to break down organic particles into particles of smaller size and vapors, removing from the stream inorganic particles which generally are of relatively high mass, and then analyzing the remaining organic particles and vapors to detect the presence of elements or compounds indicative of underlying geological conditions.

Figure 15 shows the apparatus and technique which may be used.

Figure 15: Barringer Research Ltd. Apparatus for Airborne Particulate Sampling and Analysis

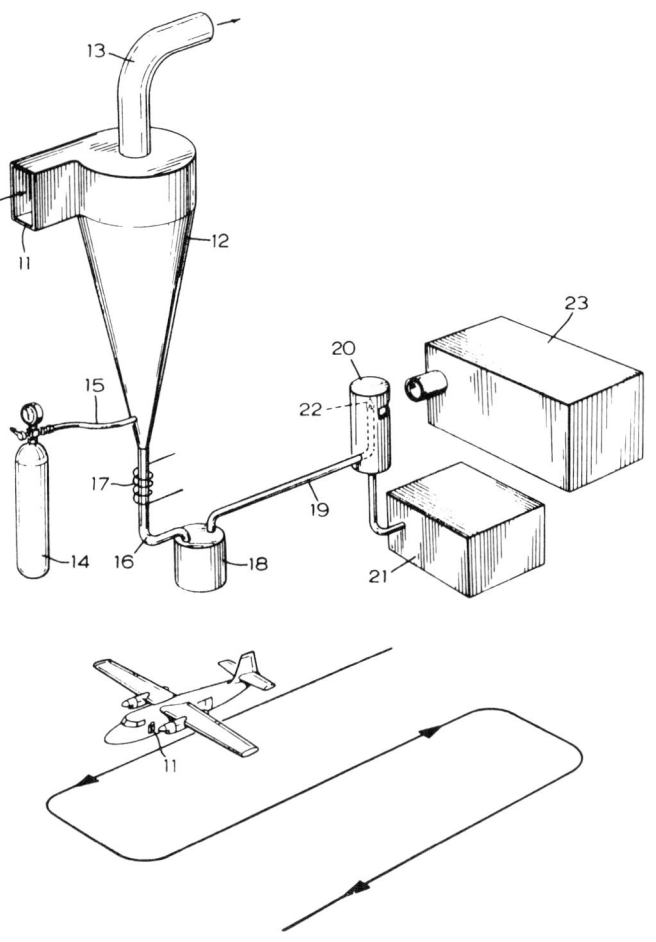

Source: U.S. Patent 3,759,617

Air containing particulates to be analyzed enter through a sampling duct **11** and thence into a cyclone **12** which causes the air to form a vortex. The particulates in the air collect on the walls of the cyclone by centrifugal action and settle towards the conical base of the cyclone under the action of gravity. Relatively clean air is discharged through an outlet duct **13**. An inert carrier gas such as argon is inserted into the cyclone in such a manner as to capture the particulates. The argon or other inert gas may be contained within a pressurized gas cylinder **14** which is connected to the cyclone by means of a length of tubing **15**.

At the point where the tubing meets the cyclone, the particulates are relatively concentrated and the air at this point inside the cyclone can be arranged to be at virtually zero pressure differential with the outside air. This condition is achieved by appropriate design of the cyclone with special reference to the relative cross-sectional areas of the cyclone inlet and clean air outlet. By inserting a nonturbulent argon stream at the conical base of the cyclone, the particulates will settle into the pool of argon so formed and will be carried through the rest of the apparatus.

In effect, the arrangement described above transfers the particulates from air into an inert gas, thus considerably simplifying the subsequent analysis of the particulates. An opening is formed at the conical base of the cyclone and a tube **16** is connected to the cyclone at the conical base for the purpose of conveying the argon stream with the particulates from the cyclone. The tubing **16** preferably is heated to a predetermined temperature such as 700°C by means of an electrical heating coil or the equivalent, for the purpose of breaking down organic particles into vapors and smaller-sized particulates.

Tube **16** is connected to a conventional dust trap **18** which serves to separate by gravity the relatively heavy inorganic particles from the relatively light organic particles and vapors. The organic particles and vapors are then directed from the dust trap via a tube **19** into a microwave cavity **20** which is connected to a microwave generator **21**. The microwave generator produces a plasma in the argon stream formed inside the microwave cavity **20**. The microwave plasma is indicated by reference **22**.

The organic vapors and particles are excited in the plasma and they will emit light which is indicative of their atomic or molecular structure, and as is well known, by spectrally analyzing the emitted light the nature of the elements and compounds in the plasma can be determined. As shown, the plasma is viewed by a spectrometer **23** which is capable of measuring the intensity of light at one or more predetermined wavelengths, depending upon the number of elements or compounds it is desired to observe.

The spectrometer has one or more output channels, depending upon the number of separate elements or compounds it is desired to measure, and electrical signals appear at the output channels in the conventional manner, from whence they may be fed to a conventional recorder.

The apparatus described above may be installed in an aircraft as shown, with the open end of the duct **11** positioned forwardly, as shown. The aircraft is flown in systematic traverses over an area to be explored, at a relatively low flying altitude and with traverse intervals varying between about one-eighth of a mile and five miles depending upon the nature of the information it is desired to obtain.

The response time of a practical system for airborne geochemical sampling should be no more than six seconds, in order to localize target areas. Assuming that the analytical equipment has an absolute detection sensitivity of between 10^{-10} and 10^{-12} g, it is necessary to concentrate the organic particulates from the atmosphere at a collection rate of about 10 m^3/min. When using cyclone collection techniques having a dust collection efficiency of at least 50%, the overall air flow capacity required is of the order of 20 m^3/min.

A technique developed by *A.R. Barringer; U.S. Patent 3,970,428; July 20, 1976; assigned to Barringer Research Ltd., Canada* is a technique for airborne prospecting in which samples of atmospheric gases, vapors or particulates are collected and contemporaneously the atmosphere near where the samples are collected is monitored to determine whether the respective samples were uplifted by an updraft of air.

Samples respectively collected during periods of updrafts are separated from the remainder of the samples, and the samples respectively collected during periods of updrafts are analyzed separately from the remainder of the samples.

The atmospheric samples collected during updraft conditions are considered to be more indicative of the nature of the underlying terrain than the remainder of the samples because they are more likely to have been transported recently from the earth's surface. For example, particles collected during a downdraft may have been in the atmosphere for many hours or even days prior to collection and may thus have been transported considerable distances from the place on the earth's surface where they originated.

GEOBIOLOGICAL PROSPECTING

The applications of geobotany and biogeochemistry in mineral exploration have been reviewed by Brooks (10).

GEOBOTANICAL PROSPECTING

In one form of geobiological prospecting, which might be called geobotanical prospecting, plants may be used as aids in prospecting for buried mineral deposits. So many factors are involved, however, that prediction of conditions under which plants will be of practical assistance is not always possible.

Many plants, by means of their extensive root systems and the absorptive ability of their roots, effectively sample many of the elements that are within reach of the roots and transfer these elements to the branches, stems, and leaves, which can be chemically analyzed. Thus, under ideal conditions, the plant has sampled the underlying soil or rock in its root zone to depths of as much as 50 feet.

The advantages to the prospector of being able to sample plants and thus to obtain information about the metals that occur at considerable depth are obvious, although problems in interpreting this information may render this method of prospecting impractical under many field conditions. For instance, some plants, because of their genetic makeup, selectively concentrate elements in their roots, stems, or leaves in higher percentages than these elements occur in the soil and rocks of the plants' environment. Whenever possible, soil and rock samples should be analyzed before concluding that a geobotanical anomaly indicates the presence of certain minerals in an area (1).

Botanical prospecting is thus a valuable adjunct to conventional prospecting methods.

GEOMICROBIAL PROSPECTING

It has been postulated that hydrocarbons emanate from a subterranean petroleum oil or gas deposit upwardly through the earth overlying the deposit and that the presence of these hydrocarbons or their reaction products in the earth is indicative of the underlying deposit. Accordingly, various methods based on the concept of emanation or migration of hydrocarbons from subterranean deposits have been proposed for locating the deposits. Among such methods are those which involve the detection of hydrocarbon-consuming microbes in the earth, it being taken that such microbes exist and thrive in the earth by virtue of their ability to utilize as nutrient, or as an energy source, the hydrocarbons migrating from the underlying petroleum deposit, and that the presence of such microbes in the earth is thereby indicative of the presence of the underlying petroleum deposit.

In one microbial method of this sort, earth samples are collected over a prospect zone and each sample is maintained in an atmosphere containing oxygen and a gaseous petroleum hydrocarbon. Any hydrocarbon-consuming microbes which may have been contained in each of the earth samples will consume the atmosphere and the rate at which the atmosphere is consumed, as measured by the decrease in pressure of the atmosphere, is regarded as a measure of the number of hydrocarbon-consuming microbes originally in each sample. The number of microbes thus determined is then related to the location of the sampling points to detect anomalous variations in the number of microbes per unit amount of earth which are indicative of the presence of an underlying petroleum deposit.

However, consumption of the atmosphere is not a specific indication of the presence of hydrocarbon-consuming microbes in the earth samples since other microbes and oxidizable organic matter present in the earth samples are capable of consuming the oxygen. Thus, anomalous variations in the number of microbes per unit amount of earth as determined by this method cannot be regarded as being a reliable indication of the presence of an underlying petroleum deposit unless the consumption of the oxygen by non-hydrocarbon-consuming microbes and oxidizable organic matter can be distinguished from the consumption of the hydrocarbon and oxygen by hydrocarbon-consuming microbes.

Consumption of oxygen by non-hydrocarbon-consuming microbes and oxidizable organic matter may be distinguished from consumption of hydrocarbon and oxygen by hydrocarbon-consuming microbes by control measurements on portions of each earth sample where the atmosphere employed is free of hydrocarbons. However, this method of distinguishing is valid only where significant differences are obtained between the consumption occurring where the atmosphere contains hydrocarbons and where it is free of hydrocarbons, and significant differences are not obtainable unless the earth sample contains extremely large numbers of hydrocarbon-consuming microbes. Another method of distinguishing involves analysis of the atmosphere to detect a decrease in the quantity of the hydrocarbon in the atmosphere but this method requires involved and expensive analytical techniques.

A technique designed to overcome these difficulties has been developed by *J.B. Davis; U.S. Patent 2,777,799; January 15, 1957; assigned to Socony Mobil Oil Company, Inc.* In this technique, earth samples taken from a prospect zone are contacted with oxygen and with a hydrocarbon, the molecules of which contain carbon 14 to form a reaction mixture with each earth sample and thereafter at least a portion of each of the reaction mixtures is analyzed for carbon 14 which has entered into a microbial metabolic process. When microbes contained in an earth sample consume a hydrocarbon, they convert it into microbial metabolism products which contain the carbons which were contained in the molecules of the hydrocarbon. The common microbial metabolism products from hydrocarbons are microbial protoplasm and CO_2. As is well known, C-14 is radioactive and any compound containing it is likewise radioactive. Thus, the detection in at least a portion of the reaction mixture of a radioactive microbial metabolism product or a radioactive material derived therefrom is a specific indication of the presence in the sample of hydrocarbon-consuming microbes. The intensity of the radioactivity is a measure of the number of these microbes in the sample.

In a technique developed by *D.M. Updegraff and H.H. Chase; U.S. Patent 2,861,921; November 25, 1958; assigned to Socony Mobil Oil Company, Inc.,* earth samples are collected from a prospect area, the samples are admixed with an aqueous inorganic salt medium, the mixtures are contacted with a petroleum hydrocarbon nutrient, and the time required for consumption of the hydrocarbon to begin is compared with the time required for known numbers of hydrocarbon-consuming bacteria to begin consumption of the same petroleum hydrocarbon. By this comparison, a measure of the number of hydrocarbon-consuming bacteria in each sample is made. Thereafter, the numbers of bacteria in each sample may be plotted relative to the sampling points in the area to determine location of any anomalously high numbers indicative of the location of a petroleum deposit.

In order to eliminate the effects arising from the ability of any bacteria in the samples, whose presence may be due to the availability of nonpetroleum hydrocarbon or other nutrient, to adapt to consuming the petroleum hydrocarbon employed and thereby give rise to the erroneous conclusion that the earth contained petroleum hydrocarbons indicative of a petroleum deposit, all samples wherein the time for initiation of consumption of the hydrocarbon nutrient employed is longer than the adaptation time are eliminated from consideration. This adaptation time is ordinarily greater than 25 days and therefore all samples which do not show initiation of hydrocarbon consumption within 25 days are regarded as being indicative of not having contained hydrocarbon migrating from a petroleum reservoir and therefore indicative of the nonproximity of a hydrocarbon reservoir.

Figure 16 shows a suitable form of test apparatus for use in such a technique. It comprises a chamber **10** provided with a filling tube **11** and a manometer tube **12**. The filling tube narrows at its end to form a tip **13** which, as shown, is closed. Prior to introduction of the earth sample **14** and the aqueous inorganic salt medium **15**, the tip is open but thereafter is closed by sealing with a flame. By means of this feature, and by making the manometer tube integral with the chamber, the use of joints, stoppers, etc., which may give rise to leaks or require lubricants in which the hydrocarbon nutrient would dissolve with consequent error in measuring the time for consumption of the hydrocarbon nutrient to begin, is eliminated. To reuse the apparatus, the tip may be resealed.

For operation, the apparatus is first cleaned and sterilized. A portion of the earth sample is weighed on sterile paper and is then introduced through the open tip by means of a sterile powder funnel. Thereafter, a measured amount of aqueous inorganic salt medium is introduced by means of a sterile pipette through the open tip. Following this, the tip is flame sealed.

The petroleum hydrocarbon nutrient is then contacted with the mixture in chamber **10**, by evacuating the chamber and permitting the gaseous nutrient mixture to enter the evacuated chamber.

Figure 16: Socony Mobil Oil Co. Apparatus for Use in Microbiological Petroleum Prospecting Method

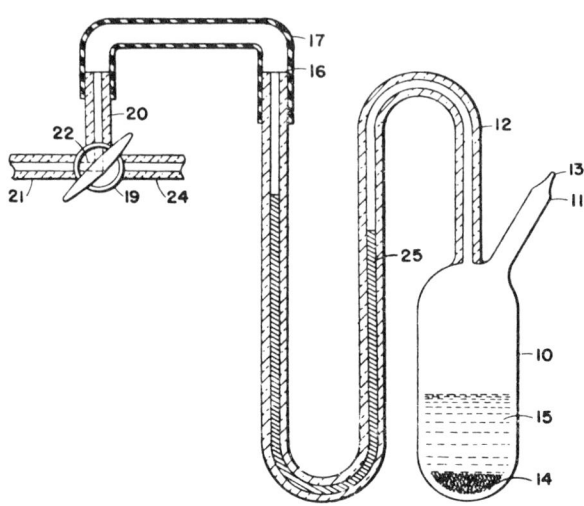

Source: U.S. Patent 2,861,921

The chamber can be evacuated by connecting the open end **16** of manometer tube **12** to tubing **17** leading to line **20** on a three-way valve or stopcock **19** provided with line **21** leading to a vacuum pump or other means for obtaining a low pressure. After a suitably low pressure has been obtained in chamber **10**, the valve **19** is turned so as to break the connection between lines **20** and **21** through channel **22** in the valve. Gaseous nutrient mixture is then permitted to enter the chamber by turning the valve so as to position the channel between line **20** and line **24** leading to a source of the gaseous nutrient mixture.

The pressure of the nutrient mixture should preferably be slightly above atmospheric to prevent entrance of air to the chamber upon disconnecting the tubing from the manometer tube. The tubing is disconnected from the manometer tube and mercury or other liquid in which the hydrocarbon nutrient is not soluble is poured into the open end **16** of the manometer tube. The apparatus is then maintained at a temperature of about 30°C as by placing in an incubator, water bath, or by other suitable means. Readings of the two levels of the mercury column **25** are made frequently and the time at which the levels begin to change at a more rapid rate than the levels in a control apparatus containing sterile aqueous medium and gas mixture only, indicating the beginning of consumption of the hydrocarbon nutrient by the bacteria in the sample in the chamber, is noted.

However, as previously mentioned, since the bacteria in the sample may adapt themselves to consume the petroleum hydrocarbon nutrient even though they may have been consuming a nonpetroleum hydrocarbon or other nutrient in the earth in place, the nutrient accounting for their presence in the earth in place, the sample is regarded as being free of bacteria significant with respect to the proximity of a petroleum reservoir if the levels of the mercury column do not change, as compared with those in the control apparatus, within 25 days.

A method developed by *J. Maddox, Jr.; U.S. Patent 2,875,135; February 24, 1959; assigned to The Texas Company,* is one in which soil samples are incubated with a hydrocarbon substrate in the presence of aqueous inorganic salt nutrient medium conducive to optimum growth of hydrocarbon-consuming microorganisms in the soil samples and the rates of hydrocarbon consumption are determined and correlated with the area being prospected as a measure of subsurface petroleum deposits.

An improvement in this particular process comprises adding a surface-active adsorbent siliceous material to the mixture of soil sample and aqueous nutrient medium in an amount equivalent to 0.5 to 5 weight percent of the soil sample-aqueous nutrient medium mixture, contacting the mixture of soil sample, nutrient medium and adsorbent material with a gas mixture comprising gaseous hydrocarbon and oxygen, incubating the resulting reaction mixture under quiescent conditions for a period of 24 to 72 hours and immediately thereafter incubating the reaction mixture under conditions of agitation.

A method developed by *D.O. Hitzman; U.S. Patent 2,880,142; March 31, 1959; assigned to Phillips Petroleum Company* is one in which an organic liquid such as methanol, ethanol or butanol is included as the sole substrate in a culture medium devoid of other sources of carbon. Thus, in order for propagation and growth to occur, the microorganisms must utilize the organic liquid as a nutrient. Aliquots of the soil sample are added to the culture medium in a culture dish, incubated, and the number of colonies of microorganisms which develop are counted. Since the hydrocarbon-consuming microorganisms are the only ones which can readily adapt to the organic liquid, the higher the number of colonies, the more indication of the presence of an oil and/or gas deposit.

A scheme developed by *D.O. Hitzman; U.S. Patent 3,281,333; October 25,1966; assigned to Phillips Petroleum Company* involves exposing hydrocarbon-consuming microorganisms in soil samples from an area under investigation to the action of certain phenolic compounds, which are normally toxic to most microorganisms, under incubating conditions for a period of time sufficient to promote the propagation and growth of the microorganisms, and then counting the number of colonies of the microorganisms which developed.

A technique developed by *W.D. Rosenfeld; U.S. Patent 2,921,003; January 12, 1960; assigned to California Research Corporation* is based on a correlation between the presence in water samples of selected liquid hydrocarbon compounds, including oxidized hydrocarbon compounds, as an indication of the possible proximity of liquid hydrocarbon accumulations, and the presence of hydrocarbon-consuming or oxidizing bacteria in these waters as a similar and analogous indication of the possible proximity of such hydrocarbon accumulations.

The types of bacteria sought include the so-called nitrate-reducing type, such as the genus *Pseudomonas,* which are primarily aerobic and which appear to be capable of oxidizing all types of hydrocarbons and oxidized hydrocarbons. The bacteria sought also include the anaerobic sulfate-reducing bacteria, such as the genus *Desulfovibrio,* which oxidize primarily the paraffinic hydrocarbons containing more than 10 carbon atoms, as well as the oxidation products of such hydrocarbons. Preferably, a given water sample is analyzed for the presence of both of the above types of bacteria, although it will be understood that the presence of either type of bacteria alone may be utilized as an indication of the possible presence of liquid hydrocarbon accumulations.

A method developed by *P.E. Oberdorfer, Jr. and D.F. Rugen; U.S. Patent 3,028,313; April 3, 1962; assigned to Sun Oil Company* is one in which a culture of methane-consuming microorganisms is placed adjacent the earth at a selected locus beneath the surface and the culture is allowed to remain there for a time sufficient to permit substantial growth of the microorganisms if any methane is present in the soil. Thereafter the culture is removed from the earth and the radioactivity of the microorganism residue is determined to ascertain the ratio of C^{14} to C^{12} in the methane consumed. Carbon derived from methane from vegetation will have a high C^{14} to C^{12} ratio while carbon derived from methane which emanated from a petroleum reservoir will have a low ratio. Hence an anomalously low ratio of C^{14} to C^{12} will serve as an indicium of the proximity of a petroleum deposit.

This prospecting method in effect utilizes the methane-consuming ability of certain microorganisms to concentrate the carbon from methane present in the soil at the selected site. When methane is present in the soil, it will continuously diffuse therefrom to the microorganism culture and a sufficiently long time, e.g., one month or more, can be allowed for a substantial amount of the methane to be consumed. Therefore, the method does not depend upon analyzing accurately for more or less trace amounts of hydrocarbon, such as is required if a soil sample itself is analyzed.

Furthermore, this method permits methane to be utilized as the hydrocarbon that indicates the proximity of petroleum by employing an analysis procedure that distinguishes between methane from such source and methane from vegetation. This is distinctly advantageous, since methane is the hydrocarbon component which will most readily diffuse from a petroleum reservoir.

A suitable form of apparatus for conducting this method is shown in Figure 17. The figure illustrates a device for holding a culture of the methane-consuming microorganisms, which can be inserted into the ground to an appropriate depth and which is so constructed that methane and oxygen from the soil can diffuse to the culture to cause growth of microorganisms. More specifically, the device comprises an elongated housing **10** formed from threaded tubing which is connected to a pointed bottom closure member **11** and a cap **12** at the top. The bottom member is constructed of a rigid porous material such as sintered metal or hard sintered resin, so that gases from the earth can diffuse through the wall and enter the housing.

The bottom member is threaded to a ring **12** and a screen **14** is held between the two pieces at shoulder **13**. The screen is used for supporting a bed **15** of a suitable absorbent for carbon dioxide, e.g., a hydrous caustic soda supported on asbestos known as Ascarite. The ring is threaded to the tube, and a membrane **17** is held between the abutting shoulders at **16**. This membrane serves to support the culture medium **18** which contains the methane-consuming microorganisms. This membrane is made of a material through which both methane and oxygen can readily diffuse. Examples of such material are polyethylene, polypropylene and other synthetic organic resins.

Figure 17: Sun Oil Co. Apparatus Using Methane-Consuming Microorganisms to Indicate Petroleum Deposits

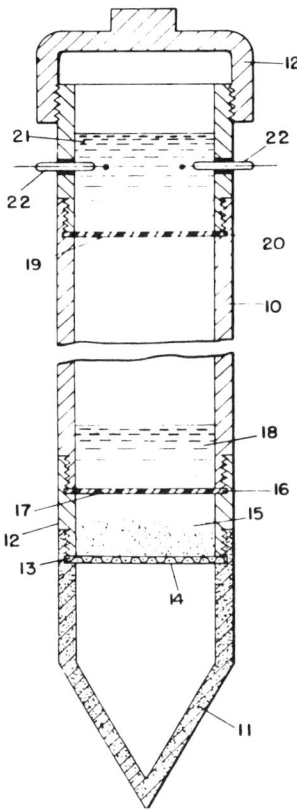

Source: U.S. Patent 3,028,313

Another membrane **19** is positioned between abutting shoulders **20** of the tube **10** and the top portion of the device. This membrane, which can also be made of polyethylene or other synthetic resins, will allow carbon dioxide to diffuse through it. The membrane is used as a support for an absorbent capable of absorbing carbon dioxide, which in this case preferably is an aqueous solution of barium hydroxide as shown at **21**. A pair of insulated electrodes **22** are positioned in packing glands in the upper wall of the device so as to be immersed in the aqueous solution. These electrodes are employed for determining any changes in the electrical conductivity of the barium hydroxide solution and thus indicate whether or not any change in the barium hydroxide concentration has occurred.

In using the illustrated apparatus for petroleum prospecting, the device containing the necessary materials is assembled as described above and then is forced into the ground at a selected site. The length of the device should be such that the porous lower portion **11** is located at a desired depth which is generally several feet beneath the earth surface, with the electrodes **22** being above ground level. By means of suitable apparatus (not shown), the initial conductivity of the barium hydroxide solution **21** is measured through the electrodes. The device is then permitted to remain in the ground so that gases contained therein can diffuse through porous member **11** and into absorbent bed **15**. Upon contact with the absorbent, any carbon dioxide present will be absorbed. Such removal of carbon dioxide in effect acts as a pumping means which aids in the diffusion of gases into the device.

After passing through the absorbent bed, oxygen and methane, if present, will diffuse through membrane **17** and be absorbed in the liquid culture medium **18**. This causes growth of the microorganism colony, the rate of which depends upon the amount of methane entering the culture medium. Carbon dioxide is formed as a by-product of the metabolism, and this product will diffuse out of the culture medium solution and be released into the chamber above it. Accordingly, carbon dioxide produced by the microorganisms will contact membrane **19** which, as previously described, is constructed so as to allow diffusion of carbon dioxide through it. The carbon dioxide thus will be absorbed by the barium hydroxide solution and will react to form insoluble barium carbonate.

Precipitation of the carbonate salt will cause a change in conductivity of the solution. Therefore, by simply measuring the conductivity of the solution from time to time, it can be determined whether or not methane has entered the device and caused growth of the bacteria without any necessity for removing the device from the ground and opening it for inspection. If after a sufficient lapse of time no growth is indicated, it is then known that proximity of petroleum is not indicated; and the device can be withdrawn and taken to another location for further prospecting.

In cases where substantial bacterial growth is shown by the conductivity measurement, the device is then removed from the ground for radioactivity tests. These tests can be made on either the precipitated barium carbonate or the microorganism residue or both. Testing of a weighed amount of the barium carbonate after drying is particularly convenient. When tests show an anomalously low radioactivity, it is known that the ratio of C^{14} to C^{12} in the metabolism products is low and therefore that at least a substantial proportion of the methane consumed by the microorganisms was derived from petroleum. This serves then as an indication that the site where the device was located is in the vicinity of a petroleum deposit.

Various modifications of the abovedescribed device can be made. For example, with the device as shown in the figure, part of the carbon dioxide resulting from the microorganism metabolic process may diffuse from the culture medium downwardly through membrane **17** and be absorbed in the absorbent bed along with any carbon dioxide which may have entered the device from the surrounding soil.

In some instances it may be desirable to convert all of the carbon dioxide produced by the metabolism into barium carbonate and determine the total amount of barium carbonate formed for purpose of quantitatively indicating the rate of growth of the microorganisms. In such cases membrane **17** can be constructed of material through which methane and oxygen can diffuse but not carbon dioxide. This will cause all of the carbon dioxide to diffuse upwardly to the barium hydroxide solution at 21. Alternatively, another barium hydroxide bath can be provided on a membrane positioned between the absorbent bed 15 and membrane 17, so that any carbon dioxide which diffuses downwardly from the culture medium will react therewith and be precipitated as barium carbonate. With this modification, the total amount of barium carbonate formed in both the upper and lower baths subsequently could be determined and utilized as a measure of the rate of microorganism growth.

This device can also be used for prospecting adjacent the floor of an ocean, sea, lake, river or the like. In such case a bag made of polyethylene or other suitable resin is filled with a liquid culture medium containing methane-consuming microorganisms and the bag is lowered on a wire or rope to the floor of the ocean. If methane is present in the water, it along with oxygen which is also present will diffuse from the water through the wall of the bag and will dissolve in the culture medium, thus causing growth of the microorganisms. The by-product carbon dioxide formed by the metabolism process will largely remain dissolved in the culture medium due to the hydrostatic pressure and will diffuse through the wall of the bag into the surrounding water.

After the bag has remained immersed for sufficient time to permit substantial growth, the bag is removed and radioactivity tests are made as previously described to determine whether at least some of the methane consumed was derived from petroleum.

A method developed by *L.R. Brown; U.S. Patent 3,033,761; May 8, 1962* involves taking earth samples at a plurality of locations within an area to be prospected; taking aliquots of each of the samples, a group of aliquots representing the sample; and subjecting each of the groups of aliquots to a test to determine the presence of microbiological entities capable of consuming at least three hydrocarbon species chosen from the group consisting of gaseous hydrocarbons higher than methane, each of the chosen hydrocarbons being tested separately on one of the aliquots.

The hydrocarbon gas will vary among the several aliquots, and may be, for example, ethane, propane, isobutane and normal butane.

This process disregards a number of criteria to which various investigators in the past have attached some importance, such as any and all tests with methane, since it is so commonly produced in swamps and other locations having no connection whatsoever with underlying petroleum deposits; the so-called "lag time" which is the period of time, usually in days, required for consumption of the hydrocarbon to begin to an appreciable extent; or the rate at which the hydrocarbons are consumed.

A method developed by *J.D. Douros, Jr. and R.L. Raymond; U.S. Patent 3,065,149; November 20, 1962; assigned to Sun Oil Company* utilizes serological techniques for determining the presence or absence of a selected type of hydrocarbon-consuming microorganism and also the concentration thereof in the sample. The method is so readily performed that it is possible for a person to make several hundred soil tests in a single day.

In practicing the method an antiserum is prepared which is specific toward any particular hydrocarbon-consuming microorganism which is known to be prevalent in soil subject to seepage of petroleum hydrocarbon gases. It is preferred that the antiserum be specific for *Mycobacterium paraffinicum,* since this type of bacteria is practically universally present in soil which is subject to continuous contact by the paraffin hydrocarbon gases. However, the antiserum can be specific for any other microorganisms that can feed on the gaseous paraffin hydrocarbons, for example, *Agrobacterium ethanicus, Bacterium hidium, Methanomonas methanica* and *Bacillus ethanicus.*

Soil samples are collected from selected locations and a small amount of each sample is allowed to incubate for a time with a small volume of the antiserum. The incubated antiserum is then filtered from the soil and serologically tested at various dilutions with a culture of the microorganism which is an antigen for the antibody of the serum. If no reaction is obtained, this means that the soil had removed the antibody from the serum and hence that the soil contained the hydrocarbon-consuming microorganism. On the other hand, if heavy coagulation is observed, the absence of the microorganism in the soil sample is indicated.

The test can be done under specific conditions that can be compared with a standardized control test and the concentration of the microorganism thereby can be ascertained. Hence an area can be mapped in terms of the quantities of the hydrocarbon-consuming microorganism present in the soil so as to indicate where a petroleum deposit is likely to be found.

The antiserum is prepared by conventional biological techniques utilizing any suitable animal such as rabbit, hamster, chicken, horse or monkey. A culture of the microorganism, preferably *Mycobacterium paraffinicum,* is grown in a hydrocarbon-containing atmosphere which hydrocarbon preferably is ethane, and the microorganism is rendered avirulent in any suitable manner such as by heat killing, formalizing, phenolizing or attenuation. The animal is then injected with the avirulent microorganism over a period of several days. Preferably the first injection is subcutaneous and is followed by three or four intravenous injections on successive days.

On about the sixth day the animal's blood will contain the desired antibodies and a quantity of it can be withdrawn. The blood is allowed to clot and stand at room temperature for at least three hours, and the serum is then decanted. It may be desirable to inactivate the serum at $56°C$ for 45 minutes in order to destroy complement since the bacterial cells may be lysed in some instances. The serum can be stored under refrigeration at $0°C$. The titer of the antiserum can be determined by dilution tests in conventional manner.

In treating the soil sample only small amounts of the soil, e.g., 0.1 gram, and of the antiserum, e.g., 2 to 5 ml, need be used. The antiserum wash and the soil sample are mixed and allowed to incubate for at least one hour at room temperature. The mixture may, if desired, be stored overnight at $4°C$. The incubated antiserum is then filtered from the soil and is ready for serological testing.

Any of the known serological tests can be used for determining the presence, absence or the amount of antibody remaining in the incubated antiserum. Tests known as neutralization, agglutination, precipitin and complement fixation can be used, with neutralization being the preferred procedure. In general these involve admixing the antiserum with a normal saline preparation of the hydrocarbon-consuming bacteria (either virulent or avirulent) in a series of standard dilutions and observing the reaction. In cases where quantitative values are desired, comparisons can be made with standard control tests in which a known number of bacteria are reacted with various dilutions of the antiserum.

A method developed by *R.J. De Falco and A.R. Brillaud; U.S. Patent 3,329,580; July 4, 1967; assigned to Sun Oil Company* involves biological prospecting for petroleum by collecting soil samples, contacting each sample with a wash liquid capable of extracting a water-soluble antigenic material characteristic of a selected hydrocarbon-consuming microorganism, and serologically testing separated wash liquid to detect such soluble antigenic material if present.

Samples of soil taken over a hydrocarbon-bearing formation will contain a much higher percentage of hydrocarbon-indicating organisms which are heat resistant, perhaps as high as 95%; while soil samples taken from a normal or "dry" area will have a much lower percentage of heat-resistant microorganisms, perhaps no higher than 25%.

Also, one critical problem encountered in microbiological prospecting arises from the unavoidable time lag between collection of the soil samples in the field and testing them for the presence of hydrocarbon-indicating bacteria, usually in a laboratory. The vegetative non-hydrocarbon-indicating forms, which are present to an extent in the soil samples taken from over a hydrocarbon-bearing formation, continue to grow in the interval between collection and analysis. Thus their growth tends to mask the presence and concentration of the hydrocarbon-indicating types of bacteria. The net effect is to increase the difficulty of isolation of microorganisms which are the indicators of hydrocarbon deposits.

A technique developed by *D.O. Hitzman; U.S. Patent 3,096,254; July 2, 1963; assigned to Phillips Petroleum Company* is based on the discovery that the isolation of soil types of microorganisms, which are indicators of hydrocarbon deposits, can be facilitated by heating the soil or soil suspension to kill off substantially all vegetative cells. There remains the more resistant forms which are predominantly hydrocarbon-indicating microorganisms since certain hydrocarbon-indicating microorganisms have forms, most commonly the spore form, which are more resistant to heat than most normal soil types of bacteria.

Although this technique will leave viable other types of heat-resistant, non-hydrocarbon-indicating spores, the soil from above an oil deposit will contain a higher proportional number of heat-resistant, hydrocarbon-indicating microorganisms, and thus indicate the location of the deposit. This method will permit soil samples to be collected; heated to prevent bacterial change in the sample by overgrowth of non-hydrocarbon-indicating microorganisms in the vegetative form; and so enable the sample to be stored for longer periods of time before testing without the soil count changing.

A method developed by *D.O. Hitzman; U.S. Patent 3,174,910; March 23, 1965; assigned to Phillips Petroleum Company* describes a process for prospecting for oil in which the microbial action of pairs of samples of earth from each of a plurality of sample points are measured for the consumption of methane in one sample of each pair as compared to the consumption of the same concentration of methane and a hydrocarbon gas or gases heavier than methane in the other sample.

GEOPHYSICAL EXPLORATION

The geophysical exploration area is an area of widespread interest and impact. Over $1 billion was spent in worldwide geophysical hydrocarbon exploration in 1976 alone, reflecting the dependency of most nations on oil and gas (the demand for which is sharply increasing as the supply is decreasing) to supply the major part of their energy needs. It is still basically a U.S. technology although foreign activity in this area has been increasing. Thus, the problem of foreign technological developments which are not patented in the U.S. is minimized (3).

Geophysical exploration, broadly, can be defined as the analysis of the earth's structure and composition using physical measurements taken at, or near, the area to be studied. Generally, the area to be studied is hidden from direct view and lies under thousands of feet of soil and rock. Such exploration is used not only in geological studies but also in prospecting for valuable natural resources, including hydrocarbons (oil, gas, coal), minerals and water.

Geophysical exploration for hydrocarbons is a relatively new area of technology that is only about 50 years old. Since its inception, methods and apparatus used in hydrocarbon exploration have been continually improved or specially modified in order to permit optimum implementation for differing exploration environments, e.g. Arctic tundra, offshore areas, deserts, forests, etc.

Geophysical prospecting combines the sciences of physics and geology to assist the prospector in exploring for both oil and mineral deposits. Familiar examples include the use of scintillation counters for detecting radioactive uranium deposits and magnetic surveys for locating iron deposits.

Five major geophysical methods—magnetic, gravimetric, electrical, nuclear, and seismic—are routinely used in mineral exploration. Application of some of these methods and techniques requires complex and costly instruments and sophisticated methods of processing and interpreting the data, but others are relatively simple and less expensive. Among the latter are the magnetic and nuclear methods and some of the electrical techniques (1).

The seismic method is primarily used to determine underground geologic structures which are indicative of adjacent hydrocarbon deposits. Electrical-magnetic and gravitational prospecting methods are primarily used in initial reconnaissance surveys where little is known of the geology and where it is desired to ascertain whether a sufficient thickness of sediments of potential interest is present. Nuclear geophysical prospecting is largely confined to determining, in boreholes, the composition of the subsurface strata (3).

SPECTRAL SENSING METHODS

Among what might be called minor geophysical prospecting methods, apart from the big five (electrical, magnetic, gravimetric, nuclear, and seismic) are techniques using various portions of the electromagnetic spectrum.

A thermal sensing technique directed at the infrared portion of the spectrum has been described by *S.W. Milochik; U.S. Patent 2,933,923; April 26, 1960* as a means of prospecting for uranium and thorium ore deposits.

According to this technique, it is possible to measure the heating effect of uranium- or thorium-bearing ore bodies located at considerable distances under the earth's surface, depending upon the size and richness of the ore body, in test bores only 5 or 6 ft deep. By measuring the temperature at the bottom of a series of such shallow bores spaced about 300 ft apart along the ground, for example, it is possible to chart a temperature curve which is sufficiently precise to indicate the temperature rises of such small amounts as might probably be due to the heating effect of uranium and thorium disintegration.

A method developed by *J.F. Grayson and P.K.H. Groth; U.S. Patent 4,093,420; June 6, 1978; assigned to Standard Oil Company (Indiana)* is based on measuring the quantity of a characteristic of light which is absorbed or transmitted by specific organic particles in geologic specimens taken at at least two different depths. This method is particularly useful in the shallow portion of wells; that is, at depths down to the order of approximately 500 ft or so from the surface. However, the benefits of this method of prospecting can also be derived at depths in excess of 500 ft.

More specifically, in this method, at least two (and preferably more) samples of subsurface rock units are taken at different vertical depths in a plurality of locations. Each such sample is processed to recover organic material present in the rocks. From this organic residue, a number of specimens of the same palynomorph taxon are selected, and a characteristic of light absorbed or transmitted by the specific organic particles is determined.

The difference in the values obtained in the same location for two sets of palynomorphs at different vertical distances (preferably same well) is called the translucency differential. This differential or the gradient (differential divided by the corresponding vertical distance between sampling points) is then plotted on a map at the location of the various points of sample collection, and a contour map or cross section is drawn. This contouring tends to produce anomalies comparable to those obtained in seismic prospecting or in magnetic or gravimetric prospecting in that they tend to form ovals above the more deep-seated mineral deposit.

Accordingly, one employs the anomalies found from the contouring of the data to aid in the location of hydrocarbons or other mineral deposits.

A method developed by *L.S. Gournay, J.W. Harrell, and C.L. Dennis; U.S. Patent 4,132,943; January 2, 1979; assigned to Mobil Oil Corporation* is one in which a radar transmitter directs a beam of microwave energy at a first frequency through the atmosphere. A gas seep in the atmosphere irradiated by the beam of microwave energy is excited to emit microwave energy at a second frequency characteristic of the particular species of gas. A radar receiver is tuned to produce video signals which are representative of the microwave energy at the second frequency.

Figure 18 shows the essentials of such a system.

Figure 18: Mobil Oil Corp. Technique for Remote Sensing of Hydrocarbon Gas Seeps Utilizing Microwave Energy

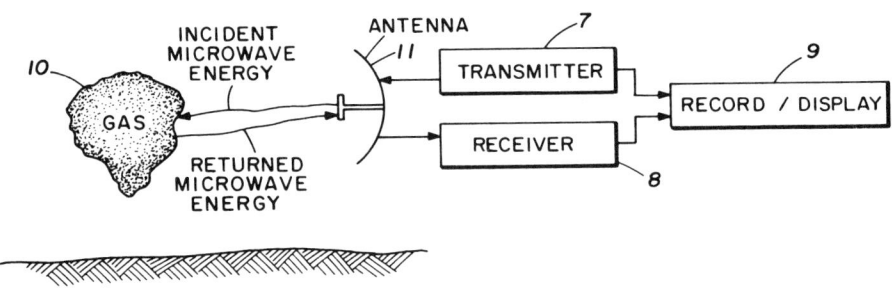

Source: U.S. Patent 4,132,943

A radar transmitter **7** and antenna **11** operate to emit a short, intense pulse of microwave energy in a concentrated beam. When a gas, such as from the subsurface gas seep **10**, is located within the path of this transmitted beam of microwave energy above the earth's surface, the microwave pulse delivers a fraction of its energy to those molecules of the gas that can absorb such energy. After the microwave energy irradiates these absorbing molecules, they immediately begin emitting some of the absorbed energy.

The frequency of the emitted energy is equal to the difference in rotational rates that were initially effected, regardless of the incident excitational frequency. This effect permits the exciting or pumping of the molecules with the radar transmitter at a frequency somewhat removed from the molecular resonance frequency. The pumping frequency may be greater than or less than the molecular resonance frequency.

In the particular system described herein, the difference between the pumping and molecular resonance frequencies may be as much as 100 MHz before system performance is adversely affected. One portion of the radar receiver **8** is tuned to the frequency of this energy emitted by the excited gas, thereby permitting the remote sensing of the gas seep without primary frequency interference from the microwave energy from the transmitter 7. Such tuned receiver produces video signals on record/display device 9 representing each gas seep detected during each sweep of the radar antenna. The amplitudes of these video signals are measured as an indication of gas concentration in the gas seeps, and the number of video signals are counted as an indication of the size of the gas seeps.

The antenna 11 is of the sector scanning type mounted in the nose of commercial aircraft for weather radar. The radar rotates through an arc of 120°, 60° on either side of the path of the aircraft. A cam arrangement rotates the antenna through its complete arc in two seconds and then reverses the movement so that the antenna is returned over the same arc in two seconds. Alternatively, a phased array antenna could be used.

REMOTE SENSING TECHNIQUES

The overall topic of remote sensing applications for mineral exploration has been reviewed in a volume edited by Smith (11).

Actually, remote sensing may be applied to a number of geophysical prospecting techniques as described later in this book under the headings of the individual techniques (see Magnetic Geophysical Prospecting, Gravitational Geophysical Prospecting and Surface Aerial Prospecting under Nuclear Geophysical Prospecting, for example).

As discussed by Smith, the newest and most intriguing example of remote sensing techniques involves the use of earth satellites rather than aircraft.

A system developed by *A. Madsen; U.S. Patent 3,043,908; July 10, 1962; assigned to Aircraft Engineering and Maintenance Co.* essentially comprises an aircraft capable of emitting a controlled pulse of ultraviolet radiant energy to an appreciable area of the earth's surface, and at a level of energy distribution such that the radiated area will exhibit a pattern of fluorescence and associated phenomena.

While ultraviolet radiation is rapidly absorbed in most forms of matter, the fluorescence of certain substances results in a luminescence which may be visually observed by personnel and equipment within the aircraft. This luminescence or fluorescence continues so long as the radiant stimulus producing it is maintained (ceasing about 10^{-8} second after the radiant excitation stops). Therefore, it is desirable to provide means to record the pattern of fluorescence for viewing and study, for example a video camera equipped with an optical recorder or viewer.

A principal application of this fluorographic survey technique is in the detection of hydrocarbon compounds, such as petroleum deposits or petroleum by-products associated with a submerged source. Specifically, it has been found that geological petroleum deposits undergo diverse conditions of diffusion and migration, through permeable faults, joints or minute fractures, principally adjacent to the edges of a deposit, and eventually exercise an influence on vegetation and cover soil at the earth's surface. Such influence is detectable as a fluorescence when radiated with ultraviolet energy in the manner herein contemplated.

In ocean areas, a similar influence is exerted at the surface of the sea by the propulsion systems of seagoing vessels so that the detection of such vessels, and particularly submerged vessels such as submarines, is made possible. In addition to hydrocarbon sources, it is contemplated that useful fluorographic information can be obtained with respect to a wide variety of other fluorescent substances on the earth's surfaces, such as various mineral deposits, etc., and that such information can be recorded from the airborne vantage point, and the information recorded or graphically presented.

A method developed by *J.S. Bradley and T.F. Moore; U.S. Patent 3,153,147; October 13, 1964; assigned to The Atlantic Refining Company* is an indirect, reconnaissance type method for locating petroliferous deposits under bodies of water by measuring at the surface of the water effects caused by seeps lifting underlying surface and subsurface waters to the surface.

Heretofore, naturally occurring seeps from petroliferous deposits have been located by various types of direct and indirect methods of exploration. The direct methods usually rely on gas analyzers or electromagnetic energy, such as ultraviolet energy, to detect the presence of hydrocarbons in the atmosphere or in earth samples. The indirect or geochemical methods rely on the detection of various indicators to determine the possible presence of petroliferous deposits. Some of the more common indicators are certain metals, H_2, CO, sulfides, salinity and hydrocarbon-consuming bacteria.

All of the known direct and indirect methods are subject to certain inherent disadvantages and limitations. This is especially true in detecting seeps in water-covered areas.

This method is based on the discovery that measuring the surface of a body of water for a preselected property affected by the seep's lifting action provides a greatly improved indirect method of exploration in water-covered areas. This improved method not only significantly increases the speed of analysis and the resulting speed of traverse, it also reduces the cost of each survey by at least an order of ten over the cheapest comparable water reconnaissance method heretofore available. Various surface water properties affected by the lifting action of the hydrocarbon seep can be used to detect the seep. Usually, the preferred method is to measure temperature changes caused by the seep lifting underlying surface waters to the surface of the body of water. However, other effects, such as resistivity changes can be used.

Figure 19 shows the seep's lifting action affecting the vertical thermal gradient of underlying surface waters. Gas bubbles **1** are shown seeping from petroliferous deposit **3** through fault **5** located in a portion of the earth **7** beneath body of water **9**. As bubbles containing methane gas and other hydrocarbons rise to the surface of body of water, they create a lifting action which moves the cooler underlying surface waters (shown in terms of isotherms displaying temperatures present in an extreme case) to the surface at bubble point or boil point **11**. The bubbles may or may not be visible.

However, this point will usually register the lowest temperature on the surface, and this effect of the cooler, underlying surface waters decreases as the waters leave the point of boil in ever increasing concentric circles.

Figure 19: Schematic of Atlantic Refining Co. Gas Lift Boil Detection Method

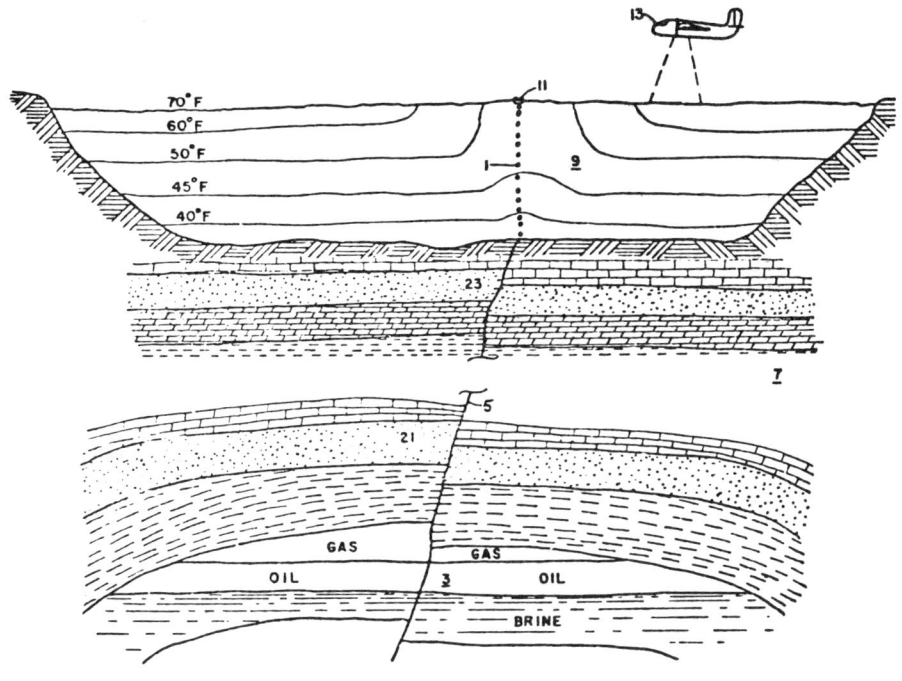

Source: U.S. Patent 3,153,147

This temperature anomaly in the vicinity of point **11** can be detected from a waterborne vehicle, not shown, or an airborne vehicle **13** mounting well-known detecting components as shown.

It is clear that as gas **1** moves from deposit **3** toward the surface of body of water **9**, it passes through sands, such as **21** and **23**, bearing mineralized subsurface waters. It is equally clear that portions of these subsurface waters and minerals dissolved or entrained therein are also moved to the surface by gas. Because of this action, minerals are added to the underlying surface waters and to the surface of the body of water and this increase in mineralization decreases the water resistivity. This decrease is readily detectable except in water already heavily mineralized, or in shallow, stationary fresh water.

In either of these two cases, the mineralization effect is usually masked. However, if the seep is large, a resistivity variation can sometimes be detected even in these waters.

A method developed by *G. Fiat; U.S. Patent 3,278,746; October 11, 1966; assigned to Ormad Systems* is one in which an aircraft is caused to travel along a systematic series of adjoining traverses, preferably at a specific altitude over the area which is to be surveyed. During such traverses, at least some specific portions of the area are continuously photographed by visible light to provide conventional strip films which may be geographically correlated with previously known objects and features of the terrain. In that manner a record of the precise position of the aircraft relative to terrain, at all times, is provided. Additionally, during such traverses of the terrain an infrared radiometric apparatus carried by the aircraft is utilized to generate first and second continuous strip photo records of the terrain.

The first photo record preferably is a recording of the infrared radiation which emanates from the terrain traversed within a shorter wavelength portion of the infrared region from about 1.0 to 5.5 microns wavelength. The second and similar (but not identical) photo record is generated by recording the radiation in a longer wavelength portion of the infrared region from 6 to 14 microns, which emanates from the traversed terrain.

Subsequently, the abovementioned infrared photographic records are developed and incrementally measured on a point-to-point basis to produce incremental information representative of the absolute intensities of infrared radiation and the differential of infrared radiation between the two bands which emanates from the various portions of the terrain.

The abovementioned incremental information is correlated with empirically predetermined information concerning the emissivities of various materials at different portions of the infrared region so that the geological nature of the individual components of the terrain may be determined solely from the records of infrared emission or from those records as they may be supplemented by maps of gamma ray activity and other records of the geological structures in the particular terrain.

A method developed by *A.R. Barringer; U.S. Patent 3,961,187; June 1, 1976; assigned to Barringer Research Limited, Canada* is a method of determining the presence of hydrocarbon seeps in the sea from an aircraft or other moving vehicle wherein an intense beam of light from an artificial source such as a laser is directed towards the sea and reflections and/or bioluminescence attributable to this beam occurring in the near surface of the sea are observed in the vehicle. Means is provided for discriminating between responses attributable to reflections from the surface of the sea and reflections and/or bioluminescence occurring below the surface of the sea.

Figure 20 shows, in the upper view the aircraft and sensing system and, in the bottom view, details of the sensing system. Referring to the figure, the apparatus may be installed in an aircraft **11** flying over the sea **12** preferably at low altitude of approximately 60 meters. Light from a pulsed laser system **13** is reflected from mirror **14** down to the surface of the sea where it penetrates through to layers of scattering **15** lying beneath the surface. Such layers of scattering which are associated with increased microbiological activity and the presence of gas bubbles, reflect the light back through a telescopic receiving system **16**.

The receiving system consists of a telescope **17** directing the received light onto a photomultiplier amplifier assembly **18** whence it is passed to an amplifier **18a**, the output of which is fed to a cathode ray tube oscilloscope **19**.

Geophysical Exploration

Figure 20: Barringer Research Ltd. Technique for Remote Sensing of Marine Hydrocarbon Seeps

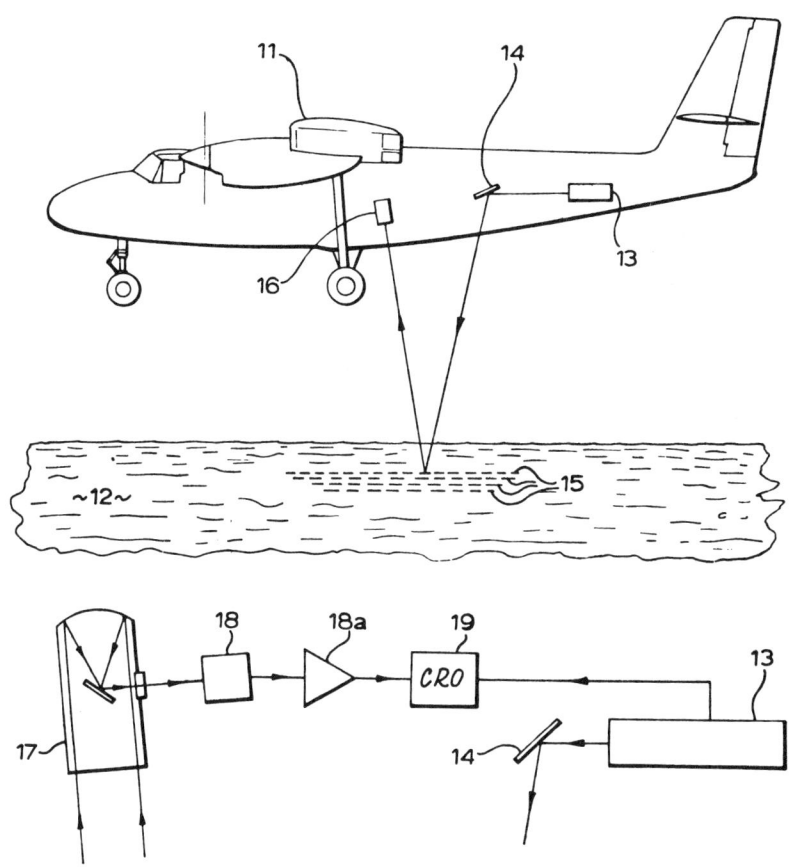

Source: U.S. Patent 3,961,187

The oscilloscope **19** is triggered from the laser system **13** and its sweep has a repetition rate that is tied synchronously to the outgoing pulses from the laser system. The face of the oscilloscope can be photographed continuously by means of a strip camera in order to produce a permanent record of the intensity of observed surface reflections.

The purpose of the oscilloscope display is to allow the signal received from the seawater surface to be differentiated from subsurface scattering and bioluminescence. This can be readily observed on the oscilloscope face and photographed. However, it will be appreciated that more sophisticated forms of recording can be employed. These include digital recording methods in which the received signal can be scanned and recorded digitally. This has the advantage of providing much greater dynamic range in the recording and allowing various types of signal processing to be employed to maximize the signal-to-noise ratio of the system. Furthermore, it is possible electronically to scan the signal and to carry out more sophisticated signal processing in real time if this is so desired.

In a typical survey operation for locating potential oil field target areas, parallel traverses over the sea are flown at fixed intervals such as one or two miles. This survey height may be as low as 60 meters in order to maximize the signal strength; however, higher altitudes may be possible depending upon the strength of the light source. Navigation is carried out by electronic means such as Doppler radar, inertial navigation or other suitable conventional techniques.

The aircraft positon is recorded during flight along with the data described above such that the two can subsequently be synchronized together. The location of anomalous subsurface zones of turbidity and bioluminescence eventually are plotted on maps of the survey area.

The images then may be analyzed to locate target areas for subsequent follow-up with marine seismic surveys. The objective is to use the method and apparatus to cover large areas at relatively high speed and low cost so that the much more costly and slower seismic techniques can be applied initially to areas of high priority selected from the sweep survey.

The depth of penetration of the light beam into the sea is a function of the wavelength and intensity of the light and the turbidity of the sea. For wavelengths of the ultraviolet range, a penetration of about 10 cm is an approximate useful working limit; for wavelengths in the visible range, the penetration of the beam would be on the order of 2 m.

A method developed by *N.K. Del Grande; U.S. Patent 4,005,289; January 25, 1977; assigned to United States Energy Research and Development Administration* is a method for locating and mapping the magnitude and extent of terrestrial heat-flow anomalies from 5 to 50 times average with a tenfold improved sensitivity over orthodox applications of aerial temperature-sensing surveys as used for geothermal reconnaissance.

The method remotely senses surface temperature anomalies such as occur from geothermal resources or oxidizing ore bodies by:

> measuring the spectral, spatial, statistical, thermal, and temporal features characterizing infrared radiation emitted by natural terrestrial surfaces;
>
> deriving from these measurements the true surface temperature with uncertainties as small as 0.05 to 0.5 K;
>
> removing effects related to natural temperature variations of topographic, hydrologic, or meteoric origin, the surface composition, detector noise, and atmospheric conditions;
>
> factoring out the ambient normal surface temperature for nonthermally enhanced areas surveyed under otherwise identical environmental conditions; and
>
> distinguishing significant residual temperature enhancements characteristic of anomalous heat flows and mapping the extent and magnitude of anomalous heat flows where they occur.

A method developed by *L.S. Gournay; U.S. Patent 4,100,481; July 11, 1978; assigned to Mobil Oil Corporation* is one in which microwave energy is radiated from an antenna transported along a traverse above the surface of the earth. Microwave energy reradiated from gas seeps is detected. Microwave energy reflected from hard targets along the traverse is also detected. Video monitors simultaneously display the detected reradiated and reflected microwave energy. In this way, hydrocarbon gas seeps are displayed in relation to topographical features along the traverse.

MAGNETIC GEOPHYSICAL PROSPECTING

Magnetic prospecting is based on the natural magnetic properties of some minerals such as magnetite. When held near a magnetite-rich rock, the needle in a compass behaves erratically because the earth's magnetic field is distorted by the local magnetic field. Minerals such as ilmenite (iron-titanium oxide), hematite, (iron oxide), and pyrrhotite (iron sulfide), are weakly to moderately magnetic, a property that can be recorded by sensitive magnetic instruments.

The common unit of measure for the strength of a magnetic field is the gamma. Where not disturbed by highly magnetic rocks, the strength of the earth's magnetic field in the conterminous United States in 1975 ranged from a low of about 48,000 gammas in Texas and Florida to a high of 60,000 gammas in Minnesota (1).

Instruments called magnetometers are used for direct detection of magnetic anomalies (that is, the distortion of magnetic minerals in crustal rocks superimposed on the earth's magnetic field). The magnetic readings over weakly magnetic host rocks may depart from local average, or background, values by 10 to 500 gammas, but over magnetic iron formations the readings may depart from background by 100 to 100,000 gammas.

The magnetometer can also be used to trace concealed rock formations that have magnetic properties differing from those of adjacent formations. For example, a prospector may know that copper is associated with an igneous rock such as quartz monozonite. If, as is often true, the quartz monzonite differs noticeably in magnetic response from the surrounding rocks, the magnetometer can be used to detect it beneath soil, talus, or other cover. Similarly, the black sand of placer deposits commonly contains grains of magnetite or ilmenite that affect the magnetometer. This instrument, therefore, can be used indirectly in the search for gold or other heavy minerals present in the black sand.

Some years ago the dip needle, a simple instrument using a pivoted magnetized needle, was often used to measure the magnetic field when great sensitivity was not required. Dip needles are no longer manufactured, as various types of more sensitive magnetometers have replaced them. These modern magnetometers can measure directly the total intensity of the earth's magnetic field at any one place or can measure the intensity in either the vertical or the horizontal direction of the field.

In a torsion magnetometer, a small magnet is attached to a fiber in such a way that the magnet twists the fiber in proportion to the intensity of the magnetic field as measured in the vertical direction. The flux-gate magnetometer is an electronic device that uses a magnetic core that becomes saturated in the earth's field to measure the strength of the field in the vertical direction (1).

The basic sensing element for a proton magnetometer consists of a container filled with a proton-rich liquid such as water or kerosene surrounded by a coil of wire. The protons within the liquid are subatomic particles that spin about rotational axes somewhat like the earth spins on its axis, completing one turn per day. The frequency with which the spin axes of the protons wobble or precess, after being excited by a current passing through the coil, is directly related to the strength of the earth's field. The frequency is measured and converted into readings in units of gammas. The proton magnetometer measures the total intensity of the earth's field rather than the intensity in the vertical or other direction.

Many commercial magnetometers range in price from $1,000 to $5,000. The less expensive ones are accurate to only about 20 gammas or less. The most expensive are accurate to about 0.25 gamma.

Magnetic surveys may be conducted either along a series of lines or in a grid pattern. The size of the area being prospected and the type of deposit being sought determines the spacing of the stations. Stations spaced 10 to 20 feet (3 to 6 m) apart may be required to locate small magnetic anomalies associated with weakly or moderately magnetic rocks, but stations spaced 100 feet (30 m) or more apart may suffice if the presence of highly magnetic rocks is suspected in a large area.

Power lines, rails, automobiles and other large magnetic objects should be avoided in any type of magnetometer survey because they create strong local magnetic fields that mask the anomalies inherent in the rocks.

Today most magnetic surveys are airborne or marine and use total-field detecting systems rather than measuring either the vertical or horizontal component of the earth's magnetic field. These surveys provide comprehensive, reliable data about regional magnetic trends. Ground magnetometer surveys are still used, however, to locate anomalies from small subsurface structures (1). The use of magnetic techniques in oil prospecting has been reviewed by Nettleton (12).

In magnetic prospecting, variations in the earth's magnetic field are measured. These variations indicate the existence of certain geologic structures in the formation lying below the earth's surface. Magnetic surveys may be carried out on the ground, under water, in the air, or within boreholes extending thousands of feet below the earth's surface.

Magnetic prospecting is one of the oldest methods of geophysical exploration, dating back in its most primitive form, to the 17th century. The basis of this technique is the fact that the earth's magnetic field at any point is a function of the geologic structure at that point.

The unit of magnetic field measurement in magnetic surveying is the gamma, as noted earlier, which is equal to 10^{-5} oersted. The magnetic field of the earth and that of the geologic bodies lying therein is expressed in terms of the magnitude and direction of the total field intensity vector T. The total field intensity vector may be resolved into its horizontal (H) and vertical (Z) intensity components. H in turn may be resolved into its X (north) and Y (east) components. The angle between H and T is known as the inclination (I) and may be defined as H = T cos I or tan I = Z/ZH. The declination, D, which is the angle between the magnetic north direction and the astronomic (true) north may be expressed in terms of the horizontal intensity (H) components X and Y as tan D = Y/X.

Although modern instruments can now measure the total field intensity, earlier measuring devices separately measured the various components of total field intensity.

Pivoted needle instruments were among the earliest devices used in magnetic exploration and included such types as the Swedish Mining Compass, the Hotchkiss Superdip, and the Thalen-Tiberg magnetometer. These pivoted needle devices, as the name implies, utilized a magnetic element in the form of a needle, positioned so as to rotate on pivots in response to an external magnetic field. Although widely used for many years, such instruments were limited in their range of application due to a lack of adequate sensitivity.

The art of magnetic prospecting experienced a significant breakthrough from 1913 through 1917 with the development by Adolph Schmidt of a vertical magnetometer. Schmidt's device, a precision-type portable magnetometer, able to measure small magnetic field variations, utilized a permanent magnet system balanced on a knife edge as its sensing element. While the prior art devices were primarily limited to searching for magnetite ore, Schmidt's instrument opened up exploration to the discovery of many other geologic structures.

Variations on Schmidt's magnetometer and other types of magnetic field measuring instruments (e.g., the magnetic torsion balance, magnetron, earth inductors and iron-induction) were developed in the ensuing years, providing a variety of instruments for magnetic prospecting. Although relatively accurate, these instruments could only be used in relatively slow-moving transport systems. Surveying with them was tedious and slow. What was needed was an instrument which could be used in rapidly moving transport systems to carry out surveys over vast regions at a rapid pace.

Just prior to World War II, Gulf Research & Development Company scientists developed a flux-gate magnetometer for use in aircraft. Such a device has been described by *V.V. Vacquier; U.S. Patent 2,406,870; September 13, 1946; R.D. Wyckoff; U.S. Patent 2,518,513; August 15, 1950; and V.V. Vacquier and G. Muffly; U.S. Patent 2,555,209; May 29, 1951; all assigned to Gulf Research & Development Co.*

Basically the flux-gate magnetometer is composed of two series-connected primary coils wound in opposition around a pair of parallel cores aligned in the direction of the earth's field. When the coils are energized the two cores are magnetized in opposite directions. Absent any external field the coils are balanced. In the presence of an external field, the field of one of the coils is enhanced while the field of the other is diminished, creating an imbalance which is detectable at the outputs of a pair of secondary coils wound around the cores.

Although used in World War II primarily for submarine detection, the airborne flux-gate magnetometer opened up new horizons for geophysical exploration. Unencumbered by the need for rotating sensing elements, the flux-gate magnetometer revolutionized magnetic prospecting. Surveys of vast regions could now be accomplished in a fraction of the time previously taken.

In the early 1950's progress in the art of magnetometers developed from different phenomena that had previously been considered. A total field proton magnetometer based on the principle of nuclear magnetic resonance was developed by *R.H. Varian; U.S. Patent 2,561,490; July 24, 1951; assigned to Varian Associates.* The magnetometer made use of the precessional frequency of oriented protons in a magnetic field which is proportional to the applied field intensity.

Another proton precession magnetometer has been described by *P.H. Serson; U.S. Patent 3,070,745; December 25, 1962; assigned to The Canadian Minister of Mines and Technical Surveys, Canada.*

Following the innovation of the proton magnetometer, other magnetic field measuring devices were developed based on nuclear resonance phenomena. These included magnetometers using alkali metal vapors as described by *H.G. Dehmelt; U.S. Patent 3,584,292; June 8, 1971; assigned to Varian Associates.*

The nuclear resonance-type magnetometer has further increased the precision and versatility of magnetic field measuring instruments. Some models are alleged to have a sensitivity of 0.005 gamma and are adaptable for use in the air, on the ground, under water or within a borehole.

Now that extremely precise magnetometers exist, different techniques are continually being developed for their actual field use. For example, *K.A. Ruddock, H.A. Slack and S. Briner, U.S. Patent 3,263,161; July 26, 1966; assigned (Slack) to Pure Oil Co. and (Ruddock and Breiner) to Varian Associates* have developed a system which employs a pair of vertically spaced magnetometers suspended from a helicopter. Spatial derivatives of the intensity of a magnetic field are measured as an indication of the depth and fall-off rate of subsurface magnetic anomalies. As Ruddock et al point out, abrupt contrasts in the magnetic susceptibility of adjacent subterranean volumes constitute sources of magnetic disturbances detectable by means of ground or aerial surveys of the earth's magnetic field. Such surveys have often been of limited value in view of difficulties in resolving and interpreting the data obtained.

For example in oil exploration, it is desirable to obtain knowledge of the structure of the igneous or metamorphic rock, known as the basement. Any abrupt change in the structure of this rock is indicative of corresponding structural changes in the overlying sedimentary layers which may be favorable to the accumulation of oil in these layers. Since the basement rock contains magnetite and is therefore magnetic, such changes in the structure of the basement rock give rise to magnetic disturbances. A condition known as basement differentiation wherein adjacent portions of the rock have markedly different magnetite concentrations also gives rise to magnetic disturbances.

Previous methods of magnetic anomaly surveying have inferred information of the depth and nature of such sources of disturbance by assuming that the disturbance arises from a magnetic body of some particular geometrical shape. The data obtained was not of sufficient accuracy to justify drilling operations until after a time-consuming and expensive ground survey could be made using acoustical methods.

Thus the principal object of the Ruddock et al system is to enable the making of magnetic surveys of greater exploratory value. Generally speaking, this is accomplished by making simultaneous measurements of total magnetic field intensity and the vertical gradient of this intensity, and then using these measurements to determine the depth and fall-off rate of subterranean magnetic disturbances.

In another development, *E.G. Zurflueh; U.S. Patent 3,490,032; January 13, 1970; assigned to Gulf Research & Development Company* employs a pair of magnetometers spaced along the direction of travel of the transport means and correlates the measurements to provide a single measurement free from time variations. Such instruments are of such increased sensitivity that it is now possible to detect anomalies much smaller than those detectable with equipment previously available.

This increased sensitivity can be used, for example, to directly detect structures in the sedimentary rocks of the earth. These sedimentary rocks, which overlie igneous and metamorphic rocks comprising the so-called basement, have very small values of magnetization compared to basement rocks. One main advantage of these high-sensitivity instruments is that they allow measurement of heretofore undetectable changes in magnetization occurring in the sedimentary rocks. This is of particular importance in oil exploration, because the great majority of all petroleum and gas deposits are found in sedimentary rocks.

A device developed by *S.H. Yungul; U.S. Patent 3,965,412; June 22, 1976; assigned to Chevron Research Company* makes use of a highly sensitive magnetometer to determine the geologic age of the segment of a formation surrounding a borehole by measuring the remanent magnetization of the formation. This enables the operator of the device to document stratigraphic boundaries occurring, e.g., during generation and migration of petroleum over a given span of geologic time such as when accumulative traps were generated.

Geologic dating of the relevant adjacent sections of strata within a borehole by conventional means is both time-consuming and costly. Use of in-hole dating equipment such as magnetometers and the like has not been successful in age-dating due to inaccuracy of the generated results. Use of cores of sediments, i.e., long cylinders of successive layers of sediment, is likewise costly and requires extensive well time to accomplish.

The direction of the remanent magnetization of an earth formation penetrated by a borehole is accurately measured utilizing a high-sensitivity and directional magnetometer positioned within the borehole so as to decrease sensitivity of the instrument to the induced magnetization of the earth's formation while maintaining high resolution for detection of the natural remanent magnetization property of the formation.

In the preferred logging posture the axis of response of the magnetometer is normal to the earth's normal field in a given azimuthal direction such that it identifies only the flux associated with the remanent magnetization in a plane normal to the earth's field. Since the response of the magnetometer is also a function of angular direction, means can also be provided for azimuthal rotation of the magnetometer in a plane perpendicular to the earth's magnetic field. Also, the azimuth for which signal intensity is maximum indicates polarity of a paleomagnetic reversal present in the adjacent earth formation which can then be used to indicate the age of the formation, if desired.

In a technique developed by *G.H. McLaughlin, H.A. Harvey, W.O. Cartier and W.A. Robinson; U.S. Patent 2,931,974; April 5, 1960; assigned to Crossland Licensing Corporation Limited, Canada* it is proposed to detect time transients of the earth's magnetic field within a frequency range of 1 to 20,000 hertz. It is the object of this technique to provide a reliable geophysical prospecting method which will locate subterranean conductors with a high degree of accuracy, yet which can be carried out with equal facility to present magnetometer methods for locating magnetic materials, without requiring any local transmitter.

Although the fields may be detected by the voltages induced in induction coils, this form of field detector is simply employed as the simplest illustration of the method. Flux-gate magnetic field detectors or other devices for measuring magnitudes of magnetic fields could be utilized by those skilled in the field of geophysics.

Current developments in magnetic prospecting involve the use of superconducting magnetic field sensors which allegedly have the versatility of flux-gate and nuclear resonance magnetometers, yet possess greater sensitivity. Such a device has been developed by *J. Nicol, S. Shapiro, and M.F. Roetter; U.S. Patent 3,829,768; August 13, 1974; assigned to The U.S. Secretary of the Navy.*

GRAVITATIONAL GEOPHYSICAL PROSPECTING

This category of geophysical prospecting comprises determining the force of gravity by direct measurement, comparison to other gravitational measurements, or determining acceleration differentials caused by gravity. The gravitation measuring systems may be directed to geophysical exploration for hydrocarbons or to prospecting for minerals.

Recent activity has been directed towards devices which measure gravitational fields by electrical measurement of relative differences in the acceleration of a plurality of masses. Particular emphasis has been placed on gravitometers capable of fine scale field measurement from high flying aircraft or earth satellites.

The use of gravitational techniques in oil prospecting has been reviewed by Nettleton (12).

The determination of gravity gradients has long been known to be a very desirable way of sampling relatively large volumes of strata because of the simple dependence of the vertical gravity gradient on the density of the strata. As an example, the determination of gravity gradients may be widely used to detect the presence and to evaluate the extent of ore bodies and oil reserves beneath the surface of the earth. Presently, vertical gradients of gravity are obtained from measurements of gravity at various levels, such as by lowering a gravimeter down a borehole.

These attempts, although encouraging, result in operations that are time-consuming and limited in resolution. Direct-reading gravity gradiometers have also been proposed. However, the practical utility of such instruments is hindered by technical difficulties which are encountered in confined spaces such as boreholes.

Other borehole or well surveying methods and apparatus have been developed over the years in an attempt to overcome the difficulties encountered in the construction of a rugged, reliable and easily operated instrument capable of measuring density of surrounding strata. These alternate methods and apparatus have been based upon the measurements of the velocity of sound or the injection of γ-rays into a stratum and determination of the γ-rays which are returned. These types of measurements suffer from the fact that their effective range beyond the borehole into which it is placed is very limited. Typically, this effective radius is not more than a few inches. In contrast, a sensitive and reliable gravity gradiometer should have an effective radius of detection in the range of several feet.

The various possible uses for a gravity gradiometer indicate the desirable characteristics one should possess. One use is in oil well logging in which the instrument is lowered into an existing well to determine the properties of the surrounding strata as a function of depth. Thus, any gravity gradiometer used in a borehole to determine the gravity gradients along the entire depth of the hole must be very compact in size, have a minimum number of external connections, and be reliable and rugged.

Another use for a gravity gradiometer is in general surface prospecting in which gravity gradients are measured and plotted over a surface area, whether the area is land or water. On land, this may be done by stopping at predetermined points and making measurements.

This same process may also, of course, be done on the water surface. However, this method of periodically stopping to determine gravity is a very expensive way of obtaining such measurements and often the terrain or sea conditions make such measurements hazardous or even impossible. This in turn has led to making gravity measurements over an area from an airplane or helicopter.

Presently available instruments are, however, influenced by the motion of the airplane or helicopter and the noise resulting from such motion detracts from their useful sensitivity. Thus, the use of a gravity gradiometer in general surface prospecting indicates that the instrument should be one which in its operation is independent of vehicle motion.

A technique developed by *P.C. Von Thuna; U.S. Patent 3,865,467; February 11, 1975; assigned to Arthur D. Little, Inc.* involves measuring gravity gradients directly. Two retroreflectors are caused to experience free flight, and the radiation reflected by these retroreflectors during free flight is directed to a detector in a manner to cause interference between the reflected beams. The lower retroreflector is made to serve as a beam splitter thus making it possible to use a single optical path and minimize the number of optical components required. The interference frequency is determined as a function of time; and the rate of change of this frequency is directly related to gravity gradient.

Such an apparatus is claimed to be in its operation independent of its motion, rugged and reliable and having a minimum number of external connections.

A technique developed by *H.D. Black; U.S. Patent 3,888,122; June 10, 1975; assigned to U.S. Secretary of the Navy* provides a system useful in measuring the fine scale gravitational field of the earth. The system utilizes two closely-spaced earth orbiting satellites employing radar altimeters, and having the same orbit. As they traverse their orbit the earth is also spinning thereby allowing the satellite pair to see a strip of the earth's surface and to accurately measure the cross-orbit component of the deflection of the vertical (the local gravitational force vector). The relative positions of the two satellites must be accurately known, but, to obtain the fine scale measurement, the absolute position of neither satellite need be accurately known.

It is well known that an instrument for continuously measuring the vertical gravity gradient with a resolution of 10 Eotvos units in an oil well environment would permit exploration geophysicists and petrology engineers to differentiate oil and water bearing strata in the vicinity of a borehole.

The principal prior art method for determining the vertical gravity gradient for well-logging purposes required a detailed gravity survey using gravitational field intensity measurements produced every two meters by a borehole gravimeter, followed by the appropriate calculation to generate the gradient. This method of determining the gravity gradient was extremely time consuming and is inherently incompatible with continuous well-logging practice.

Other techniques for generating the gravity gradient have been developed for airborne geophysical use and for space flight.

However, none of these known techniques are suitable for an instrument capable of continuously measuring the vertical gravity gradient in a borehole environment due primarily to the extreme sensitivity to vibration and acceleration of the associated instruments. Furthermore, the speed of response and common mode rejection of the earth's gravitational field in such systems is severely limited due to inherently small dynamic range and insufficient linearity of the sensor.

A device developed by *S.W. Buck; U.S. Patent 3,926,054; December 16, 1975; assigned to The Charles Stark Draper Laboratory, Inc.* was designed to overcome these problems and to provide a continuous reading gravity gradiometer for use in a borehole environment.

Two dynamically balanced pendulums (DBP's) or alternatively, rotate-to-null specific force detectors, each including a pendulous sensor, are configured with their respective sensitive axes being parallel to a common gradiometer sensitive axis and with the sensors being separated by a predetermined distance along the gradiometer sensitive axis. A coupling means rotationally couples the sensors about their sensitive axes.

According to one embodiment, e.g., the two sensors are pendulous integrating gyros (PIG's). These sensors are connected by a coupling shaft of predetermined length with the longitudinal axis of the coupling shaft corresponding to the sensitive axis of the gradiometer. The gradiometer sensitive axis passes through the intersection points of the input and output axes of both PIG's. The two PIG's are further constrained to have substantially identical pendulosity-to-wheel angular momentum ratios.

The first DBP (DBP-1) is arranged in an accelerometer configuration (PIGA) including its PIG, associated servo electronics (including a signal generator) and a torque motor for rotating the coupling shaft and the coupled PIG's about the sensitive axis. The first PIG and associated servo electronics and torque motors are configured to null the gravitational acceleration-induced angular displacement of the first PIG input axis relative to the sensitive axis by rotating the coupled PIG's about the sensitive axis. The resultant precession rate of the shaft and coupled PIG's about the sensitive axis at null is proportional to the gravitational force sensed at the first PIG.

The second DBP (DBP-2) is arranged in a torque-to-balance loop configuration including its PIG, associated servo electronics (including a signal generator) and PIG torquer for supplying torque about the second PIG output axis. The second PIG and associated servo electronics and torquer are configured to null the angular displacement of the second PIG input axis from the sensitive axis by torquing the second PIG about its output axis. Since the second PIG is constrained by the coupling shaft to precess about the gradiometer sensitive axis at the same rate as the first PIG, the torque-to-null signal generated by the second PIG servo electronics is proportional to the difference in the gravitational force sensed at the first and the second PIG's.

Where the sensitive axis is aligned with the vertical axis, this signal is directly proportional to the vertical gravity gradient along the length of the coupling shaft. In this configuration, the vertical gravity gradient may be continuously measured with a high speed of response and substantial common mode rejection of the earth's gravitational field due to the inherently large dynamic range and linearity of the gyro sensors. Because measurements are made with respect to inertial space, rotational stability of the outer case is not required.

A device developed by *J.N. Preston; U.S. Patent 3,991,625; November 16, 1976;* is a gravity-measuring device for water and mineral detection. The device consists of an electric motor and generator whose moving parts are connected to each other by torque-transmitting means so that the torque of the electric motor and the resistance of the electric generator operates to lift either the electric motor or generator for a distance against the force of gravity.

The distance is determined by a mercury-type electric switch and the differences in gravity at different places above the earth are measured both by a motor ammeter and a generator ammeter to detect minerals and water beneath the surface of the earth.

Another device developed by *J.N. Preston; U.S Patent 4,022,064; May 10, 1977* consists of an electric generator powered by an electric motor that is activated by any source of electricity so that the torque of the electric motor reacting against its stator assembly and the action of the stator of the generator resisting the magnetic force of its rotor both act to electrically indicate the differences in acceleration of gravity and other accelerations when electrically measured by electric current measuring means.

Electrically measuring the minute difference in gravity intensities at different places on the earth's surface by such sensitive means indicates the existence of nonexistence of water and minerals of various densities beneath the earth's surface.

A device developed by *S. Stauber; U.S. Patent 4,023,413; May 17, 1977; assigned to Wyler AG, Switzerland* is one for measuring accelerations, particularly components of accelerations due to gravity. It comprises a stator with first and second spaced apart electrodes. A deflecting part is suspended between the first and second electrodes and it is displaceable under the influence of gravity in a direction toward one or the other electrodes and in so doing produces an electric signal which is proportional to the displacement amount. The signal is connected to means for indicating the magnitude of displacement as a measure of the acceleration acting upon the device.

A method developed by *H.B. Hunt; U.S. Patent 4,068,160; January 10, 1978; assigned to Texas Pacific Oil Company, Inc.* is a method for determining a set of surface locations lying along a predetermined line of survey beneath which there is a relatively high probability of existing subsurface mineral bodies exhibiting relatively high density and relatively low magnetic susceptibility.

It is carried out by selecting for membership in the set those surface locations lying in regions where a local minimum of magnetic field intensity substantially correlates to a local maximum of gravitational field intensity; plus those surface locations lying within regions of local topographic irregularity where the gravitational field intensity substantially directly correlates with the surface elevation; and those surface locations lying within regions where a steep gradient between a local maximum and an adjacent local minimum of the magnetic field intensities substantially directly correlates to a steep gradient between a local maximum and an adjacent local minimum of the gravitational field intensities.

Within the field of geophysical prospecting, it is well known that valuable information pertaining to the characteristics of subsurface features can be obtained by measuring the variations of the earth's magnetic and gravitational fields as a function of surface location. As a consequence, various well known apparatus have been developed which are capable of determining with a high degree of accuracy the relatively low magnitude variation in the earth's magnetic and gravitational fields resulting from differences in subsurface structure underlying different surface locations.

For example, the vertical component of the magnetic field intensity may be measured using the Askania Model GSZ Magnetometer or the total magnetic field intensity may be measured using the geoMetrics Model G-816 Magnetometer, while the gravitational field intensity may be measured using the Worden Prospector or Master model gravity meter or the LaCoste Romberg Model G gravity meter.

In general, such apparatus is utilized to determine the relative magnetic field intensity and relative gravitational field intensity at each of a plurality of predetermined surface locations selected along a predetermined line of survey. The method of the process may then be utilized to determine a master set of the predetermined surface locations beneath which there is a relatively high probability of existing subsurface mineral bodies having relatively high density and relatively low magnetic susceptibility.

For example, the method of the process may be used advantageously to detect near surface fluorspar deposits in a sedimentary rock structure when the density of the fluorspar deposit is greater than that of the surrounding sedimentary rocks and the magnetic susceptibility of the fluorspar deposit is less than the magnetic susceptibility of the surrounding sedimentary rocks.

Figure 21 shows a diagrammatical representation of the preferred embodiment of the method as it could be practiced using either analog or digital automatic data processing equipment or the like.

Figure 21: Texas Pacific Oil Company Method for Detecting Subsurface Mineral Bodies

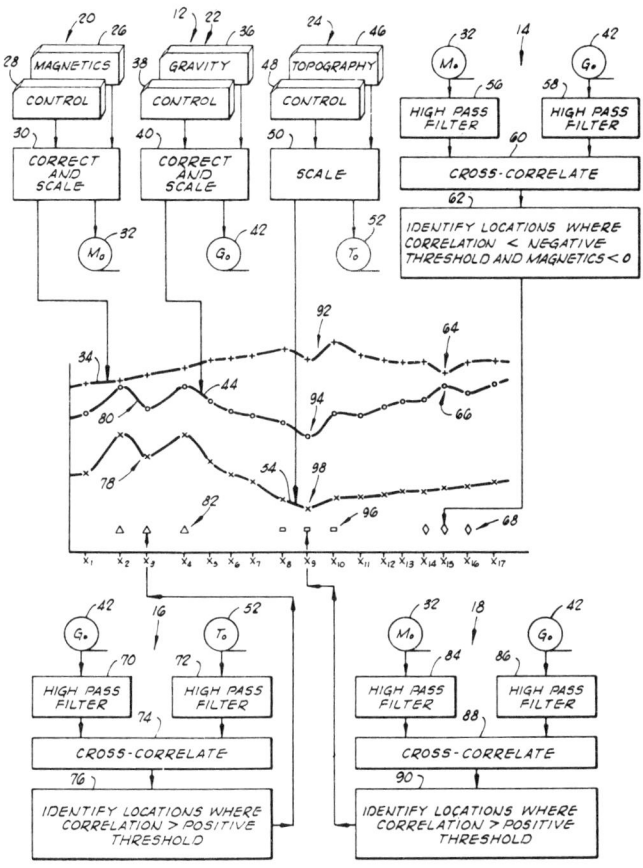

Source: U.S. Patent 4,068,160

If desired, the method may be practiced by a skilled geophysicist by allowing the automatic data processing to perform only those functions for which it is best adapted, such as the operations comprising the data preparation phase **12**, with the geophysicst performing the specific identification function. For convenience of reference, the preferred embodiment of the method can be considered as consisting of a data preparation phase **12**, a vein detection phase **14**, a bed detection phase **16**, and a fault detection phase **18**. The data preparation phase **12** can be further subdivided into a magnetics preparation phase **20**, a gravity preparation phase **22** and a topographical preparation phase **24**.

In the magnetics preparation phase **20**, a raw magnetics data file **26** is constructed in a conventional manner by recording the magnetic field intensity measured at each of the predetermined surface locations on a medium acceptable to the automatic data processing equipment, such as punched cards. However, it is well known that the raw magnetic field intensities contain components attributable to physical phenomena other than near surface structure. In general, it is preferable to remove these undesired components. To this end, a magnetics control information file **28** is constructed in a manner similar to the magnetics data file, with the control information file containing such information as the diurnal variations of the earth's magnetic field during the period that the raw magnetic field intensities were obtained, the ambient temperature at each predetermined surface location at the time the associated magnetic field intensity was obtained, and the normal correction factor attributable to the normal variation of magnetic intensity over the earth's surface, each of these factors being well known in the art.

The magnetics data file **26** and the magnetics control information file **28** are thereafter utilized by the automatic data processing equipment in the magnetics correction process **30** under the control of an appropriate program constructed to operate in a conventional, well known manner. Preferably, the automatic data processing equipment will also be programmed to adjust the amplitude scale of the magnetic field intensities to a preselected range or scale. After correction, the magnetic field intensities **32** should then be made available for use in subsequent phases by being recorded on a suitable medium such as magnetic tape. If desired, a two dimensional representation of the corrected magnetic field intensities, such as the magnetic plot **34**, may be constructed utilizing conventional plotting equipment.

In the gravity preparation phase **22**, a raw gravity data file **36** is constructed in a conventional manner by recording the gravitational field intensity measured at each of the predetermined surface locations on a medium acceptable to the automatic data processing equipment. As is well known in the art, the raw gravitational field intensities contain components attributable to physical phenomena other than near surface structure.

In general, it is preferable to remove these undesired components. To this end, a gravity control information file **38** is constructed in a manner similar to the magnetics control information file **28**, except that the gravity control information file **38** contains such information as the latitude and relative surface elevation at each predetermined surface location, the latter factor being multiplied in a well known manner by a regional elevation correction having a free air correction factor to compensate for the vertical decrease of gravity with an increase in elevation and a Bouguer correction factor to compensate for the attraction of material between a predetermined reference elevation and the particular surface location, each of these factors being well known in the art.

The gravity data file **36** and the gravity control information file **38** are thereafter utilized by the automatic data processing equipment in a gravity correction process **40** under the control of an appropriate program constructed to operate in a conventional, well known manner. Preferably, the automatic data processing equipment will also be programmed to adjust the amplitude scale of the gravitational field intensities to a preselected range or scale.

After correction, the gravitational field intensities **42** should then be made available for use in subsequent phases by being recorded on a suitable medium, such as magnetic tape. If desired, a two dimensional representation of the corrected gravitational field intensities, such as the gravity plot **44**, may be constructed utilizing conventional plotting equipment.

In the topographic preparation phase **24**, a raw topographic data file **46** is constructed in a conventional manner by recording the surface elevation at each of the predetermined surface locations on a medium acceptable to the automatic data processing equipment. Although the raw surface elevation information is generally of sufficient accuracy, it is preferable that the amplitude scale of the surface elevations be limited to a preselected range or scale. To this end, a topographic control information file **48** is constructed in a manner similar to the magnetics control information file **28**, except that the topographic control information file **48** contains information relating to the preselected range or scale.

The topographic data file **46** and the topographic control information file **48** are thereafter utilized by the automatic data processing equipment in a topographic scaling process **50** under the control of an appropriate program constructed to operate in a conventional, well known manner to adjust the amplitude scale of the surface elevations to the preselected range or scale. After scaling, the surface elevations **52** should then be made available for use in subsequent phases by being recorded on a suitable medium, such as magnetic tape. If desired, a two dimensional representation of the scaled surface elevations, such as the topographic plot **54**, may be constructed utilizing conventional plotting equipment.

In the vein detection phase **14**, the magnetic field intensities **32** and the gravitational field intensities **42** are utilized to select for membership in the master set a first subset of surface locations at which a local minimum of the magnetic field intensities **32** substantially directly correlates to a local maximum of the gravitational field intensities **42**. More particularly, the magnetic field intensities **32** and the gravitational field intensities **42** are utilized by the automatic data processing equipment in respective high pass filter processes **56** and **58** under the control of appropriate programs constructed to emphasize those variations in magnetic and gravitational field intensity attributable to near surface anomalies of rather narrow extent. Such filtering operations may be performed utilizing such conventional techniques as time or frequency domain filtering, least squares polynomial approximations or moving averages.

The filtered magnetic field intensities and the filtered gravitational field intensities are then utilized by the automatic data processing equipment in a first cross-correlation phase **60** under the control of an appropriate program constructed to correlate the set consisting of the filtered magnetic field intensities with the set consisting of the filtered gravitational field intensities to produce a set consisting of the first correlation value for each of the predetermined surface locations.

The automatic data processing equipment may then utilize the set of first correlation values in a first identification phase **62** to identify for membership in the first subset those surface locations where the first correlation value is less than a predetermined negative threshold indicating that a local anomaly of the magnetic field intensities (one such anomaly being indicated by reference number **64**) negatively correlates with a local anomaly of the gravitational field intensities (one such anomaly being indicated by the reference number **66**). In order to eliminate from consideration those surface locations at which a local maximum of the magnetic field intensities negatively correlates with a local minimum of the gravitational field intensities, the automatic data processing equipment should further identify those surface locations where the first correlation value was determined to be less than the predetermined negative threshold and where the filtered magnetic field intensities are less than zero indicating a local minimum thereof. This first subset of surface locations may then be identified as the two-dimensional representation in an appropriate manner, if desired, such as the diamond-shaped marks **68**.

In the bed detection phase **16**, the gravitational field intensities **42** and the surface elevations **52** are utilized to select for membership in the master set a second subset of surface locations consisting of those surface locations lying within each region of local topographic irregularity where the gravitational field intensity substantially directly correlates with the surface elevation, indicating a subsurface density greater than expected. More particularly, the gravitational field intensities **42** and the surface elevations **52** are utilized by the automatic data processing equipment in respective high pass filter processes **70** and **72** under the control of appropriate programs constructed to emphasize those variations in gravitational field intensity attributable to near surface anomalies of rather narrow extent and local topographic irregularities, respectively.

The filtered gravitational field intensities and the filtered surface elevations are then utilized by the automatic data processing equipment in a second cross-correlation phase **74** under the control of an appropriate program constructed to correlate the set consisting of the filtered gravitational field intensities with a set consisting of the filtered surface elevations to produce a set consisting of the second correlation value for each of the predetermined surface locations.

The automatic data processing equipment may then utilize the set of second correlation values in a second identification phase **76** to identify for membership in the second subset those surface locations at which the second correlation value exceeds a predetermined positive threshold, indicating that a local topographic irregularity (one such irregularity being indicated by reference number **78**) positively correlates with the local anomaly of the gravity field intensities (one such anomaly being indicated by reference number **80**). This second subset of surface locations may then be identified on the two-dimensional representation in an appropriate manner, if desired, such as the triangle-shaped marks **82**.

In the fault detection phase **18**, the magnetic field intensities **32** and the gravitational field intensities **42** are utilized to select for membership in the master set a third subset of surface locations consisting of those surface locations lying within each region where a steep gradient between a local maximum and an adjacent local minimum of the magnetic field intensities substantially directly correlates with a steep gradient between a local maximum and an adjacent local minimum of the gravitational field intensities.

More particularly, the magnetic field intensities **32** and the gravitational field intensities **42** are utilized by the automatic data processing equipment in respective high pass filter processes **84** and **86** under the control of appropriate programs constructed to emphasize those variations in magnetic and gravitational field intensity attributable to near surface anomalies of rather narrow extent.

The filtered magnetic field intensities and the filtered gravitational field intensities are then utilized by the automatic data processing equipment in a third cross-correlation phase **88** under the control of an appropriate program constructed to correlate the set consisting of the filtered magnetic field intensities with the set consisting of the filtered gravitational field intensities to produce a set consisting of the third correlation value for each of the predetermined surface locations.

The automatic data processing equipment may then utilize the set of third correlation values in a third identification phase **90** to identify for membership in the third subset those surface locations where the third correlation value exceeds a second positive threshold, indicating that a local anomaly of the magnetic field intensities (one such anomaly being indicated by the reference number **92**) positively correlates with a local anomaly of the gravitational field intensities (one such anomaly being indicated by the reference number **94**).

This third subset of surface locations may then be identified on the two-dimensional representation in an appropriate manner, if desired, such as the rectangle-shaped marks **96**. In general, the ability of the fault detection phase to detect actual subsurface faulting may be improved by correlating the set of surface elevations with the set of magnetic or gravitational field intensities to detect local topographic anomalies (one such anomaly being indicated by the reference number **98**) which correlates with the identified magnetic and gravitational anomalies.

A gravity related technique has been developed by *A.R. Geiger; U.S. Patent 4,121,464; October 24, 1978.* To locate potentially hydrocarbon-bearing subsurface formations, tiltmeters measure the time of arrival and the apparent direction of a lunar-induced earth tide at an array of points on the earth's surface. The measurements are combined to determine the shape of the tidal wave in a region of interest. The shape is indicative of subsurface viscosity in the region. Subsurface formations having an abnormally low viscosity are considered potentially hydrocarbon bearing.

ELECTRICAL GEOPHYSICAL PROSPECTING

In electrical prospecting, electrical properties of earth formations are measured to provide a clue to the nature of the geologic structure situated below the surface. Electrical surveys, in one form or another, may be carried out on the ground, under water, in the air or within boreholes extending thousands of feet below the earth's surface.

Numerous electrical exploration techniques have been developed, yet they have had limited application because of their lack of effectiveness at extreme depths below the earth's surface. Electrical methods can be divided into two groups: those which depend on naturally occurring fields and currents within the earth, such as the self-potential, telluric and magnetotelluric methods; and, those which depend on the artificial introduction of currents and fields in the earth, such as resistivity, potential profile, induced polarization, and electromagnetic methods. Electrical exploration, like magnetic prospecting, may be carried out on the ground, in the air, under water and below the earth's surface.

Aboveground electrical surveys are used more for mineral deposit investigation than for oil prospecting. However, the adaptability of certain electrical techniques to a borehole environment presents a useful tool in the discovery of hydrocarbon deposits within the formations surrounding a well bore. While certain electrical techniques are limited in their operational procedures, electromagnetic processes are adaptable to airborne, ground and borehole exploration.

Some historical notes on electrical geophysical prospecting have been recorded by Accaud and Martin (13).

Most electrical prospecting is based on the fact that various minerals and rocks offer differing degrees of resistance to the flow of electricity. The common unit of measure for electrical resistivity is the ohm. The electrical resistivity of common rocks can vary from several thousand ohms for some igneous and metamorphic rocks to only a few ohms for shales and clays. The resistivity of most common sulfide minerals such as chalcopyrite (copper-iron sulfide) and galena (lead sulfide), but not including sphalerite (zinc sulfide), is only a fraction of an ohm. The resistivity of graphite and carbon is also very low. If the individual grains in a sulfide orebody are in good electrical contact with each other, the entire orebody may offer a very low resistance to the flow of electricity when compared with the surrounding rocks. Such bodies are refined by geophysicists as conductors (1).

A large variety of techniques may be used in searching for conductors that are orebodies. Many of these techniques require very expensive and complicated instruments and large field crews to make the measurements, and mathematical computations must be applied to the data before interpretation. A few techniques, however, use only moderately expensive equipment that is easily operated by one or two persons and that requires little or no use of mathematics.

NATURALLY OCCURRING FIELD TECHNIQUES

Self-Potential Field Techniques

Electrochemical processes sometimes cause orebodies to act as natural batteries. This causes the surface of the earth over the top of the orebody to have a negative electrical potential with respect to the surrounding area. In the self-potential method two nonpolarizing electrodes placed on the surface of the earth are connected to a sensitive millivoltmeter, and the difference in potential is measured.

Ideally, one electrode is left in a fixed position, and the other electrode is moved along lines or a grid to measure the variations in self-potential. Price of the equipment ranges from $1,000 to $2,000. No calculations have to be performed on the data unless more than one location is used for the fixed electrode.

A number of sources other than mineral deposits have variations in self-potential. Unless other evidence also indicates the presence of an orebody, caution should be used in considering the importance of a self-potential anomaly (1).

In the search for mineral bodies, the self-potential technique dates back to the early part of this century and involves the measurement of naturally occurring electrical potentials on the ground or within a borehole.

A very early patent on the application of this technique to "location of ores in the subsoil" was that by *C. Schlumberger; U.S. Patent 1,163,469; December 7, 1915.*

A somewhat later patent by *C. Schlumberger; U.S. Patent 1,913,293; June 6, 1933* describes an electrical process whereby the porous strata, traversed by drill holes, may be studied by means of potential measurements effected inside the drill hole; and thereby their thickness and depth may be determined as well as the pressure existing therein.

The process consists in studying the various strata with reference to their porosity and their pressure, by taking measurements of the electrical potentials existing at various depths in the uncased part of the drill hole, after filling the drill hole with water.

The process is based upon the phenomenon known as electrofiltration, i.e., the occurrence of an electromotive force whenever an electrolyte flows through a porous dielectric, for instance when water filters through a layer of sand.

Such an electromotive force is known to depend both on the chemical constitution of the liquid under filtration and on the dielectric constituting the filter. Its order of magnitude is proportional to the pressure which causes the filtration. In most cases the sign of this electromotive force is such that the electric current which it causes to flow possesses the same direction as the liquid in movement. This electric current gives rise by ohmic effect to differences of potential, and the measurement of these differences constitutes the basis of the process.

Telluric Techniques

The telluric current method relies on the measurement of potential differences at the earth's surface caused by telluric current flow. Telluric currents are naturally occurring variable currents which permanently circulate in vast sheets within the outermost layers of the earth's crust. The source of such currents is believed to be astronomic in nature, caused primarily by electromagnetic phenomena of ionospheric origin. The distribution of these telluric currents is determined by the resistive nature of the formations through which they flow. Thus, any resistive anomaly will cause a disturbance in the telluric current flow pattern which can be detected at the earth's surface by measuring the potential gradients associated therewith.

Much of the early work on this technique was done in the 1930s by M. Schlumberger who established the basic technique of comparing the telluric current potential at different stations. One early disclosure was given by *M. Schlumberger; U.S. Patent 2,034,447; March 17, 1936; assigned to Compagnie Generale de Geophysique, France.*

Further details were given by *M. Schlumberger; U.S. Patent 2,284,990; June 2, 1942.* As shown in Figure 22, in field operation telluric measurements are obtained at different stations, at a predetermined frequency or bandwidth and compared over a selected period of time.

Figure 22: Telluric Current Prospecting

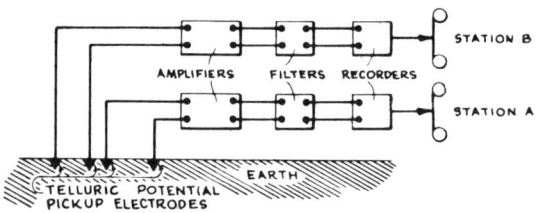

Source: U.S. Patent 2,284,990

Later, *G. Kunetz; U.S. Patent 2,586,667; February 19, 1952; assigned to Compagnie Generale de Geophysique, France* expanded on this technique.

Then, *S.H. Yungel; U.S. Patents 3,188,558; June 8, 1965; and 3,188,559; June 8, 1965; assigned to California Research Corporation* introduced a collinear three-electrode array and Fourier analysis techniques into the measurement of telluric potentials, which contributed to the development of telluric prospecting.

A method developed by *S.J. Pirson and J.E. Pirson; U.S. Patent 3,943,436; March 9, 1976* is a method of exploration for deposits of oil, gas and of other minerals in the earth including geothermal energy, which is based on the existence of electrotelluric currents that are generated spontaneously by such deposits because of the geochemical modifications caused by their presence within rocks in the proximity of such deposits, which method consists in measuring the magnetic perturbations created by the electrotelluric currents in the normally exisitng earth magnetic field.

When such electrotelluric currents exist, closed line-integrals of the earth magnetic field performed at or near the earth's surface do not vanish and the residual values of such integrals are a direct function of the magnitude and of the polarity of the electrotelluric current flux densities generated by the underground mineral deposits sought.

The concept of magnetoelectric exploration and its subsequent verification by field tests have established that the vertical electrotelluric currents generated in the earth by the fuel cell effect of oil and gas accumulations, of sulfide ore deposits and of geothermal heat sources give rise to anomalies in the static earth magnetic field.

Figure 23 gives a generalized representation of the physico-chemical modifications generated by an oil and gas accumulation within a stratigraphic trap in the overlying sediments and it indicates the paths taken by the vertical electrotelluric currents generated by the oil field subsurface fuel cell.

The fundamental mechanism of oil and gas migration and accumulation which is implied here, and which is also generally accepted by the petroleum geology profession, is that formation waters which were expelled from shales under the compacting effect of the weight of the rock overburden and which contained highly disseminated hydrocarbon globules or hydrocarbons in solution because of the elevated formation water temperatures at depth, escaped vertically upward above the loci of oil and gas entrapment, i.e., structural or stratigraphic traps.

Such expelled waters were filtered out of their suspended hydrocarbons; oil and gas pools were thus formed. In their vertical escape paths, the waters released the hydrocarbons still in solution in a gradual manner as temperature declined at shallow depths, and these hydrocarbons rendered the overburden rocks more reduced than when they were originally laid out, thus forming a chimney or funnel of reduced rocks which extended to the surface of the earth.

When surface rocks were exposed to weathering agents, the reduced nature of the upper part of the chimney may have been partially oxidized. However, the escape of compaction waters and of hydrocarbons takes place continuously during geologic time and thus the fuel cell is being replenished continuously.

Figure 23: Diagram Showing Electrotelluric Currents Generated by Underground Petroleum Deposit

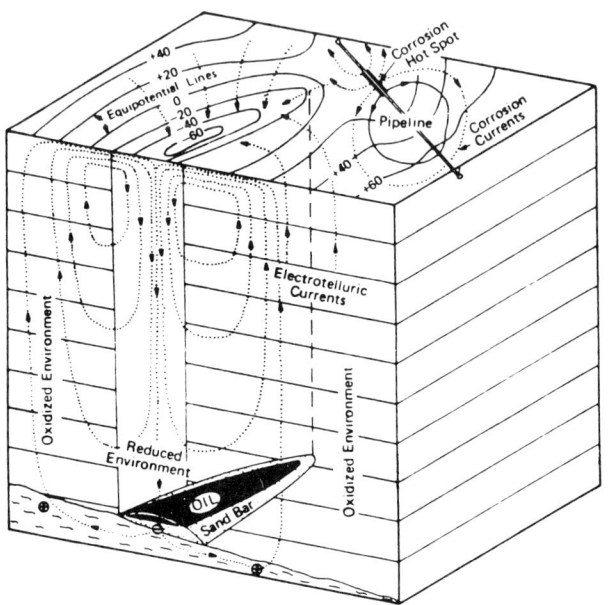

Source: U.S. Patent 3,943,436

A permanent and durable physico-chemical contrast is thus created above oil and gas pools which is manifested in many ways, but for this purpose, this contrast is mostly observed in the near-surface rocks by their ability to generate electric potentials and currents as if from a giant fuel cell, the source of potential being distributed on the external surface of the subsurface chimney.

This fuel cell generates current flow lines closed on themselves, a few of which are represented in the figure. Normally, the direction of the earth currents, or electrotelluric currents so generated is as shown over this stratigraphic trap. Such currents and their distribution in the earth last indefinitely during geologic time, or as long as the fuel cell is being resupplied with fuel in the form of escaping hydrocarbons in solution or in suspension in the compaction-expelled waters or by molecular gas diffusion from the accumulation of oil and gas.

Magnetotelluric Techniques

L. Cagniard in the early 1950s proposed to compare variations occurring in the electric telluric field in a given area over a given period of time with variations occurring in the magnetic field of the earth in that area. This technique, which is based on the fact that magnetic field variations occurring at the earth's surface are not independent of variations in direction and intensity of telluric currents, became known as the magnetotelluric method.

An early disclosure of this technique was by *L. Cagniard; U.S. Patent 2,677,801; May 4, 1954; assigned to Centre National de la Recherche Scientifique, France*. As pointed out by Cagniard, it is preferable, in general, to record the respective variations of the electric and of the magnetic fields at the same station. This offers the following advantages:

> Suppression of the fixed station, generally called a base-station, where it would otherwise have been necessary to carry out a permanent recording of the magnetotelluric elements while stationing at different points of the ground to be explored;
>
> Possibility of appreciating ipso facto the simultaneity of phenomena, the coincidence or phase shift, due to the fact that they are recorded on the same film. An accurate synchronic marking for two instruments distant from each other could evidently be obtained by conventional technical means, but it would mean unnecessary complications;
>
> A higher standard in the correlations, i.e., a higher precision for the checking of the linear relationships between magnetotelluric components.

A subsequent development by *L.P.E. Cagniard; U.S. Patent 3,514,693; May 26, 1970; assigned to Centre National de la Recherche Scientifique, France* provides a method for use in submarine magnetotelluric surveying in which measurements of the telluric and magnetic fields are made only on the sea bed, thus excluding interference due to waves, tidal currents, and the like.

The method includes placing two terminal electrodes on the sea bed, separated by a distance of the order of a kilometer, placing a magnetic sensing device on the sea bed, orienting the device to sense the variations in the desired magnetic component, and recording the telluric and magnetic data produced. Also disclosed is a magnetometric means adapted to rest in a stable manner on the sea bed, comprising essentially a frame supported by weighted feet and three platforms adapted to move in relationship to each other, the magnetometer being carried on a vertical rod adapted to rotate in relation to one of the platforms.

The magnetometric means is used in conjunction with a pair of telluric electrodes which are positioned on the sea bed. The magnetometer and electrodes are connected to recording means which are positioned at the surface of the water.

ARTIFICIALLY PRODUCED FIELD TECHNIQUES

Most electrical prospecting methods, unlike those discussed above, rely on the use of an artificial source of energy.

Resistivity Techniques

The resistivity method is among the oldest of these types, employing direct, commutated or low frequency alternating currents to determine the resistivity distribution of the subsurface formations. Since the earth's resistivity is not uniform, apparent resistivity is determined rather than true resistivity.

In practicing the resistivity technique, an electric current is passed into the earth formations between a first pair of electrodes embedded in the earth while the resulting potential is measured between a second pair of embedded electrodes. The electrodes are usually arranged in a straight-line configuration. Electrode arrays often differ with regard to electrode spacing and position of the current electrode pair in relation to the potential pick-up electrode pair. Early basic configurations included current electrodes, symmetrically spaced along a line outside the potential electrodes. Such an arrangement was described by *C. Schlumberger; U.S. Patent 1,719,786; July 2, 1929.*

A later-developed configuration utilized the current electrode pair spaced a considerable distance from the potential electrode pair as described by *G.C. Summers; U.S. Patent 2,644,130; June 30, 1953; assigned to Socony-Vacuum Oil Company, Incorporated.* As pointed out by Summers, it has been determined in electrical prospecting methods that the pulses must be of relatively high power, at frequencies variable in the range of 2, 4, 8, 16 cycles per second, more or less.

It has heretofore been proposed to generate such signals by the opening and closing of circuit breaker contacts driven by a variable speed motor. However, when the contacts must perform the work of making and breaking the high current flow, a certain amount of pitting and erosion takes place and the device does not remain precise over an appreciable period. More importantly, any arc forming upon the opening of the contacts represents current flow after the separation of the contacts. Hence, the length of each pulse represents an uncontrolled variable.

It has also been proposed to utilize gas discharge tubes of the Thyratron type. However, the wave shape of the pulses from such tubes is not rectangular and with alternating current successive pulses are not alike except where the ratio of the frequency of the alternating current source to the frequency of commutation is a whole number.

In still other systems, an electric commutator comprising gas discharge tubes capable of delivering rectangular pulses has been proposed. In such systems the termination of a positive pulse coincides with the initiation of a negative pulse. Such systems do not provide for the spacing of the successive pulses.

In carrying out the technique in one form thereof, there is provided a method and system for generating pulses of relatively high power which are rectangular in shape, which are spaced one from the other, and all of which are of the same shape. More particularly, there is provided, in circuit with electrodes for applying current pulses to the earth load, a circuit-reversing device which is operated during the intervals between successive pulses.

Accordingly, its operation does not affect the character or timing of the pulses carried thereby. The current flow is derived from a source capable of producing rectangular pulses under the control of an electric valve having a grid for controlling the conductivity thereof. The negligible current demand of the grid circuit makes feasible the use of a mechanical commutator in that circuit for controlling the impulses though the initiation of each impulse may be mechanical while the length thereof is electrically determined.

In a further preferred form of the technique, the length of each pulse is determined by an artificial transmission line having a plurality of capacitors arranged to be charged through circuits individual to one or more capacitors and to be discharged through a separate circuit forming a part of the artificial transmission line. That line may be connected to the earth electrodes either through a mechanical circuit-reversing device or through an electronic device.

The power pulses, because identical and spaced one from the other, make possible the detection of electrical signals with such accuracy that important subsurface information as to the character of subsurface strata may be obtained.

Various operational procedures may be applied to these arrays. For example, the entire electrode array, while maintaining a fixed separation between electrodes, may be moved laterally, in toto, along the ground to provide a horizontal profile of subsurface resistivity variations as described by *C. Schlumberger; U.S. Patent 1,719,786; July 2, 1929.*

Alternatively, the spacing between current electrodes or voltage electrodes may be varied to provide an indication of resistivity variation with depth as described by *H.M. Evjen; U.S. Patent 2,342,626; February 29, 1944; assigned to Nordel Corporation.* Such a system is illustrated in Figure 24.

The earth current circuit comprises a source **10** of direct current, such as a generator or battery. The source **10** is connected in series circuit with a line **11**, the field winding **12** of a dc generator **25**, an ammeter **13** and a line **14**. Lines **11** and **14** are connected to segments **15** and **16** respectively of a commutator **17** which is driven by suitable means to be described.

Alternating current is taken from the commutator **17** by means of lines **18** and **19** which are connected respectively to suitable ground electrodes **20** and **21** herein referred to for convenience as current electrodes. The current electrodes **20** and **21** are spaced a convenient distance apart, which may be on the order of several thousand feet, and when the commutator **17** is driven at a suitable speed, an alternating current having a frequency corresponding to the speed of rotation of the commutator, passes through the earth circuit. This current may be measured by the ammeter **13** and may be adjusted within desired limits by suitable regulation of the source **10**.

The armature **26** of the dc generator **25** is connected by lines **27** and resistances **38** and lines **38a** across a potentiometer **39** having variable contacts **40** and **41**. The generator **25** is driven at a constant speed by a constant speed motor **28** of any suitable type. The motor **28** may be energized from the generator **10** if desired.

The contact **40** is connected by a line **42** to a galvanometer **43** and then through a line **44** to a commutator **46**. The contact **41** is connected by a line **47** to the commutator **46**. Segment **45** of the commutator **46** is connected by a line **50** to an earth potential electrode **51**. Segment **48** of the commutator **46** is connected by a line **52** to the movable contact **53** of a potentiometer **54** which is connected across a battery **55**. The midpoint of this battery **55** is connected by a line **56** to an earth potential electrode **57**.

Figure 24: Nordel Corporation Scheme for Resistivity Profile Measurement

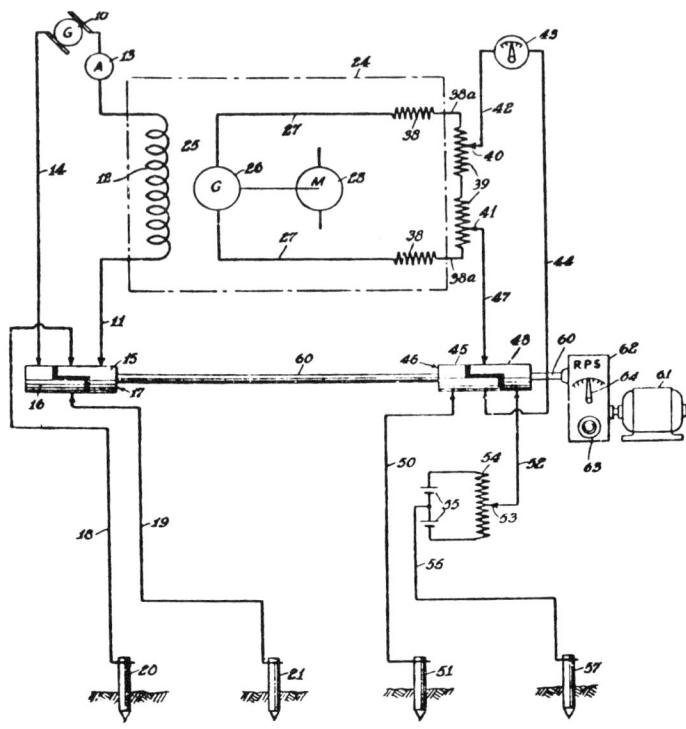

Source: U.S. Patent 2,342,626

The commutators **17** and **46** are shown as mounted on a common shaft **60** and as driven in synchronism by a motor **61** through a suitable control box **62** which may comprise a gear box and a rheostat for adjusting the motor speed. A control handle **63** is provided for controlling the motor speed as desired. The speed of rotation of the commutator may be indicated by a speed indicator **64**.

The galvanometer **43** is preferably of the highly damped type so that it is influenced only by the direct current component and not by any ripple or other alternating current component in the circuit under measurement.

In the operation of this system, the current electrodes **20** and **21** are located at selected points in the earth and the potential electrodes **51** and **57** are located at points within the field of influence of the earth currents produced between the current electrodes.

The source **10** is adjusted to pass a predetermined amount of current through the earth circuit, as indicated by the ammeter **13**. The frequency of alternation of the current is regulated by making suitable adjustment of the control handle **63**.

If the measurements are to be taken with direct current, the commutator **17** remains stationary. For low frequency alternating current, the commutator is slowly rotated at the selected speeds as indicated by the speed indicator **64**. The potential generated by the generator **25** is proportional to the field excitation which, in turn, is proportional to the current flowing in the earth current circuit. A potential drop is thus developed across the potentiometer **39** which is at all times proportional to the current flowing in the earth current circuit.

With the circuit operating in this manner, suitable adjustments are made of the contacts **40** and **41** to produce a zero reading on the galvanometer **43** which indicates that the average voltage drop across the intermediate portion of the potentiometer **39** between the contacts **40** and **41** exactly balances the potential which is picked up by the earth potential electrodes **51** and **57** after the latter has been rectified by commutator **46**.

The potential picked up by the potential electrodes corresponds in frequency to the frequency of the earth current. The alternating potential is rectified by the commutator **46** which operates in synchronism with earth current commutator **17** to supply a direct voltage across the galvanometer **43** and to the potentiometer **39**.

In the abovedescribed system, a definite component of the picked-up potential is utilized. The measuring potential is in phase with the earth current whereas the picked-up potential has, in general, a phase shift. By means of commutator **46** a component of this potential, having a definite phase relation to the current, is selected. The phase relation of this component to the current is determined by the phase relation between commutators **17** and **46**.

In the accompanying figure, a phase angle of 0° is shown, but it is to be understood that any desired phase angle may be used and that this angle may be changed as desired.

The effect of any residual earth current, which would tend to produce a direct potential across the electrodes **51** and **57**, is eliminated by suitable adjustment of the potentiometer **54**. This adjustment may be made, for example, while no current is flowing through the earth circuit so that the only effect on the galvanometer **43** represents the effect of the residual earth current.

Since the earth current impresses a direct current potential across the probe (or potential) electrodes, any unbalance between this potential and the potential taken from potentiometer **54**, will appear as an alternating potential after commutation by commutator **46**. This will make the galvanometer kick from one side to the other in synchronism with the commutation. Any drift in the natural earth potential may thus be immediately perceived, and can be compensated by readjusting potentiometer **54** until the kicking of the galvanometer is eliminated.

A filter may be included in the measuring circuit if desired in order to eliminate ripples of fluctuations from various causes such as, for example, those produced by the action of the commutators.

The potentiometer **39** may be suitably graduated in convenient units and when the balance is obtained, the reading corresponding to the point of adjustment is noted. The frequency is also noted from the indicator **64**.

The potential developed across the potentiometer **39**, being dependent upon the total value of the earth current at the instant that the measurement is being taken, automatically compensates for any variations in electrode resistance or in contact resistance and eliminates any necessity for making simultaneous measurements of the earth current and the induced potential.

This measurement is repeated with different frequencies until a complete potential spectrum of the area between electrodes **51** and **57** is obtained. These electrodes may then be moved to another location and the measurements repeated to obtain a potential spectrum at various frequencies in the manner above pointed out. The various spectra thus obtained may be interpreted to show the characteristics of the earth's strata in the area under investigation. It is to be understood, of course, that the potential electrodes may be variously positioned with respect to the current electrodes **20** and **21** and that the positions thereof may be changed in accordance with the depth which is under investigation at any particular instant.

It will be noted that in the abovedescribed system only a single reading need be taken and the necessity for mathematical calculations is avoided. The system is accordingly well adapted to field use and may be operated by unskilled persons. The data thus compiled may be interpreted by geological physicists in accordance with well-known principles.

The size of the electrodes may vary over wide limits. In practice, metal stakes about 1" or 2" in diameter and 3 or 4 ft in length have been found satisfactory. The distance between the current electrodes may vary widely, for example, from 200 to 5,000 ft, depending upon the depth to be investigated. The potential of the source **10** may likewise be adjusted as desired. Voltages of 100 to 200 volts and earth currents of 1 to 2 A have been found satisfactory.

The resistivity method has also been adapted to well logging by supporting the electrodes on a logging instrument adapted to be moved within a borehole. In this manner the resistivity of the formations along the borehole can be determined. There are numerous types of electrode configurations used with such instruments, and over the years the electrode arrays and the circuitry have become complex. These instruments include extensions of the so-called focusing electrode arrangements as described by *R.G. Piety; U.S. Patent 2,347,794; May 2, 1944; assigned to Phillips Petroleum Company.*

Current may be forced to flow horizontally for some distance into the formations surrounding the borehole, by means of guard electrodes situated on each side of the current-emitting electrode as described by *P. Threadgold and G.S. Thomas; U.S. Patent 3,262,050; July 19, 1966; assigned to The British Petroleum Company Limited, England,* for example.

Another such device has been developed by *H. Janssen; U.S. Patent 3,068,401; December 11, 1962; assigned to PGAC Development Company* and is shown in Figure 25.

There is illustrated an apparatus for electrically logging a well or borehole **10** in order to determine the characteristics of earth formations **11** penetrated or traversed by the borehole. It will be understood that the borehole **10** may contain drilling fluid with mud suspended therein which generally remains in the hole after the removal of the drilling equipment although such fluid has not been illustrated in the drawing.

Figure 25: PGAC Development Company Electrical Logging System

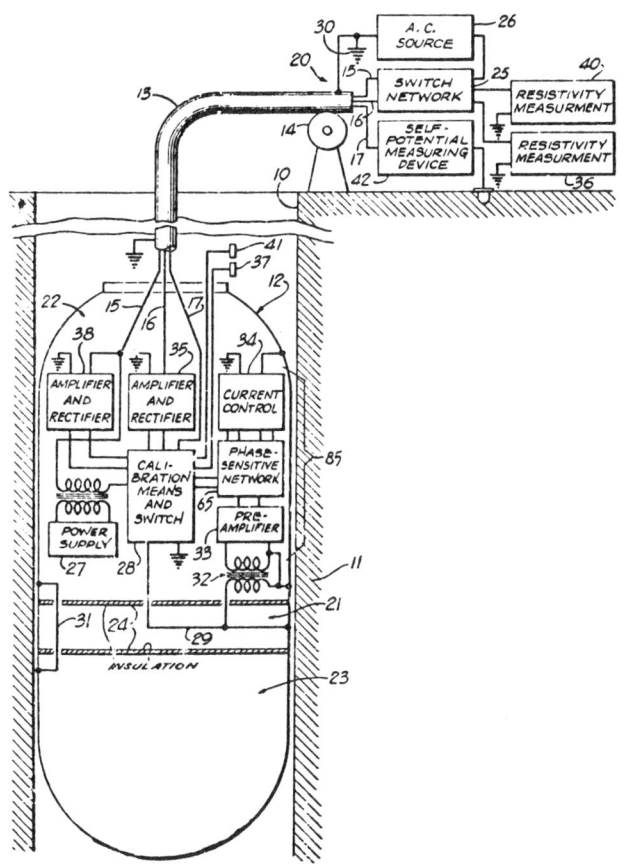

Source: U.S. Patent 3,068,401

Investigation apparatus or downhole equipment, indicated generally by the reference character **12**, is carried upon the lower end of a multiconductor cable **13** for movement up or down within the borehole **10**. To effect the raising and lowering of the apparatus **12**, the cable **13** is trained over a motor-driven sheave **14** or the like at the earth's surface and may be wound upon a suitable take-up reel (not shown) in conventional manner.

In the form illustrated, the cable **13** contains three conductors **15, 16** and **17**, each of which terminates at one end in the investigation apparatus 12 and at the other end is connected to surface equipment designated generally by the reference character 20.

The downhole equipment 12 comprises a housing which carries on its outer surface a current or measuring electrode 21 and a pair of elongated guard or screen electrodes 22 and 23 respectively disposed above and below the measuring electrode. The screen electrodes 22 and 23 are interconnected by conductor 31 and are electrically insulated from the measuring electrode 21 by means of suitable insulation indicated by the reference character 24. All three of the electrodes are exposed to the fluid in the borehole.

The conductor 15 is adapted to be connected at its upper end through a manually operated switch network 25 to a suitable source of alternating current 26 which applies a logging or measuring current preferably having a frequency of 400 cycles to the measuring electrode 21.

The switch network 25 may be selectively rendered effective either to send the described logging current down the cable, or, alternatively, to supply a switching signal to the apparatus 12 in order to calibrate the surface equipment 20 preparatory to the initiation of a logging operation. The source 26 is so designed that, when the switch network 25 is in its logging or operating position, a constant current is furnished to the downhole equipment 12. This current is used to drive a power supply 27 which supplies filament and B+ signals to electrical components of the subsurface equipment.

The described measuring current is also delivered through a calibration and switch network or circuit 28 and through a signal connector 29 to the measuring electrode 21 from which it passes through the borehole fluid to the formations 11. The circuit for this 400-cycle measuring current is completed through a ground connection 30 at the earth's surface connected to one side of the source 26.

To obtain a resistivity measurement of the formation or stratum lying adjacent the measuring electrode 21, the difference of potential existing between the electrode 21 and a remote reference point, as for example, the remotely positioned electrode 37 carried on the downhole apparatus at a point spaced some distance from the screen electrode 22 is applied through the calibration and switch circuit 28 to an amplifier and rectifier 38, where the signals are converted to a slowly varying dc signal which is passed through cable conductor 15 to the surface equipment 20.

At the surface equipment the signals appearing upon conductor 15 are passed through the switch network 25 to a recording galvanometer 40 which provides a continuous indication of the magnitude of these dc resistivity signals as a function of borehole depth as the downhole apparatus 12 traverses the borehole.

The borehole 10 is effectively plugged electrically in order to prevent the dispersion of the measuring current along the mud column or the borehole fluid lying adjacent the borehole formations by supplying current to the screen or guard electrodes 22 and 23 of sufficient magnitude to maintain these electrodes at substantially the same potential as the measuring electrode 21.

As indicated above, measuring systems of this kind exhibit a focusing effect in which the measuring current flows from the measuring electrode **21** laterally in a thin sheet or disk extending perpendicular to the borehole and having a thickness which is a function of the height of the electrode **21**. This measuring current flows laterally into the borehole formations to an extent which is determined by the potential and length of the screen electrodes **22** and **23**.

A large difference of potential between the measuring electrode **21** and the electrically connected screen electrodes **22** and **23** would result in a change of the focusing of the measuring current and, hence, would introduce a substantial error in the determination of the true resistivity of the formation lying adjacent the electrode **21**.

To prevent such a potential difference, the gap between the measuring electrode **21** and the screen electrode is bridged by a transformer **32** having an extremely low primary impedance.

The secondary of the transformer **32** supplies excitation signals for an electronic regulator for supplying current to the screen electrodes **22** and **23**. This regulator is indicated generally by the reference numeral **85** and includes a preamplifier **33**, a phase sensitive network **65** and a regulated current control circuit **34**.

The proportion of the screen control current to the measuring current depends upon the ratio of the specific resistivity of the stratum lying adjacent the electrode **21** to the adjoining strata lying opposite the screen electrodes **22** and **23**. Specifically, the ratio of screen current to measuring current increases when a thin stratum is being investigated, when the specific resistivity of the mud is relatively low or when the ratio of the specific resistivity of the investigated stratum is relatively high in comparison to the resistivity of the adjoining strata. Current ratios on the order of 10^5 and higher are frequently obtained, and since the available current from the current control circuit **34** is limited, the achievement of such a high proportion can be obtained only by keeping the measuring current relatively low.

Since it is impossible to obtain linear amplification over a range covering several decades or 10 to 1 ratios of input signal, the amplifier **38** may saturate when very high formation resistances are encountered. Thus, about one-tenth of the signal input to the amplifier **38** is diverted to an amplifier and rectifier circuit **35** which converts the ac signals to a slowly varying dc signal for passage through conductor **16** to the surface equipment.

At the surface, the signals appearing upon the conductor **16** pass through the switch network **25** to a measuring circuit **36** where they are recorded as a function of borehole depth simultaneously with the apparent resistivity measurements provided by the measuring instrument **40**. The indications provided by the device **36** may thus be used for high resistance formations while the indications provided by the device **40** may be used to determine the resistivities of the relatively low resistance formations.

Natural earth potentials are recorded simultaneously with the apparent resistivity measurements by supplying the signals picked up by a remotely positioned electrode **41** through the calibration and switch circuit **28** and through the cable conductor 17 to the surface equipment. The differences of potential existing between electrode 41 and a ground or surface electrode are then recorded by a self-potential measuring device 42 simultaneously with the signals recorded by the devices 36 and 40. The electrode 41 is preferably located at some distance from the elongated electrode 22 due to the fact that the presence of a long metallic body in the borehole adversely affects the self-potential signals present in the vicinity of the long electrodes.

Other electrical methods utilizing the introduction of current into the ground include equipotential line survey as described by *S.R. Phelan; U.S. Patent 2,314,597; March 23, 1943* and potential drop-ratio methods as described by *J.J. Jakosky; U.S. Patent 2,192,404; March 5, 1940.*

A scheme developed by *J.M. Forgotson; U.S. Patent 3,820,390; June 28, 1974* relates to methods of recognizing the presence of hydrocarbons and associated fluids in reservoir rocks (strata) below the surface of the earth. A determination is made of the relationship between the porosity and the water (salt water or fresh water) in the pore spaces of the reservoir rocks. When certain relationships exist between the porosity and the water, the presence of recoverable hydrocarbons and associated fluids and non-water-miscible substances is indicated.

Measurements are made of the electrical resistance (or electrical conductivity) of the reservoir rocks at particular depths to facilitate the interpretation of the contents of the reservoir rock. Such measurements are particularly made to aid in the determination of the percentage of water in the pores of the reservoir rocks.

Measurements are also made of the transit time of a sonic wave through the rocks exposed to the borehole as a means of measuring the porosities of the reservoir rock exposed to the borehole. These measurements also facilitate an interpretation of the percentage of water saturation in the reservoir rocks at particular depths. Measurements are also made to determine the percentage of shale in the reservoir rocks at particular depths.

Induced Polarization Methods

In the 1950s and 1960s a field procedure known as the induced polarization method gained widespread recognition in the field of electrical prospecting. Induced polarization is the term used in referring to the phenomenon which occurs with respect to certain types of conducting minerals, located within a formation zone, in the presence of an externally applied current source.

In passing an electric current through the zone an electrochemical barrier in the form of an opposing voltage is produced at the interface between a mineral body and the surrounding electrolyte. (The mineral is said to be polarized at the interface.)

In order to overcome this polarization effect an added voltage known as the overvoltage is necessary to drive the current across the barrier. When the current is turned off, the overvoltages which were set up decay in time. Observation of these voltages provides an indication of the nature of the minerals within a particular zone of investigation. In addition to time domain studies, induced polarization measurements can also be made in the frequency domain by comparing the effect of the overvoltages on alternating currents of different frequencies.

An early disclosure of the induced polarization method was that of *A.A. Brant and E.A. Gilbert; U.S. Patent 2,611,004; September 16, 1952; assigned to Geophysical Exploration Company*.

As pointed out by Brant and Gilbert, electrical methods of exploration may be considered as indirect in principle as they involve a critical interpretation of the data obtained in order to arrive at a conclusion with respect to the type, nature and character of the subsurface conditions, strata, deposits, etc.

The accuracy of interpretation of electrical prospecting results depends upon the ease with which interfering factors may be eliminated. While the electrolytic polarization method is adapted for oil exploration, due to the polarization effects arising from the porosity differences (percent electrolyte) of the formations concerned, it is not satisfactory for the detection and identification of scattered mineral, metallic and carbon particles. These substances are generally found in nature as sulfides and a practical method and apparatus for the detection of such substances must inherently involve means for ascertaining what may be termed effects of the second order of refinement.

The process for geophysical exploration is based upon an electrochemical effect identified as the overvoltage effect or electrical double layer charging effect occurring at the surfaces of metal particles in an electrolyte.

H.O. Seigel, having done considerable work with the induced polarization method, extended this procedure to include measurement of the magnetic field created by the applied polarization current, before and after termination of current flow, as an indication of the induced polarization characteristics of the medium under investigation. This technique was described by *H.O. Seigel; U.S. Patent 3,210,652 October 5, 1965*.

Magnetic field measurements were later applied to frequency domain induced polarization studies as described by *H.O. Seigel; U.S. Patent 3,745,445; July 10, 1973; assigned to Scintrex Limited, Canada*. In this technique a polarized condition in the medium being explored is created by conductively passing primary electric currents of repetitive wave form through the medium at two distinct frequencies in the active induced polarization range which differ from each other by at least a factor of 2, maintaining the ratio of the electric current amplitudes of the wave forms at the distinct frequencies substantially constant, measuring the magnitude of a magnetic field component primarily due to current flow in the medium at each of the distinct frequencies and comparing the magnitudes to obtain an indication of the presence of regions of anomalous induced polarization.

Figure 26 shows the current producing and detecting means involved in the practice of this technique. The numeral **1** refers to a power generator capable of producing a primary electric current wave form containing as constituents one or more sinusoidal wave forms of frequencies in the range of 0 to 100 Hz, the relative amplitudes of which constituents are accurately maintained and whose frequencies differ by at least a factor of 2 from one another.

Composite wave forms with frequencies such as 1, 2, 5, 10 and 20 Hz have been found effective in detecting areas of anomalous polarization, for example, zones of sulfide mineralization. Because of the superposition of the multiple sinusoidal wave forms, possibly harmonically related, the resultant wave form may have a complex form, e.g., a square wave form comprising the fundamental frequency and all odd harmonics, a sawtooth wave form comprising the fundamental and all odd harmonics in lesser amounts, a wave form comprising the fundamental and at least one harmonic. It will be understood that a half wave rectified wave form is considered as a repetitive wave form (e.g. half wave rectified square).

Figure 26: Scintrex Limited Induced Polarization Technique Showing Current-Producing Means (Above) and Detecting Means (Below)

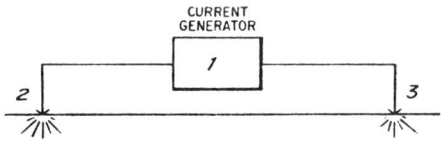

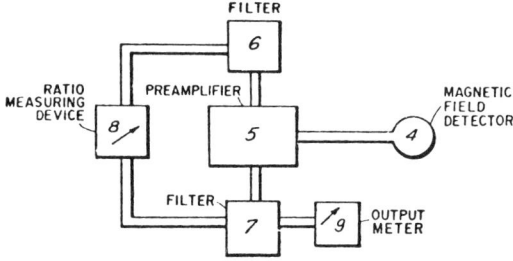

Source: U.S. Patent 3,745,445

Where continuous measurements are not required, it will be adequate to produce the current wave form of different frequencies consecutively rather than concurrently.

It will be important to maintain the ratio of current amplitudes at the different frequencies constant to and preferably within 0.1%. This may be achieved through the known use of saturable reactors or other current stabilizing devices.

The current output of current generator **1** is passed into the ground through the contact points **2** and **3**, which are usually well-grounded metal electrodes.

Numeral **4** is a magnetic field detector which is so oriented as to detect primarily the magnetic field due to passage of current in the ground. This detector may be a multiturn induction coil, a flux gate or alkali vapor magnetometer or other magnetic field sensitive element, with a sufficiently broad pass band to receive signals at all the frequencies being transmitted by current generator **1** above. Alternatively two or more similar magnetic field detectors may be used, one for each transmitted frequency, in place of the single detector.

The output of the detector 4 is fed into a low noise, distortion-free preamplifier 5. Items 6 and 7 are narrow band filters (e.g. phase lock amplifiers) which select, from the output of the preamplifier, signals of two specific transmitted frequencies. The outputs of these two filters are compared in a ratio measuring device 8. One output amplitude is displayed on an output meter 9. These two quantities (ratio and amplitude) may also be recorded on suitable analog or digital recorders if continuous measurements are being made.

If intermittent measurements are being made, a single variable frequency filter may be employed in place of 6 and 7 and the magnetic field amplitudes at the two or more frequencies being transmitted will be measured consecutively and then compared. This technique works on the basis that in areas of polarization which are higher than normal for the general environment in the area, there will be an increase in current with frequency, so long as the frequencies used are in the active induced polarization range.

With change in the frequency of the primary current the current paths through the earth would change, resulting in a corresponding change of magnetic field amplitude with frequency in the vicinity. It is this variation of magnetic field amplitude with frequency that can be taken as an indication of induced polarization characteristics of the medium.

A method developed by *K.L. Zonge; U.S. Patent 3,967,190; June 29, 1976* is a method for establishing data related to the complex resistivity of selected subsurface earth deposits whereby increased discrimination and identification of ore deposits are established. The method utilizes induced polarization in the earth with establishment of both time domain and frequency domain data and correlation of such data with an analog equivalent circuit of the earth section to derive a value closely approximating the complex resistivity of the selected earth substructure thereby to provide more discriminating data, enabling determination of the percentages and types of sulfide ore deposits.

A method developed by *A.A. Kaufman; U.S. Patent 4,114,086; September 12, 1978; assigned to Scintrex Limited, Canada* is a method of geophysically prospecting a polarizable medium by measuring induced polarization effects. The medium is inductively polarized by a time-varying current in an induction coil that varies with respect to time unidirectionally to maintain a unidirectional current flow in the medium for a first period and then does not vary with respect to time for a second period and the induced polarization effects caused by the unidirectional current flow in the first period are measured during the second period.

Electromagnetic Methods

Electromagnetic methods involve the creation of an electromagnetic field (called the primary field) which penetrates the earth's formations causing currents to be induced in conducting bodies present within the formations.

One electrical prospecting technique employs very low frequency (VLF) radio signals or electromagnetic fields transmitted from several powerful stations that operate in the range of 15 to 25 kHz. The primary fields from these stations penetrate into the earth to depths on the order of 30 to 300 ft (10 to 100 m) and cause electrical currents of the same frequency to flow in the earth. The current flowing in the earth causes secondary fields that can be observed at the earth's surface.

In areas where the electrical resistivity of the earth is uniform, the primary and secondary fields at the surface are also uniform and are oriented in the horizontal direction. When the resistivity of the earth is not uniform, however, the currents tend to concentrate along low resistance paths such as may be provided by orebodies. This causes disturbances in the secondary fields at the surface that can be measured and used to predict the presence and location of orebodies.

The measuring instruments consist basically of one or more induction coils, which are used to sense the VLF magnetic field, a VLF radio receiver and a readout device. In some instruments the tilt of the VLF magnetic field from the horizontal plane is measured; in other instruments, the amplitude of the magnetic field in the horizontal direction is measured.

Measurements are made along lines or a grid in the same manner that magnetic surveys are conducted. Ordinarily a station spacing of 50 to 100 ft (approximately 15 to 30 m) is adequate. Power lines, pipelines, metal fences and other large metallic objects, even if they are not steel, should be avoided because they act as conductors and cause false anomalies.

Another type of electromagnetic method uses a local transmitter consisting of a battery-powered source of alternating current and an air- or metal-cored induction coil that serves as the antenna. The operating frequencies range from about 400 to 4,000 Hz. The separation between the transmitter and the receiver varies from about 200 to 800 ft (approximately 60 to 240 m).

The receiver for this technique functions in about the same way as the VLF receiver to measure the tilt of the alternating magnetic field from horizontal; the presence of low resistivity orebodies distorts the fields that are observed at the surface. In general this technique has a greater depth range than the VLF method and often provides data that are easier to interpret than data from the VLF method. Disadvantages are that two persons are needed to make measurements and that survey lines must be cleared in advance to work in wooded terrain.

Both electromagnetic techniques described above respond to the presence of rocks bearing significant quantities of graphite or carbon, to water-filled shear zones, and to overburden, particularly when clays or saline waters are present. Where geologic evidence indicates the possible presence of an orebody, magnetometer measurements may help discern anomalies likely to represent valuable mineral deposits (1).

Two categories into which electromagnetic methods are often divided include: (a) the method in which the primary field source is held stationary while the receiver system is moved over the area of investigation; and (b) the method in which both the transmitter and receiver are moved.

By varying the frequency of the applied field or the separation between transmitter and receiver, responses from different depths below the surface may be obtained. In the development of these techniques, Sundberg, as far back as the 1920s, provided early pioneering work in ground methods of electromagnetic prospecting. These developments have been described by *K. Sundberg; U.S.*

Patents 1,678,489; July 24, 1928; and 1,748,659; February 25, 1930, as well as by K. Sundberg and E.D. Lindblom; U.S. Patent 1,820,953; September 1, 1931; all assigned to AB Elektrisk Malmletning, Sweden.

However, the electromagnetic methods are not confined to ground prospecting, but find considerable application in airborne use (wherein both transmitter and receiver are simultaneously moved). Over the years airborne systems such as the one shown in Figure 27 have been developed to deal with the problems of supporting the transmitter and receiver coils with respect to each other and the aircraft.

Figure 27: Airborne Electromagnetic Prospecting

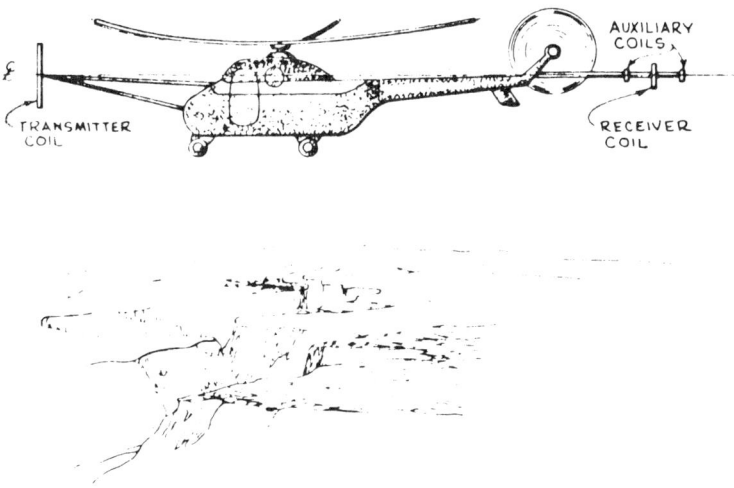

Source: U.S. Patent 3,108,220

Disclosures of techniques for airborne magnetic prospecting include those by M. Puranen, A.A. Kahma and V. Ronka; U.S. Patent 2,929,984; March 22, 1960; assigned to Canadian Airborne Geophysics Limited, Canada. Such a system is shown schematically in Figure 28.

In this figure, **1** is the ac generator for frequency f_1 and **2** the generator for f_2. These generators supply the transmitting coil **3** which sets up primary field shown by arrows **P**. This field, upon encountering an orebody such as **4**, which may lie beneath an overburden **5**, creates a secondary field shown by arrows **S**.

Receiving coil **6** receives the second field of frequency f_1 while receiving coil **7** receives the secondary field f_2. Frequencies f_1 and f_2 are separately recorded as shown in the drawing, reference coils **8** and **9** detecting frequencies f_1 and f_2 respectively and feeding these reference signals for recording of the phase shift of f_1 and f_2 secondary signals.

Figure 28: Canadian Airborne Geophysics Limited Electromagnetic Survey Scheme

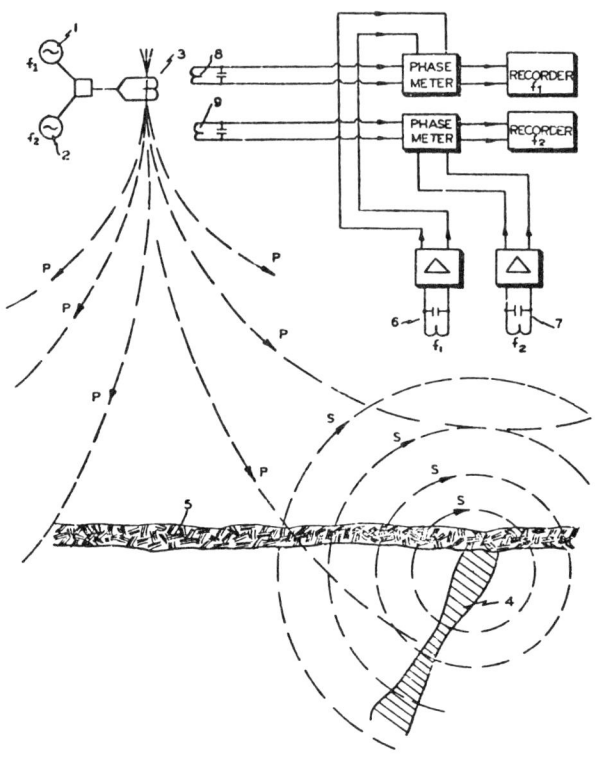

Source: U.S. Patent 2,929,984

In operation, each receiving coil will be in a circuit whereby the phase angle may be measured and recorded for each transmitted frequency. The existence of a phase shift at either frequency is an indication of the presence of a conductive body, and a comparison of the phase angle at each frequency will give an indication as to the class or size of the material involved.

Another description of an airborne electromagnetic technique is that by *V. Ronka; U.S. Patent 3,042,857; July 3, 1962*. Such a scheme is shown in Figure 29.

An aircraft **1** is shown towing a bird **2** by means of a tow cable **3**. In the terrain below, under the overburden **4**, is a steeply dipping conductive orebody **5**. Within the aircraft an alternating current generator **6**, operating at, say, 1,000 cycles per second, energizes a transmitting coil **7** mounted at the front of the aircraft and having a horizontal axis. The coil **7** sets up a primary electromagnetic field which passes through both the orebody **5** and a receiving coil **8** carried by the bird behind the coil **7**.

Figure 29: Airborne Electromagnetic Prospecting Operation

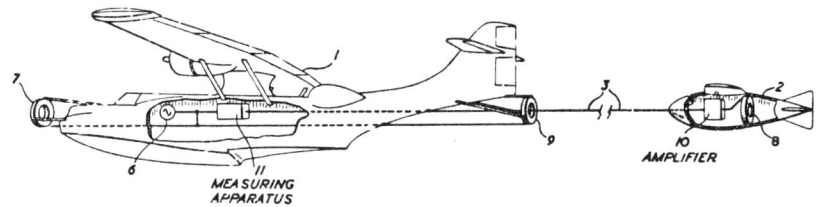

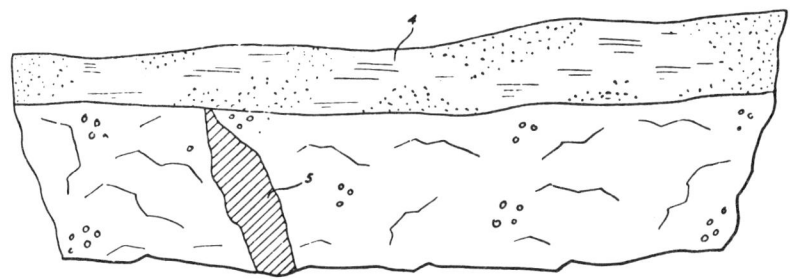

Source: U.S. Patent 3,042,567

In the example illustrated, there is provided a second transmitting coil **9** carried at the rear of the aircraft and also energized by the generator **6** and setting up at the receiving coil **8** a compensating electromagnetic field the characteristics of which are described below.

The primary field from the transmitter **7**, modified by the field from the transmitter **9**, causes eddy currents in the orebody **5** and these, in turn, set up a secondary electromagnetic field also passing through the receiving coil **8**. Thus, there is induced in the receiving coil **8** a first signal caused by the primary field from the transmitter **7**, a second signal caused by the compensating field from the transmitter **9**, and a third signal caused by the secondary field from the orebody. (These signals, and the fields causing them, of course, produce resultant or net signals or fields which are the signals or fields actually detected, but it is convenient to consider the individual signals and fields as separate parts of these observed resultants.)

The sum of these signals is amplified by an amplifier **10** and carried along the tow cable **3** to measuring apparatus **11** in the aircraft.

A technique developed by *K.A. Ruddock; U.S. Patent 3,108,220; October 22, 1963; assigned to Varian Associates* involves an airborne, electromagnetic arrangement for establishing the existence of a subsurface conducting ore zone.

An airborne electromagnetic prospecting technique described by *G.H. McLaughlin, H.A. Harvey, W.O. Cartier and W.A. Robinson; U.S. Patent 3,015,060; December 26, 1961; assigned to Nucom Limited, Canada* was designed to eliminate the major source of error caused in the prior art airborne methods caused by the relative movement of the aircraft and bird.

Another important object is to devise a method and means of prospecting which will eliminate the necessity of the previously required long towing cables connecting the bird to the aircraft whereby a helicopter may be employed and the equipment required to carry out the method may be trailed in close proximity to the helicopter to allow the topography of the area being investigated to be closely followed and the distance of the detecting equipment from the ground to be closely ascertained.

By enabling the use of a helicopter it will be understood that large areas can be quickly explored and when indications of a conductor anomaly are detected, the helicopter may be maneuvered slowly in the vicinity of the anomaly or caused to hover over a particular location to provide extensive and reliable information concerning the anomaly, eliminating the prior problems of trying to correlate the results achieved from the aircraft with the ground positions.

In this regard it is to be noted that it has not heretofore been possible to safely fly a bird of any appreciable size with a helicopter as the bird becomes dangerously unmanageable and it is a particular object of this technique to provide a bird which can be safely used with a helicopter and requires a relatively short towing cable.

A method developed by *A.R. Barringer; U.S. Patent 3,852,659; December 3, 1974; assigned to Barringer Research Limited, Canada* is one in which a primary electromagnetic field having a complex waveform is generated and is radiated toward the earth.

Means are provided for receiving electromagnetic signals which are responsive to either the resultant of the primary field and secondary fields emanating from the earth, or else to such secondary fields alone. A plurality of stored reference waveforms corresponding to known geological structures is matched against the received signals, and the optimum match is indicated.

Another method developed by *A.R. Barringer; U.S. Patent 3,950,695; April 13, 1976; assigned to Barringer Research Limited, Canada* is a method of detecting conductive bodies in an area wherein a primary field having a complex waveform, such as a frequency-modulated, rapidly swept signal is generated and is radiated toward the area.

A receiving system is provided for sensing the primary and secondary fields. Computerized signal processing techniques are used to obtain the best match between the waveforms of the received signals and those of stored reference waveforms which correspond to responses of known conductive bodies.

Figure 30 shows an aircraft carrying the receiving and transmitting apparatus as well as a block diagram of the principal components of the signal processing system.

Electrical Geophysical Prospecting

Figure 30: Barringer Research Limited Geophysical Prospecting Method Utilizing Correlation of Received Waveforms with Stored Reference Waveforms

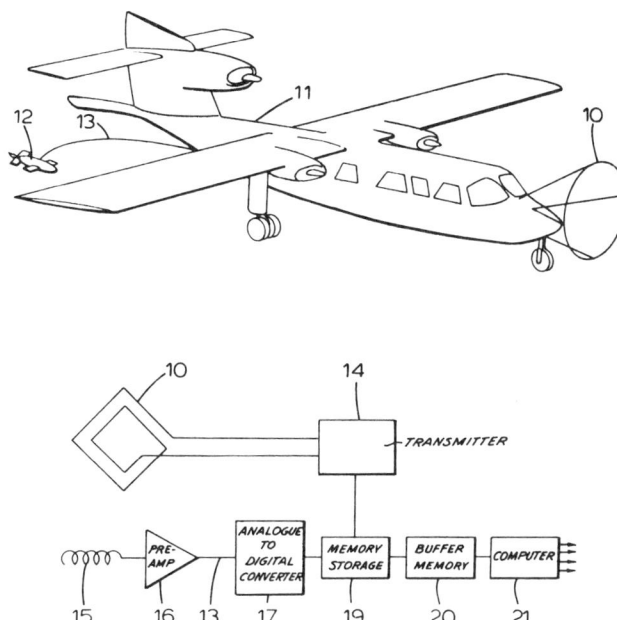

Source: U.S. Patent 3,950,695

Referring to the above figure, a primary electromagnetic field is generated by passing current produced by a transmitter **14** through a multiturn loop **10** which is installed on an aircraft **11** or other vehicle. The power delivered by the transmitter **14** should be at least about 1 kW.

The primary field and the secondary fields reradiated from conductive bodies in the underlying terrain are picked up in one or more receiving coils **15** towed in a finned, torpedo-shaped vehicle, hereinafter referred to as a bird, which is indicated by reference number **12**. The bird **12** may be equipped with small wings if desired to provide it with some lift to raise it to a convenient towing position behind the aircraft where it will not be caught by trees when flying at low altitude.

It is usual to place the receiving coils in orthogonal relationship to each other with one of the coils in maximum coupling with the primary field generated by the loop **10**. The receiving coils are connected to preamplifiers, the output of which is carried to signal processing equipment in the aircraft **11** via electrical conductors within tow cable **13**.

The primary field is detected by the receiving coil **15** in the form of its derivative since the coil **15** is sensitive to the rate of change of the magnetic flux.

A receiving coil **15** in the bird **12** is designed to have sufficient bandwidth to faithfully follow the waveform of the primary field and is connected to a preamplifier **16**. The output of the preamplifier **16** is fed to an analog-to-digital convertor **17** in the aircraft via electrical conductors in the tow cable **13**.

The analogue-to-digital convertor **17** is connected to a scanning memory storage **19**, which synchronously scans the repetitive waveform and stores it in a digital buffer memory unit **20**. The buffer memory **20** is unloaded at periodic intervals such as twice per second and is digitally matched in a small computer **21**, against a family of stored waveforms specific to typical overburden and orebody responses.

In borehole investigation, another electromagnetic method used is the so-called induction logging system. In this system a logging tool, containing spaced transmitter and receiver coils, is moved along the borehole to provide a continuous profile of downhole characteristics, e.g., conductivity or susceptibility. One such system has been described by *C.B. Aiken; U.S. Patent 2,220,070; November 5, 1940; assigned to Schlumberger Well Surveying Corporation.*

Another such system has been described by *H.-G. Doll; U.S. Patent 2,582,314; January 15, 1952; assigned to Schlumberger Well Surveying Corporation.*

A system developed by *G.O. Buckner, Jr.; U.S. Patent 3,112,443; November 26, 1963; assigned to Halliburton Company* is particularly applicable to induction systems wherein eddy currents are induced in the ambient formations and the effect of such eddy currents is indicated for showing either formation conductivity or resistivity.

An analog computer provides compensation for deviations of apparent conductivity from true conductivity and the same is applicable not only to induction logging systems, but electrical logging systems wherein resistivity or conductivity determinations are made without magnetic induction.

A device developed by *G.R. Atwood and D.J. Dowling; U.S. Patent 3,555,409; January 12, 1971; assigned to Texaco, Inc.* comprises a borehole logging tool that includes air core transmitter and receiver coils, plus auxiliary air core nulling coils.

A compensated magnetic induction circuit is employed for nulling a major portion of the receiver coil signals in order that the changes caused by magnetic property change in the surrounding borehole may be readily detected. The auxiliary nulling coils are pancake-shaped and are located outside the fields of the transmitter and receiver coils; but, they are located on the logging tool so that they are subject to the same ambient conditions as are the transmitter and receiver coils.

All of these electromagnetic methods rely on the use of continuous electromagnetic waves as the energy source. However, the use of transient pulses as an energy source has also been developed. In this method, the electromagnetic waves produced by short, abruptly terminated transient current pulses applied

Electrical Geophysical Prospecting

to a loop of wire or grounded electrodes are caused to penetrate the formations. The resultant voltage induced in a pick-up coil is measured as an indication of the presence of a conducting body as described by *W.J. Yost; U.S. Patent Reissue 24,464; August 22, 1958; assigned to Socony Mobil Oil Company, Inc.*

A.R. Barringer; U.S. Patent 3,105,934; October 1, 1963; assigned to Selco Exploration Company Limited, Canada has adapted a similar technique for use in an airborne system.

Other electromagnetic methods include observation at a receiver of the effects of subterranean features on the transmitter signal. One such method has been described by *L.W. Blau; U.S. Patent 2,268,106; December 30, 1941; assigned to Standard Oil Development Company*.

Another such method has been described by *D.B. Daniel; U.S. Patent 4,010,413; March 1, 1977; assigned to Geo-Nav, Inc.* It is a system for investigating subterranean formations by analyzing the propagation of low frequency radio transmissions.

Commensurate signals of different frequencies are received from a transmitting station and are processed into a phase-comparable form. The phase-comparable signals are then compared in phase to provide a representative signal indicative of a subterranean formation. A system utilizing at least three transmitted signals is disclosed to provide specific information related to the depth of formations.

A system utilizing signals from a plurality of transmitting stations is disclosed to afford improved formation illumination. Also, a mobile form of the system is disclosed which incorporates a position-information system for combined use.

Another electromagnetic method involves the use of radar techniques for mapping subsurface interfaces, as described, for example, by *C.A. Donaldson; U.S. Patent 2,657,380; October 27, 1953*.

R.M. Morey; U.S. Patent 3,806,795; April 23, 1974; assigned to Geophysical Survey Systems Inc. has disclosed a geophysical survey system for determining the character of the subterrain by analysis of reflections from electromagnetic pulses radiated into the ground.

The system repetitively radiates into the ground a short-duration electromagnetic pulse having a rise time on the order of 1 nanosecond. The antenna which radiates the pulse into the ground is employed to receive the reflections of the pulse. The received signals are coupled through a transmit-receive network to a receiver which permits the input signal waveform to be reconstructed from a sequence of samples taken by the receiver. The system is capable of generating a profile chart indicating the magnitudes of the reflected signals and the depths at which the reflections occurred.

Such a system is shown schematically in Figure 31. Operation of the system is governed by a controller **1** which emits a trigger pulse to a pulse generator **2**. In response to the trigger pulse, a pulse of the requisite shape and duration is generated and is coupled through a T-R network **3** to an antenna **4**. The antenna is a broad-band device which is impulse-excited to radiate electromagnetic energy into the ground. In practice, the antenna is elevated 2" to 4" from the surface of the ground and the antenna is mounted on a carriage that can be pulled over the ground to permit a continuous profile of the subterrain to be obtained. The system, except for the antenna, is carried in a vehicle which tows the carriage carrying the antenna. A flexible transmission line is employed to couple the T-R network to the antenna to permit the carriage to easily ride over uneven or rough ground without the restraint imposed by a rigid connection.

Figure 31: Geophysical Survey Systems Inc. Scheme Employing Electromagnetic Impulses

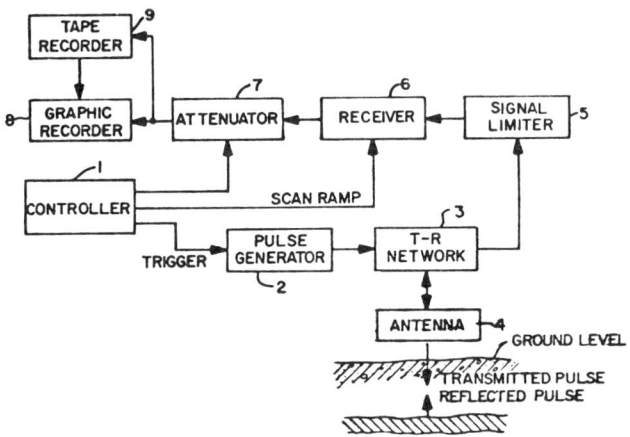

Source: U.S. Patent 3,806,795

Inasmuch as the system is intended to be employed for shallow depths where the maximum depth of interest may, for example, be 40 ft, and where reflective interfaces or reflective objects may be within 4 or 5 ft directly under the antenna, the antenna **4** is employed for the reception of reflected pulse energy. Because the energy is propagated vertically into the ground, the reflections tend to be vertical or very nearly so. However, the greater the depth, the greater is the probability that more of the wave energy will reflect at angles which cause that reflected energy to miss the antenna.

Thus, in addition to the attenuation due to the longer path length, a smaller portion of the energy reflected from the deeper depths reaches the antenna. Because the depths are shallow and the path lengths are short, the scattering which occurs at the higher frequencies (i.e., 300 MHz and above) becomes of lesser significance to the greater detail which is obtained with short wave lengths.

Although it is known that the frequencies of 30 MHz and above do not penetrate the soil to the depths achieved with the longer wave lengths, nevertheless, the process utilizes the higher frequencies because adequate penetration for shallow depths is obtained at the higher frequencies with moderate amounts of power, even where the frequency spectrum includes frequencies up to 400 MHz.

Energy reflected from a discontinuity such as the soil-rock interface shown is incident upon the antenna **4**. The received reflected energy is coupled through the T-R network and through a signal limiter **5** to a receiver **6**. The receiver may include an oscilloscope of the sampling type whose sweep is synchronized with the transmitted pulse by a scan ramp signal emitted by the controller **1**. The signal limiter prevents the receiver from being overdriven by the large signals that result from reflective objects or interfaces close to the antenna.

Of course, as the electromagnetic waves penetrate deeper into the ground, the energy is attenuated and signals reflected back to the antenna become increasingly weaker with the depth of the reflector. The output of the receiver is coupled to a variable attenuator **7** which provides maximum attenuation during and immediately following transmission, with the attenuation being gradually reduced to provide minimum attenuation at the end of the receiving interval.

The attenuator **7** is of the type whose attenuation can be varied by an electrical signal and in the process the attenuator is controlled by a signal emitted by controller **1**. The output of the attenuator may be applied to a graphic recorder **8** to obtain a profile chart. In that profile chart, depth is indicated in feet along the vertical axis and the distance over which the antenna was towed is indicated along the horizontal axis. The strength of the received signal is shown by the density of the markings.

A borehole electromagnetic survey method has been described by *W.T. Holser, R.R. Unterberger and S.B. Jones; U.S. Patent 3,286,163; November 15, 1966; assigned to Chevron Research Company*. More particularly, it relates to a method of mapping sides of a salt body by applying electromagnetic radiation from a well bore in a salt body to measure the distance to the side of the salt body at any desired level.

It is a particular object of this technique to map the sides of a salt body so that the interface between the salt body and the sedimentary formations may be located exactly at a known depth and with this knowledge to direct the drilling of oil wells into the sedimentary formation close to this interface.

In carrying out the method, a borehole is deliberately drilled into the salt body. The borehole is extended to a depth where electromagnetic radiations, such as those at radio, microwave and infrared frequencies, can be transmitted in known elevation and azimuth directions from the borehole to the interface between the salt body and the sedimentary formations. The time of travel from the well bore to the interface and back to the well bore is then recorded in accordance with the depth, azimuth, and elevation position of the transmitter and receiver, to map three-dimensionally the underground contour or configuration of the salt-sediment interface.

Figure 32 schematically indicates the use of the method to map, at depth, the location of the side wall of salt dome **10**. The purpose of such mapping, of course, is to locate and direct a proposed well bore, such as **12**, into sedimentary beds **14**, **16** and **18**. As shown, these beds are normally tilted upwardly by the intrusion of salt dome **10** through the beds after they have been laid down horizontally. The salt at the ends of beds **14**, **16** and **18** lying against wall **20** of the salt dome, together with the overlying impermeable beds, such as **15**, **17** and **19**, respectively, form traps for gravity segregation of petroleum as indicated by the reservoirs denoted at **22, 24** and **26**. Obviously, if the exact horizontal and vertical location of side wall **20** is known for each of the beds **14, 16** and **18**, well bore **12** can be directed as shown by the deviations **121** or **122** to encounter reservoirs **22, 24** or **26**.

Figure 32: Chevron Research Company Method for Mapping a Salt Dome

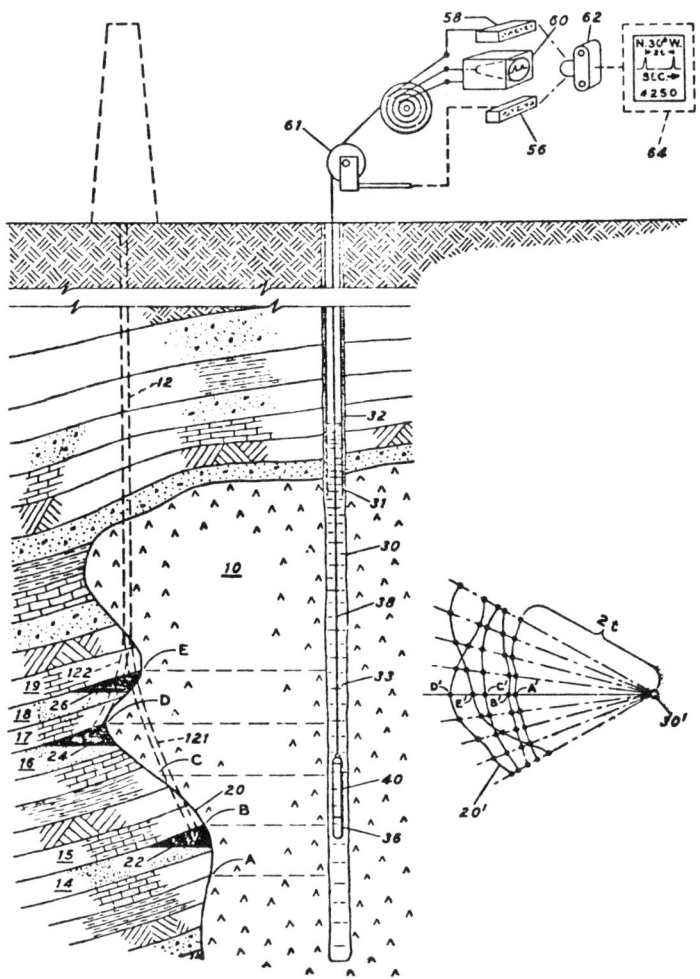

Source: U.S. Patent 3,286,163

To map the location of side wall **20** at depth, a borehole **30** is deliberately drilled into the main body of salt dome **10**. The exact location of this borehole is not critical, but it will generally be selected by normal surface exploration methods, such as gravity or seismic survey. It is most desirably located so that it is well within the main body of salt dome **10**. Suitably, the minimum distance will be at least a few hundred feet from the well bore to the side wall **20**. Well bore **30** may be drilled in any manner to the salt. Preferably then it should be cased with pipe **31** and cemented, as at **32**, and the salt drilled dry, as by air-drilling. However, the well bore into the salt body can also be drilled by conventional rotary methods using saturated salt water, or oil if the drilling fluid is replaced at the measuring levels by a suitable low-loss fluid, as indicated at **33**.

In the form of apparatus used, well bore **30** is indicated as being drilled large enough to permit a full range of azimuth motion for an antenna, or director-receiver unit **36**. To position electromagnetic wave antenna **36**, electrical signals are transmitted and received over cable **38**, which supports sonde **40**, that houses the downhole equipment.

The surface recording equipment for mapping the salt dome side wall distance to the borehole includes three indicators: for depth, **56**; for azimuth, **58**; and for distance, **60**. Depth indicator **56** shows the mapping depth of antenna **36** in well bore **30**. Each map depth is measured by pulley **61**; in turn, the position of pully **61** is shown on indicator **56**. The azimuth position of antenna **36** may be in polar coordinates, but in the process example, it is indicated as a gyroscope position indicator **58**. The distance from borehole **30** to side wall **20** at each mapping depth is then indicated by the time between transmission and reception of pulses of electromagnetic energy at antenna **36**. The travel time is suitably shown on oscilloscope **60**. By physically associating depth indicator **56**, gyro indicator **58** and oscilloscope **60**, the information on all three units can be simultaneously photographed by camera **62**.

The results of such a simultaneous photograph produce a plate of the type indicated as **64**. Plate **64** indicates compass heading, the two-way travel time **2t** in microseconds for electromagnetic signals, and the depth at which antenna **36** was located. With the **2t** and a conversion factor for time-to-distance for the transmission of electromagnetic energy, the exact position of each point at five different levels, such as **A'**, **B'**, **C'**, **D'** and **E'**, can then be plotted as shown in the detail at the lower right of the figure. The connecting lines between each map point, as indicated, represents the location of the interface, or side wall **20**, relative to well bore **30'**.

In the area of electromagnetic borehole investigation there appears to be a trend toward increasing the frequency of the energizing fields. Recently, considerable work has been done utilizing frequencies from the low megacycle to microwave region of the electromagnetic spectrum.

A method developed by *R.K. Warren; U.S. Patent 3,866,111; February 11, 1975; assigned to Exxon Production Research Company* is a method for exploring for subsurface electrically conductive bodies in which an existing electrical power transmission line is used as a source or transmitter of electromagnetic energy departed to the ground.

The electromagnetic energy induced in the ground by the transmitted electromagnetic energy is measured at spaced-apart, horizontally aligned stations in transverse, parallel and vertical directions relative to the line of stations and plotted against the station locations. The presence and location of a subsurface conductor body capable of producing measurable induced electromagnetic energy is determinable from resolution and shape of the plotted curves.

Such a method is shown schematically in Figure 33. An electrical transmission line **9** departs electromagnetic energy, indicated by the solid curved lines **10**, to the ground (the earth's surface being indicated at **30**) by induction and ground leakage.

Although shown as being positioned above ground transmission line **9** may be located underground or on the surface of the ground. The energy **10** causes current to flow in a subterranean conductive body **11** remotely located from transmission line **9**. The distance between body **11** and transmission line **9** is dependent upon many variables involving the conductance of surrounding ground material **12**. The current in body **11** radiates electromagnetic energy, indicated by the dashed lines **13**, at the same frequency as the current flowing in electrical transmission line **9**.

Measurement of the electromagnetic energy **13** at various station locations is used to determine the location and estimate the depth **14** of conductive body **11**.

Data analysis is accomplished by plotting the measured values in profile and plan views. The plan view gives a strike direction and an indication of geologic structure.

Figure 33: Exxon Production Research Company Method of Mineral Exploration by Detecting Electromagnetic Energy at Power Line Frequency

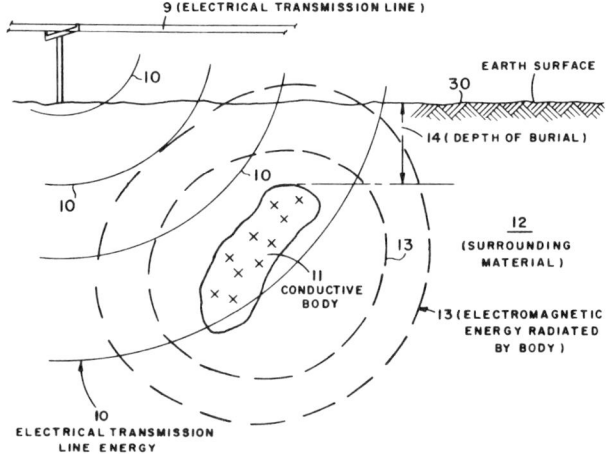

Source: U.S. Patent 3,866,111

NUCLEAR GEOPHYSICAL PROSPECTING

Nuclear geophysical prospecting can be defined as the utilization of nuclear radiation in studying underground deposits of ores, minerals, oil, gas or water. In its simplest form, the variations in the natural radioactivity of the earth are measured and correlated with known standards. This radioactivity occurs primarily from naturally radioactive elements such as: uranium, radium, thorium, and hydrocarbons containing carbon-14 substances.

A recent review of the use of nuclear methods in mineral exploration is that edited by Morse (14).

There are two main types of nuclear prospecting. One utilizes naturally-occurring radioactive elements in the earth, simply measuring their radiation. A second type of nuclear prospecting involves introducing a source of radiation into the earth, usually via a borehole, and measuring the effects produced by that radiation.

The radiometric technique is applied primarily to prospecting for uranium deposits whereas the source-reaction technique is applied to hydrocarbon prospecting, although it may be used in mineral prospecting as well.

In the radiometric technique, use is made of the fact that naturally occurring radioactive elements such as uranium or thorium break down to decay to other elements or isotopes by emission of subatomic particles. Gamma rays (similar to x-rays but of higher frequency), alpha particles (nuclei of helium atoms), and beta particles (electrons) are most commonly emitted during this process.

The portable Geiger and scintillation counters, which detect differences in the intensity of radioactivity, have been widely and effectively used in recent years for prospecting for uranium and thorium deposits. While sensitive to very small differences in amounts of radioactive elements in rocks, these instruments do not show what element produces the radioactivity. These distinctions can be made by chemical analysis of a sample of the radioactive rock. The Geiger counter is a tube filled with a gas such as helium, argon, or krypton. A high-voltage wire extends into the central part of the tube. When gamma radiation or beta particles pass into the tube from a radioactive source, some of the rays collide with gas molecules and produce electrically charged particles that are then attracted to the central wire and produce electrical pulses. The electrical pulses can be translated into dial readings of counts per minute. Scintillometers use crystals of certain compounds, such as sodium iodide, which emit flashes of light when struck by radiation. A photoelectric cell sees the flashes of light and counts the number of flashes per unit of time. Such scintillometers are more sensitive than Geiger counters.

When using radiation counters in the field, the prospector commonly walks over the terrain while listening to counts on earphones or watching the dial on a counter. Radioactive deposits may produce readings that are 10 to 100 times as great as background readings. If the deposits are covered by even a few feet of overburden, however, the radiation cannot be detected. When a portable counter is used, the information should be interpreted with caution until it is verified by adequate sampling and chemical analysis. Prices for portable counters range from $150 to more than $600 for Geiger counters and $1,100 to $1,300 for scintillometers. A good-quality Geiger counter probably will cost at least $350.

Exploration for uranium has changed markedly over the past quarter century. The simple ionization chambers and Geiger counters of the 1940s have been superseded by sophisticated instruments of great reliability and sensitivity that are capable of discriminating among uranium, thorium, and potassium. The prices of portable gamma-ray spectrometers range from $5,500 to $8,000. Highly sensitive gamma-ray spectrometers for vehicle and aircraft mounting generally cost $40,000 or more; surveys by this type of instrument are commonly obtained from geophysical service companies (1).

Research on the movements of radon, helium, and other daughter products of uranium is likely to produce new or improved tools and methods to detect concealed uranium deposits. No doubt uranium will be more costly and more difficult to find in the future because most of the easily discovered deposits in the world have been found.

There are two main physical arrangements for nuclear geophysical prospecting: surface, and well or borehole.

SURFACE CONTACT PROSPECTING

The surface type is carried out by physical contact with the surface; that is, by a person walking or riding in a land vehicle with suitable measuring instruments as described by *D.J. Belcher, T.R. Cuykendall and H.S. Sack; U.S. Patent 2,781,453; February 12, 1957; assigned to Cornell University and G. Swift, R. Monaghan, D.E. Barkalow and C.G. Denny; U.S. Patent 3,341,706; September 12, 1967; assigned to Dresser Industries, Inc.*

In the surface contact type, earth samples can be taken and measured as described by *W.M. Stratford, C.F. Teichmann, and G. Herzog; U.S. Patent 2,562,961; August 7, 1951; assigned to Texas Company and G.H. Milly; U.S. Patent 3,609,363; September 28, 1971; assigned to Geomet Mining and Exploration Company.*

In this latter technique prospecting, particularly prospecting for uranium, thorium and other radioactive ore deposits may be carried out by monitoring a gaseous decay product which diffuses through the earth's structure and becomes wind-borne.

A scheme developed by *E.P. Howell and O.J. Gant, Jr.; U.S. Patent 4,017,731; April 12, 1977; assigned to Atlantic Richfield Company* detects radioactive gaseous decay products emanating from a buried deposit of uranium or other radioactive ore which migrate upwardly through the earth and are exhaled into the atmosphere. These products may be trapped at the surface in a series of low profile, dome-shaped plastic shelters 5 to 20 feet in diameter. Radiant energy permeating each shelter cover heats the soil beneath the enclosed surface area to accelerate the escape of gas through the soil. The air confined within the shelters is continuously recirculated over the exposed surfaces of a plurality of highly adsorbent discs. After equilibrium decay conditions are reached, the discs are analyzed to determine the radioactivity of the discs, which is proportional to the concentration of radioactive gas in the vicinity.

A method of detection of the presence of such deposits consists in drilling holes in a geologically favorable area, evacuating a quantity of gas from the substrata region under consideration, and thereafter measuring by known techniques the quantity of a particular radioactive isotope present in the gas. This method suffers from the drawback, however, that even in the vicinity of a radioactive mineral deposit the gas sample collected from a given well may be a poor indicator. It is known that the fractures or channels characteristic of crystal rock provide high-speed, directional transport paths for radioactive gas, for example, radon-222. However, a single borehole may conceivably extend to a depth of several hundred feet without intersecting more than a small number of such microfractures.

Conversely, such a borehole may fortuitously intersect a very large number of such paths. In either case, the readings taken from such boreholes may not be representative of the average gas concentration in the vicinity and thus may not accurately predict the location and probable contour of a buried radioactive deposit such as uranium.

Still another surface exploration technique for locating buried ore bodies typically involves placement of small inverted cups in shallow covered holes on the order of 2 to 3 feet deep in the vicinity of interest. Sensitive film may be positioned within the cups and left for a period of time to arrive at equilibrium conditions. For the same reasons as discussed above, the sampling of gas concentration in the soil in accordance with such a method may not predict with consistent accuracy the presence or contour of buried ore bodies.

The technique here is designed to overcome the cited difficulties in prior art methods. A sampling dome and its use in relation to a mineral deposit are shown in Figure 34.

There is illustrated a gas collection shelter **10** resting on the surface of the earth **11**. The shelter **10** consists generally of a rigid, easily-disassembled frame structure **12**, preferably of lightweight tubular aluminum, which supports a removable outer cover **14**.

Figure 34: Atlantic Richfield Co. Sampling Dome for Detection of Radioactive Emanations from Mineral Deposits

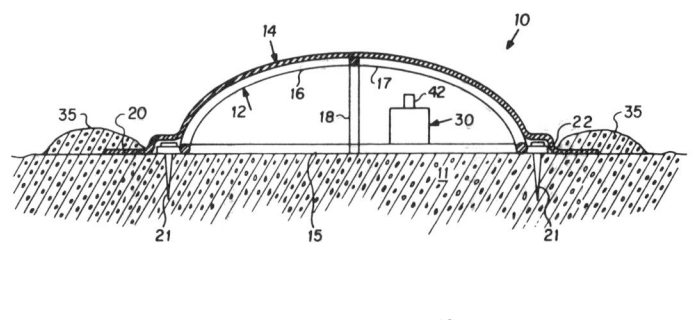

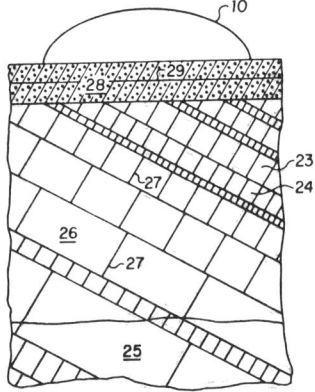

Source: U.S. Patent 4,017,731

The frame structure **12** is composed of a generally circular base **15** forming the perimeter of the shelter **10** and smoothly curved upper support members exemplified by characters **16, 17,** and **18,** suitably interconnected to establish the dome shape of the shelter **10**. It is believed, that improved results may be achieved in most instances by providing a minimum diameter for the shelter **10** of between 5 and 20 feet. The height of the shelter **10** is optional, but ideally it should not be more than a few feet above the surface of the earth to give the shelter **10** a relatively low profile. This lessens wind resistance and thus adds to its stability over long periods of emplacement.

The cover **14**, which is preferably of a lightweight, flexible, translucent, vinyl plastic such as polyvinylchloride, is provided with a perimeter skirt **20** extending outwardly of the base **15** contiguous with the earth's surface. To avoid upset, it is desirable to secure the frame structure **12** firmly against the earth's surface. For this purpose, a plurality of fasteners **21**, such as bridge nails, may be driven downwardly into the earth through corresponding brackets **22** extending outwardly from and secured to the base **15**. After the cover **14** is in place, loose earth **35** may be piled on top of the skirt **20** so as to substantially impede entry into the shelter **10** of external windblown contaminants and to confine gases exhaled over the enclosed surface area for long periods of time.

As shown in the lower view in the figure, the collection shelter **10** is positioned at the surface above a buried uranium ore deposit **25**, and further, the overburden includes a rock formation **26** characterized by a plurality of microfractures **27** and a top soil layer **28** containing an intermediate groundwater level **29**. If for example the rock formation is of sedimentary origin, as is the case over a substantial portion of Continental land masses and over the Continental Shelf, it is laid down in successive strata or layers, such as layers or beds **23** and **24**.

If these layers are of differing thickness, the microfractures which extend perpendicularly between their respective bedding planes may typically be separated by substantially different intervals. Radioactive gas, for example radon-222, emanating from the buried deposit **25** migrates upwardly through the rock formation **26** largely by transport through these microfractures or along the bedding planes between successive layers **23** and **24**. Therefore, it is hypothesized that the concentration of radon gas exiting under these circumstances into the atmosphere over a given area at the surface depends in large measure upon the presence or absence of gas exit paths from the rock formation into the portion of the subsurface soil layer **28** lying vertically beneath a given surface area and upon the cross-sectional area of these paths.

As the size of the given area is increased, the variation in effective area of available escape paths between surface areas substantially equidistant from the deposit will decrease. Investigation of the subsurface rock in a location of interest will make it possible to estimate the pattern and character of microfractures **27** or bedding planes along which radon transport will occur, depending upon the geology of the vicinity. This may involve, for example, examination of fracture spacing in rock outcrops or along creek bottoms. By providing the shelter with a diameter of sufficient size to insure coverage of a plurality of adjacent radon escape paths from the rock formation, the possibility becomes vanishingly small that any shelter directly above the buried ore deposit will fail to indicate a substantial concentration of radon gas. For example, if the average spacing of adjacent radon escape paths is approximately 2 feet, a shelter of at least 5 feet in diameter would stand a high probability of covering some of such paths.

In operation, a series of shelters **10** may be spaced apart, for example, at intervals of several hundred feet, generally in the same direction. Significantly lower concentrations of radon gas found in given shelter **10** will indicate that the contour limit of the underlying ore body **25** is approached. Clearly as the size of the area covered by a shelter **10** is increased, the accuracy of this sampling technique is enhanced, the purpose always being to insure that consistently high ratings will occur near the ore body giving rise to the gas emanations and consistently low readings will occur elsewhere. Enlargement of the shelter diameter to twenty feet would insure a high degree of correspondence between the number of covered escape paths spaced apart as in the above example, consequently minimizing variation in gas concentration in successive shelters equally close to the buried deposit.

Owing to the known dispersion characteristics of an isotope such as radon-222, a closer spacing of shelters **10** would be employed in prospecting for a deposit estimated to be relatively shallow or if the lithology of the subsurface indicates a pattern of possible migration paths of substantially higher density. Conversely, a larger spacing would be employed for ore bodies **25** estimated to be at a greater depth or having an overburden of low permeability rock.

The material of the cover **14** is selected such that it is highly permeable to infrared radiation. In addition, the shelter **10** is shaped to avoid any sharp corners tending to shadow the enclosed area of soil or to cause a reflection loss. Consistent with these design features, the shelter **10** may be formed alternatively of a section of preshaped rigid or inflatable plastic, thus eliminating the need for the frame **12**. The enclosed surface will experience heating by a greenhouse effect, and the subsurface soil temperature should be increased significantly by an amount up to 30° or 40°F each day and to a depth of 4 to 5 feet.

It is recognized that a certain amount of radon gas will be trapped at the ground water level **29** by interfacial tension with the air above. Heating the soil at the surface in the manner outlined will therefore reduce this tension and free a greater percentage of gas trapped near the surface. In this manner, variations in gas dispersion into the atmosphere which might otherwise be expected with changes in pressure or moisture will be reduced.

After securing the shelters in an area of interest, they are allowed to remain in place for a sufficient period of time to allow any stray airborne radon entering the shelters **10** to decay and to allow the radon gas reaching the surface from the buried deposit **25** to reach decay equilibrium. This period will be preferably about two or three half-lives for radon-222, a half-life being approximately 3.8 days or to provide a total time of about 7 to 10 days. Some gas will of course flow through the soil under the edges of the cover **14**, but this volume will be small compared to the total enclosed volume of the shelter.

As radon gas is exhaled into the atmosphere within the shelter, its decay products tend to attach themselves to particulate matter or impurities in the air. This air is continuously recirculated through a collector **30** positioned within the shelter.

The gases from the collector exit at **42**. The collector contains activated charcoal discs. After decay equilibrium conditions are reached, the discs may be removed and the radioactivity evidenced, thereby determined by known means such as a scintillation counter or other known types of radiation detectors measuring alpha-radon decay or gamma-daughter product decay.

The utility of the shelters is not limited to the collection of radon gas nor to the detection of uranium deposits. For example, the shelters and the collector may be employed in collecting helium, hydrogen sulfide, methane, and other gases. The collection of any soil gas or any gas-borne particulate matter which is amenable to adsorption in the manner described, is enhanced by the techniques outlined herein. While the collector is preferred, one can alternatively position a sensitive film plate (and/or thermoluminescent dosimeters) in each shelter so as to detect alpha or gamma radiation. Also, it may be preferable to pump the contents of each shelter directly to a remote location so long as due consideration is given to the half-life of any radioactive decay product to be analyzed.

However, since pumping to a remote location may tend to pull external air through the soil into the shelter, it may be advisable to collapse each shelter as the contained air is drawn into an external filter system. The method and apparatus may be easily adapted to an offshore environment by constructing the shelter over a frame **12** of floatable material and anchoring the shelter in position with respect to some fixed offshore location. To help free radon gas from the water, it will not be difficult, by means well known to the art, to continuously spray quantities of the water into the confined atmosphere within each shelter **10**.

A technique described by *W.J. Ward, III; U.S. Patent 4,064,436; December 20, 1977; assigned to Terradex Corporation* is one for reducing or removing the background noise caused by thoron gas (^{220}Rn) in uranium exploration conducted by the detection of radon gas (^{222}Rn) emanating from the ground. This is accomplished by the use of a number of alpha particle detectors, each of which is disposed in a protective enclosure. A permselective membrane, which permits, but selectively retards, the passage therethrough of gases is disposed in the path to be traversed before such gases can reach the alpha particle detector. The retarding influence of the membrane should be sufficient to make the concentration of thoron inside the enclosure small relative to the concentration of thoron outside the enclosure. The influence of the membrane on radon should be negligible, i.e., the radon concentration inside and outside the enclosure should be substantially equal.

The placement of such a device is shown in Figure 35. The cup **10** is placed in excavation **11** so as to rest on the soil **12** at the bottom thereof. Typically a board would be used to cover the hole with this in turn being covered with dirt from the excavation thereby burying the cup.

Figure 35: Terradex Corporation Device for Reducing Noise in Uranium Exploration

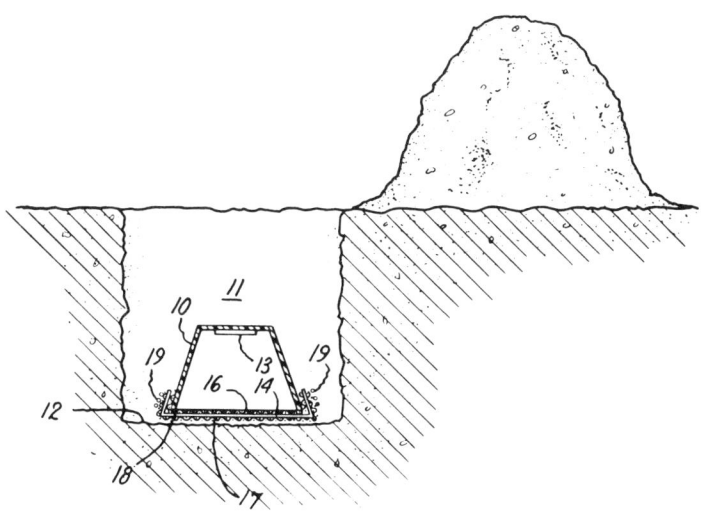

Source: U.S. Patent 4,064,436

This procedure would be repeated with a number of such cups in some desired pattern, the cups remaining buried for the test period. Mounted within each cup is a piece, or sheet, **13** of alpha particle track detector material, preferably cellulose nitrate, to measure the emanation from the ground below of the gaseous radon isotopes ^{220}Rn and ^{222}Rn. The ^{222}Rn gas is a decay product of uranium and, therefore, the detection of such emanations would be an indication of the presence of uranium in the earth. The other alpha particle emitting gas, ^{220}Rn, is a decay product of thorium and, hence, tracks induced thereby in track detector **13** constitute an unwanted background signal, when uranium is being sought.

In order to reduce or remove this unwanted background caused by alpha particle emissions from thoron gas entering the mouth of cup **10**, instead of permitting gases leaving the soil through surface **12** to enter directly into the internal volume of cup **10**, the nonporous permselective membrane **14** is disposed between surface **12** and sheet **13** (or other alpha particle detector). The sides and top (in the inverted position) of the cup are imperforate in order to prevent the short-circuiting by soil gases into the cup without passing through the membrane. Thus, any soil gases reaching the sheet of track detector material **13** must first pass through layer **14** and be subjected to the selective transit periods peculiar to each gas depending upon its particular permeation rate.

A zone, or volume, at least 6.0 centimeters thick must remain in the cup between the upper surface of layer **14** and the underside of the surface of sheet **13** so that in passing through this zone of air, the alpha particles are slowed sufficiently so that they can be detected by this form of alpha particle detector.

In the arrangement shown, membrane **14** and protective screens **16, 17** are turned over lip **18** of the cup and held in place as by a rubber band, wire or string **19** so that the mouth of the cup is closed off thereby.

The nonporous permselective barrier is typically a polymer membrane (preferably made of organopolysiloxane or a copolymer containing organosiloxane units) sufficiently impermeable to thoron so that the concentration of thoron inside the cup is a small fraction i.e., less than 20%, of the thoron concentration outside the cup during the test period. Also, the permeability of the membrane to the ^{222}Rn must be great enough so that the ^{222}Rn concentration inside the cup reaches a level which will be a large fraction of the concentration outside the cup. Preferably the concentration inside the cup will be 90% by volume or more of the concentration of the ^{222}Rn outside the cup. Silicone rubber nonporous membranes in the thickness range 2 to 12 mils and silicone/polycarbonate copolymer nonporous membranes (55% silicone by weight) in the thickness range 1 to 5 mils are examples of nonporous polymer membranes fulfilling these requirements.

The protective screens **16, 17** are preferably woven polymer (e.g., polyester or nylon monofilament) screen cloth having mesh openings in the range of 400 microns and an open area ranging from about 45 to 55%.

It has been found that by selecting the nature of the membrane material to be employed and the thickness of the membrane in accordance with the guidelines given herein, the ^{220}Rn signal reaching within 6.0 centimeters of detector **13** (the active air space) can be reduced as desired without materially affecting the ^{222}Rn signal.

A technique developed by *P.E. Felice; U.S. Patent 4,156,138; May 22, 1979; assigned to Westinghouse Electric Corp.* is one in which underground uranium deposits are located by placing wires of dosimeters each about 5 to 18 mg/cm^2

thick underground in a grid pattern. Each dosimeter contains a phosphor which is capable of storing the energy of alpha particles. In each pair, one dosimeter is shielded from alpha particles with more than 18 mg/cm² thick opaque material but not gamma and beta rays and the other dosimeter is shielded with less than 1 mg/cm² thick opaque material to exclude dust. After a period underground, the dosimeters are heated which releases the stored energy as light. The amount of light produced from the heavily shielded dosimeter is substracted from the amount of light produced from the thinly shielded dosimeter to give an indication of the location and quantity of uranium underground.

Such a device is shown in Figure 36. The upper view shows the placement of the device in a hole in the ground and the lower view shows a detail of the dosimeter.

Figure 36: Westinghouse Electric Corporation Device for Locating Underground Uranium Deposits

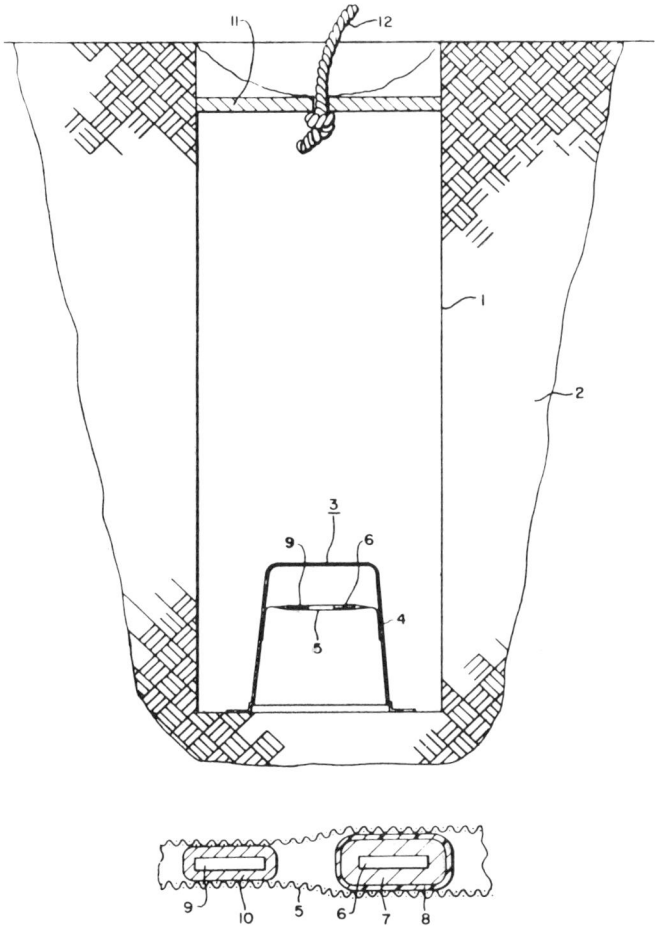

Source: U.S. Patent 4,156,138.

In the drawings, a hole **1** excavated in ground **2** houses dosimeter housing **3** which consists of an inverted cup **4** to which metallic screen **5** has been stapled or otherwise affixed. To the screen are affixed two alpha-particle sensitive dosimeters each about 5 to 18 mg/cm^2 thick, one **6** shielded with more than 18 mg/cm^2 aluminum or other material to exclude alpha-particles but not beta and gamma ray background radiation, and sealed in 2 mil polyethylene **8**, and the other **9** contained in 0.030 mil aluminum **10** which permits the passage of alpha, beta and gamma particles but excludes dust particles. A cover **11**, secured to rope **12**, protects the dosimeters.

SURFACE AERIAL PROSPECTING

In the aerial type of prospecting, the variations in the natural radioactivity can be measured directly or atmosphere samples can be periodically or continuously taken and the radioactivity of these samples measured.

Air-borne radioactivity prospecting has been employed heretofore by flying an aircraft across the surface of the earth while detecting radiation with a single radiation detector unit located within the aircraft. By such a procedure, it has been possible to obtain an indication of the extent of mineral deposits within a relatively large area beneath the aircraft, for example a strip of the earth 150 feet wide being examined from a helicopter 150 feet above the earth. Since the ore body detected may lie at any point within the 150 foot width of the strip, its location by subsequent surveying on the surface of the earth has been difficult. The search may be narrowed by reducing the width of the flight lanes, but this requires considerably more traverses of the aircraft across the surface and obviously increases the expense and time for conducting a survey.

A technique developed by *R.P. Mulligan; U.S. Patent 2,904,691; September 15, 1959; assigned to Texaco Inc.* overcomes the difficulties by simultaneously flying at least two radiation detectors above the surface of the earth at positions spaced from one another, detecting radiation emanating from the earth with the detectors, recording measurements of the detected radiation, and comparing and correlating the measurements of detected radiation with one another and with the areas of the earth's surface seen by the respective detectors to determine the location of mineral bodies more exactly than has been possible heretofore.

In one embodiment of the method, a self-propelled aircraft carrying a radiation detector is flown across the earth at one level while towing a second radiation detector at a lower level but in substantially the same vertical plane. Other embodiments involve flying more than two detectors at successively lower levels; and flying two or more detectors at positions spaced laterally from one another either in the same or different horizontal planes. If desired, the self-propelled aircraft need not carry a detector, in which case the two or more detectors are towed.

Figure 37 shows the technique in operation and indicates, from top to bottom, the aerial survey vehicles, a plan of the traverse over the ore deposit area, and strip charts showing the instrumental records produced.

Figure 37: Texaco Inc. Aerial Prospecting Method

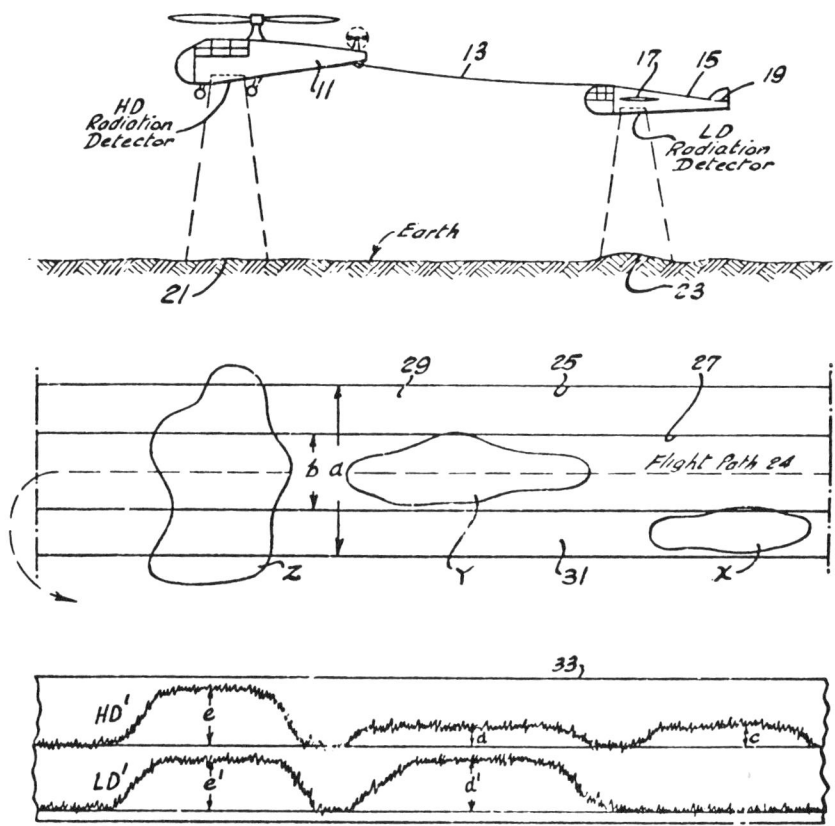

Source: U.S. Patent 2,904,691

As shown, a self-propelled aircraft **11** carrying a radiation detector **HD** is flown across the surface of the earth at a selected height, such as 150 feet for example, while towing behind it on a cable **13** a streamlined glider **15** carrying a second radiation detector **LD** in substantially the same vertical plane but at a lower level such as 100 feet above the surface of the earth. The relationship between the positions of the aircrafts **11** and **15** is maintained substantially constant by means of adjustable vanes **17** and **19** on the glider **15**.

The radiation detector **HD** at any given point in the survey observes a relatively large area **21** of the earth, while the detector **LD** observes a considerably smaller area **23**. As the aircrafts fly along a straight flight path **24** these observed areas merge continuously to form long narrow lanes, as shown more in detail in the plan of the traverse over the ore deposit area in the center of the figure.

The wide lane **25** of width **a** is observed by the radiation detector **HD** while the narrow lane **27** of width **b** is observed by the radiation detector **LD**. Both radiation detectors are arranged in the same vertical plane so that the narrow lane **27** is centered within the wide lane **25** and divides the latter into three narrow adjoining strips **27, 29** and **31**.

One of the advantages of this procedure is that when the radiation detector **HD** provides an indication of an anomaly in the area of lane **25**, the location of the ore producing this anomaly can be determined quite accurately by observing the measurement of radioactivity with the radiation detector **LD**. If **LD** shows an anomaly in radioactivity it indicates that the source is probably located within the narrow strip **27** rather than within either of the side strips **29** and **31**. Conversely, when the detector **LD** shows relatively little evidence of an anomaly in radioactivity it indicates that the source must lie either in the strip **29** or the strip **31**. Such indications make it possible for a ground party to locate the ore much more quickly and economically than if the entire wide lane **25** had to be prospected on the ground, and more quickly and economically than if a series of separate aircraft flights had to be made along narrow lanes similar to **29, 27** and **31**. Consequently, one flight across the surface of the earth accomplishes as much as three flights heretofore.

Another important advantage of the process is that it is possible to determine the general shape and extent of an ore body producing an anomaly. For example, ore bodies **X, Y** and **Z** are shown of different shapes and at different locations within the purview of radiation detector **HD**. If a single aircraft carrying only a single radiation detector unit were flown along the flight path **24**, it would observe the relatively wide lane **25**, and upon passing over each of the ore bodies, **X, Y** and **Z** would detect their presence but not their precise location and shape. As shown, a strip chart **33** would bear a single trace **HD'** showing changes in radioactivity above the ore bodies, but the ore bodies could be in any part of the wide lane.

Interpretation is made much more accurate and complete by employing the second radiation detector **LD** at a lower level to show variations in radioactivity by the trace **LD'** on chart **33** as the detector passes above the ore bodies **X, Y** and **Z**.

The change in radioactivity above the ore body **X** is recorded by a hump of magnitude **c** in the trace **HD'** whereas the trace **LD'** remains unchanged, thus indicating that the ore body **X** lies in either the strip **29** or **31** rather than in **27**. Next, when the detectors pass above the ore body **Y**, the trace **HD'** assumes a hump of magnitude **d** whereas the trace **LD'** assumes a hump of much greater magnitude **d'**, consequently showing that the ore body **Y** is located predominantly within the central lane **27** and is of a length corresponding to the length of the humps **d** and **d'**.

Next, when the detectors fly above the large ore body **Z**, both traces **HD'** and **LD'** exhibit humps of relatively great magnitudes **e** and **e'**, the magnitude of **e** being considerably greater than the magnitude of **d**, thus showing that the ore body at least extends along the entire width of the lane **25** and is of a length corresponding to the lengths of the humps **e** and **e'**.

It is evident that with detailed information of the type described above the work of the prospector is lightened greatly and the cost of prospecting for ores is greatly reduced.

For carrying out the method described above, gamma ray radiation detector units of conventional types may be employed, for example, radiation detectors employing Geiger-Mueller tubes, or scintillation counters employing gamma ray sensitive crystals such as thallium-activated potassium iodide or calcium tungstate.

This technique may be used for detecting positive anomalies caused by gamma rays emanating from radioactive minerals such as uranium and radium ores, or by detecting negative anomalies caused by substantially nonradioactive ores such as iron, copper, lead, zinc and the like.

A method developed by *C.H. Johnson, Jr. and R.S. Foote; U.S. Patent 3,825,751; July 23, 1974; assigned to Texas Instruments Incorporated* is one in which an airborne gamma ray detection system is flown over the surface of the earth at a predetermined height. The energy spectrum of gamma radiation from the surface of the earth is detected and recorded at locations spaced along the travel path of the system. The gamma radiation emanating from uranium decay products borne by the atmosphere of the earth is also sensed and recorded with sufficient regularity to detect naturally occurring variations therein. The gamma radiation data detected from the surface of the earth is then corrected in order to reduce the inaccuracies created by the gamma radiation sources present in the atmosphere.

BOREHOLE PROSPECTING

Well or borehole prospecting is often called well or borehole logging because a log or record is made of the geological formations or strata through which the well passes. Borehole nuclear prospecting developed primarily around the continuing search for oil and natural gas.

Recent problems with the supply and price of energy as well as projected shortages have spurred increased research and patent activity in both surface and borehole nuclear geophysical prospecting technology.

An early patent on nuclear borehole prospecting was that by *S.A. Scherbatskoy; U.S. Patent 2,219,273; October 22, 1940; assigned to Well Surveys, Incorporated.* Figure 38 shows the form of apparatus used.

The device consists of a main housing **1** containing the detecting and amplifying devices and suspended on the lower end of a cable **2** which serves both to support the housing and furnish electrical connections between the housing and the surface apparatus.

Within the casing **1** is an ionization chamber which consists of an hermatically sealed compartment **3**, containing a pair of spaced electrodes **4** and **5** in an atmosphere of nitrogen under a pressure of around 300 psi. The inner electrode **4**, as shown in the figure, is in the shape of a vertical rod, and the outer electrode **5** has the shape of a cylindrical tube with a radius of about 2". Both electrodes are preferably of iron although other metals can be used. Other gases and other pressures can also be used.

Connected across the electrodes of the ionization chamber is a battery **6** having a potential of around 150 volts. The negative side of the battery is connected to the outer electrode and the positive side is connected through a resistor **7** to the inner electrode. The resistor **7** preferably has a relatively high resistance, for example, of around 10^{12} ohms. The voltage drop across this resistance is used to operate an alternating current amplifier **8** also contained in the housing **1**, and this is accomplished by connecting one end of the resistor directly to the amplifier and the other end of the resistor to the amplifier through a secondary resistance **9** which may, for example, be of the order of 10^{11} ohms.

Figure 38: Well Surveys, Inc. Process for Nuclear Well Logging

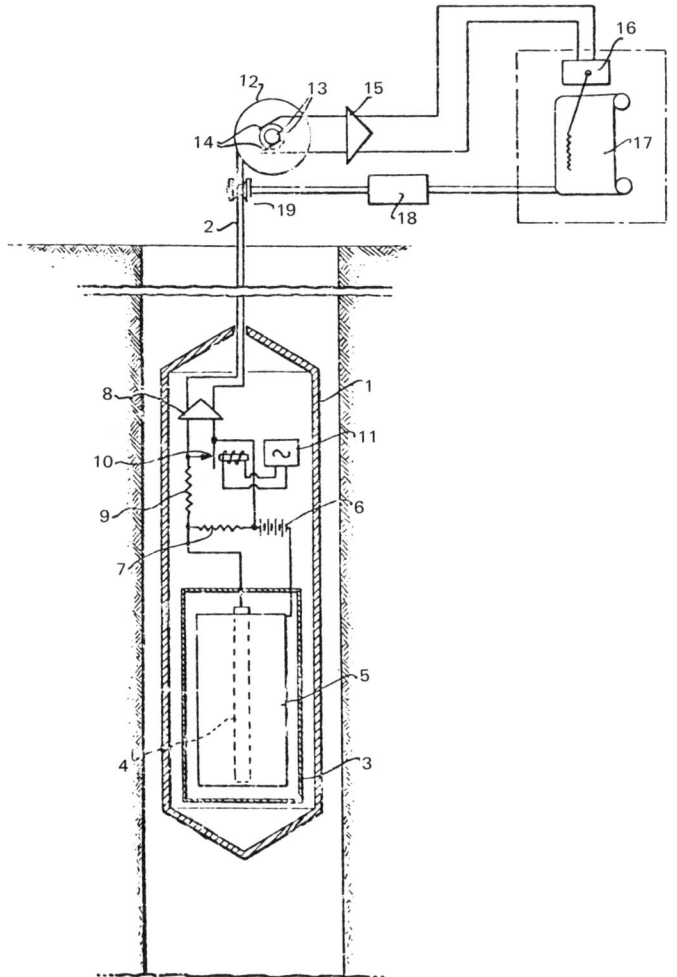

Source: U.S. Patent 2,219,273

In order to interrupt the current to the amplifier so as to permit the amplifier to function as an alternating current amplifier rather than as a direct current amplifier, a magnetically operated contacting device **10** is placed directly across the input terminals of the amplifier, and this contacting device is operated preferably at a frequency of around 100 cycles by a source of alternating current **11**, which may be an oscillator, or a buzzer and battery, or any other combination of electrical elements that will generate the necessary alternating current. A preferred form of contactor **10** contains a vibrating reed with a tungsten contact face and a stationary contact element which touches the vibrating element during part of the cycle.

During the intervals of contact between the vibrating element and the stationary elements the voltage applied across the input terminals of the amplifier **8** is zero. During the intervals when the contactor is open, the voltage across the amplifier will be derived from the output of the ionization chamber. The above-described cycle of operations in the contactor results in alternating voltage applied across the input terminals of the amplifier, the magnitude of which voltage is proportional to the steady current derived from the ionization chamber. The output of the amplifier is connected to the cable **2** through which it passes to the surface of the ground.

The purpose of the secondary resistor **9** is to prevent the potential across the main resistor **7** from falling to zero when the contactor is closed. With the secondary resistor in the circuit, the potential across the main resistor will be substantially maintained although the potential across the input of the amplifier is reduced to zero or substantially so.

The upper end of the cable is wound on a drum **12** which is rotated by means not shown to raise and lower the exploring device in the ground. The drum carries slip rings **13** to which the conductors of the cable are connected. By means of brushes **14** the currents from the cable are taken from the slip rings and applied to a further amplifier **15**, the output of which is connected to an alternating current voltmeter **16** which acts in conjunction with a recorder **17** to record the various measurements. The recorder is driven through connection **18** and a spool **19** which contacts the cable as it goes in and out of the drill hole and thus the recording made by the recorder plots the various measurements in direct correlation with the depths at which they are taken. Other methods of correlating the recorder operations with the movement of the exploring device up and down in the well bore may be used, for example, a Selsyn transmission system may be connected between the cable drum **11** and the recorder.

Another method for borehole prospecting has been described by *R.E. Fearon; U.S. Patent 2,390,433; December 4, 1945; assigned to Well Surveys, Incorporated.*

NATURAL GAMMA RAY LOGGING

As recently as 40 years ago, the state of oil exploration consisted of drilling oil wells until a recognized oil-bearing formation was reached or until the oil under natural pressure gushed out of the well. When oil from this formation or pool was exhausted, the well was either drilled deeper to another possible oil-bearing formation or abandoned. Often oil-bearing formations, under low pressure but retrievable by pumping, were missed in drilling through to the formation used.

The problem resulted primarily from the inability to electrically log a well cased for production of the first usable oil pool. Gamma radiation, naturally present, provided a tool for doing this because such radiation will pass through the steel well casing. Thus, a gamma ray measuring instrument, usually a Geiger counter, passed through the well can be used to determine where further oil-bearing formations may exist. This technique is called a natural gamma ray log.

One technique for natural gamma ray logging has been described by *G.L. Hassler; U.S. Patent 2,197,453; April 16, 1940; assigned to Shell Development Company.*

A method developed by *J.W. Merritt; U.S. Patent 2,947,870; August 2, 1960* is a method of exploring for and locating subsurface hydrocarbon deposits. It comprises the following steps: (1) measuring and recording the surface radiation intensity of the radioactive constituents of the earth at a plurality of stations in a selected area to obtain for each of such stations a radiation value; (2) obtaining a representative sample of the soil at each of these stations; (3) determining and recording the relative water-soluble mineral retentivity characteristics of these samples; (4) applying to the original radiation values a correction based on the variation from station to station in the retentivity characteristics in order to obtain a corrected radiation value at each station; and (5) charting the corrected values to visually reveal the corrected radiation pattern.

When other oil-bearing formations are detected in a cased well, the casing is merely perforated by a perforating gun and the oil is removed as described, for example, by *S.A. Scherbatskoy; U.S. Patent 2,957,083; October 18, 1960; assigned to PGAC Development Company.*

In surface prospecting by means of a vehicle in contact with the earth's surface, the following types of logging can be used, in addition to a natural gamma log:

(1) gamma-gamma for density of the substances near the surface as described by *P. Kehler; U.S. Patent 3,846,631; November 5, 1974; assigned to Applied Invention Corporation.*

(2) neutron-gamma for specific identification of elements present at or near the surface as described by *F.E. Senftle, A.F. Hoyte, and P. Martinez, Jr.; U.S. Patent 3,463,922; August 26, 1969; assigned to the U.S. Secretary of the Interior;* and

(3) neutron-neutron for moisture content of the soil at or adjacent the surface as described by *B.F. Wack; U.S. Patent 3,428,806; February 18, 1969; assigned to Electricite de France, France.*

A technique developed by *R.S. Foote; U.S. Patent 3,919,547; November 11, 1975* is based on the discovery that the relative distribution of surface-or near surface-detected gamma radiation indicative of the presence of potassium-40 bears a definite relation to the horizontal periphery of a subterranean petroleum-bearing deposit. More particularly, it has been discovered that surface areas, wherein the gamma radiation indicative of the presence of potassium-40 is low, relative to adjacent surface regions, characteristically define the horizontal periphery of subterranean petroleum-bearing deposits.

The relative intensity of surface radiation indicative of potassium-40 is determined through the use of surface vehicles, underwater vehicles or airborne vehicles traversing a region of earth's surface while sensing the intensity of emitted gamma radiation. The measurements are taken at periodic intervals during a plurality of horizontally spaced, substantially parallel linear traversals over a region of the earth's surface. Simultaneously, gamma radiation emanating from the atmosphere above the sensing apparatus is measured and recorded. The measurement of radiation in the atmosphere is conducted with sufficient regularity to detect naturally occurring variations in the gamma radiation sources present in the atmosphere. The detected gamma radiation from the surface is then corrected to reduce the effect of atmospheric-borne sources of radiation.

A system developed by *J.A. Murphy; U.S. Patent 4,059,760; November 22, 1977* is a prospecting system for hydrocarbons using emanoradiation measurements of the neutron, beta, and gamma rays, and radon gas emitted at the earth's surface and atmosphere from the earth basement complex and, the overlying sedimentary deposits of carbonaceous rocks, shales and sandstone that are impregnated with uranium, thorium and potassium derived from the eroded earth basement complex materials. A number of check points are measured and the radiation levels are plotted to form a georadiograph. A comparison between the background level and the check point levels is used in determining the contour data basically of the earth's stratosphere, altered in its variation by hydrocarbons, radiation active substances, and/or mineral deposits.

A technique developed by *I.R. Supernaw, D.M. Arnold and A.J. Link; U.S. Patent 4,071,755; January 31, 1978; assigned to Texaco Inc.* is one for the in situ evaluation of the organic carbon content of earth formations penetrated by a well borehole. The energy spectrum of natural gamma radiation occurring in earth formations penetrated by a well borehole is observed in energy regions corresponding to uranium, potassium and thorium. Quantitative evaluations of the relative abundances of these elements are made by comparing the observed spectra with standard gamma ray spectra. The relative abundances of these elements may then be interpreted in terms of the organic carbon content of earth formations by comparison with predetermined relationships found to exist therebetween.

Figure 39 is a schematic showing the application of this technique.

A fluid-filled well borehole **10** is shown penetrating earth formations **11** and having suspended therein a well logging sonde **12**. The well logging sonde is suspended in the borehole by means of a conventional armored well logging cable **13** having one or more conductors for supplying power to the downhole instrumentation and for conducting signals from the downhole sonde to the surface instrumentation. The fluid-tight hollow body member or sonde contains a naturally occurring gamma ray detecting system comprising a scintillation crystal **15**, which may either be a sodium or cesium iodide thallium activated crystal or the like. The scintillation crystal is optically coupled to a photomultiplier tube **16** for producing output electrical signals representative of natural gamma radiation occurring in the earth formations in the vicinity of the borehole.

As is well known in the art, gamma radiations from the earth formation, impinging upon the scintillation crystal produce light flashes therein whose intensity is proportional to the energy of the gamma ray, causing the scintillation. The photomultiplier senses the scintillations in the crystal and produces electrical pulses whose height or voltage level is proportional to the intensity of the light flashes produced in the scintillation crystals. Thus electrical voltage signals in the form of pulses are produced as output from photomultiplier tube **16**. These voltage signals are amplified in an amplifier **17** and are transmitted to the surface on a conductor of the well logging cable.

158   Geophysical and Geochemical Techniques for Exploration

Figure 39: Texaco, Inc. Method for In Situ Evaluation of the Source Rock Potential of Earth Formations

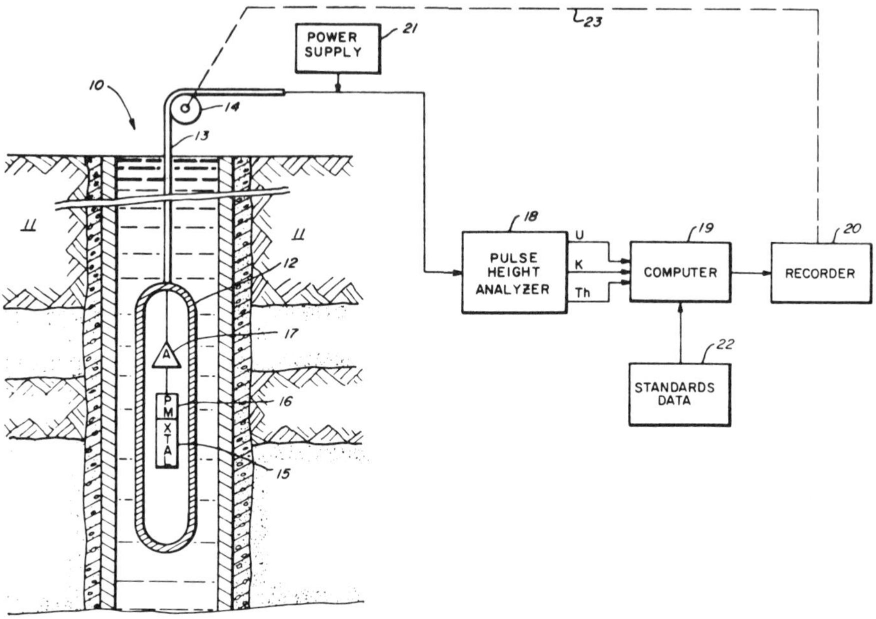

Source: U.S. Patent 4,971,755

The downhole sonde is suspended in the borehole **10** by means of a sheave wheel **14** which may be electrically or mechanically coupled (as indicated by the dotted line **23**) to a recorder **20**. Thus, depth information concerning the depth location of the well logging sonde **12** which is provided by the sheave wheel may be supplied to a recorder for recording output signals from the downhole instrumentation as a function of borehole depths.

The voltage pulses, representative of the energy of the naturally-occurring gamma radiation in earth formations **11** in the vicinity of the borehole are decoupled from the well logging cable **13** conductor at the surface and supplied to a pulse height analyzer **18** which may be of conventional design as known in the art. The pulse height analyzer functions to sort and account for the naturally-occurring gamma radiation from the earth formations as a function of energy. In the case of the process, the pulse height analyzer is provided with at least three energy windows or bands of energy acceptance for separating gamma radiation occurring from the radioactive decay of radioactive isotopes of potassium, uranium and thorium.

For the purpose of separating gamma radiations occurring from radioactive potassium, uranium, and thorium isotopes in earth formations, the pulse height analyzer is provided with at least three energy windows corresponding to these elements.

The potassium energy window is chosen to extend from approximately 1.36 to approximately 1.60 meV. The uranium energy window is chosen to extend from approximately 1.60 to approximately 1.95 meV. Finally, the thorium energy window is chosen to extend from approximately 2.40 to approximately 2.86 meV in the gamma ray energy spectrum.

Thus, the pulse height analyzer 18 provides output signals representative of the number of counts occurring in the energy window characteristic of the radioactive decay of the radioactive potassium, uranium and thorium atoms in the earth formations 11 penetrated by the well borehole 10. These output signals are supplied to a computer 19 which may comprise for example, a small general purpose digital computer such as the model PDP-11 provided by the Digital Equipment Corporation. This small digital computer may be programmed according to techniques known in the art to perform the function of spectrum stripping or curve fitting of the gamma ray spectral information provided from the downhole sonde 12.

The spectrum from the downhole instrumentation is compared against spectral standards, supplied from a standard spectrum data source 22 in which the gamma spectrum of known standard elements may be quantitatively compared with that of the unknown earth formations penetrated by the well borehole. Coefficients representative of the fraction of the gamma ray spectrum caused by the standards postulated to be contained in the downhole earth formations may thus be derived. This information is then supplied as an input to the recorder 20 where it may be recorded as a function of borehole depth as previously discussed.

Power for the operation of the downhole instrumentation contained in the sonde is supplied from power supplies contained therein (not shown). The source of this power is surface power supply 21 which supplies electrical energy via conductors of the well logging cable 13 to the downhole power supplies. This electrical power is then converted into the proper voltage levels to power the photomultiplier tube and associated circuits in the downhole sonde in a conventional manner.

Thus, a prospective area of exploration can be analyzed by performing a natural gamma ray spectral logging operation through the earth formations penetrated by an exploratory borehole. The gamma radiation occurring in the earth formation penetrated by the borehole is quantitatively analyzed for its uranium, potassium and thorium content and one or more of these signals may then be used in conjunction with calibration curves to determine from this uranium, potassium or thorium content the percentage of these shale formations which is organic carbon. The organic carbon content of the shales is thought to be indicative of the source rock activity or source rock potential of the shales penetrated by the exploratory well borehole.

GAMMA-GAMMA LOGGING

The main problem with natural gamma ray logging is that often false indications are obtained, that is, gamma ray patterns may indicate an oil formation situated behind a well casing when none exists.

A technique was developed to supplement, and, in some cases, replace gamma ray logging which uses externally applied radiation other than natural radiation. When gamma rays from a source such as, radium, cobalt-60, or cesium-137, are directed onto a specific substance, the gamma radiation will penetrate and backscatter from that substance in direct relation to its density. This indication of density can be used to locate oil deposits. This technique is known as gamma-gamma logging and usually involves passing an instrument containing a gamma ray source and a gamma ray detector through a well.

One gamma-gamma logging device has been described by *B.J. Kalb and D.H. Wise; U.S. Patent 2,365,763; December 26, 1944.* This device is shown in various views in Figure 40.

Figure 40: Gamma-Gamma Logging Device

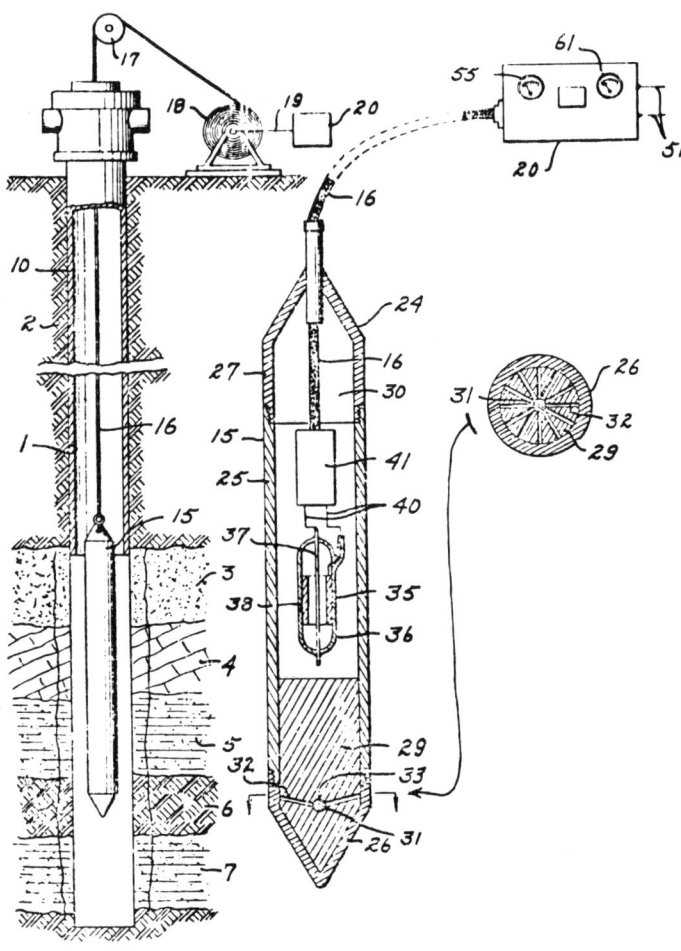

U.S. Patent 2,365,763

The well bore **1** is shown as traversing the overburden **2** and successive superimposed strata **3, 4, 5, 6** and **7** of sand, shale, water sand, rock and oil sand respectively. The upper portion of the bore is shown as provided with a casing **10** although it is to be understood that the entire length of the borehole may be cased and the survey conducted inside of such casing or drill stem without departing from the scope of the process. Likewise, concrete may surround the borehole either alone or in combination with the casing.

The exploring unit generally designated as **15** is attached to a conductor cable **16** which passes over the sheave **17** to the drum **18** suitably mounted near the mouth of the borehole. The cable **16** is connected through suitable slip rings at the drum to a cable **19** and then to the indicating instrument **20**.

With the construction thus generally described, it is believed apparent that the exploring unit **15** may be moved within the well bore and that current may be conducted between such unit and the indicating instrument **20** through the conductor cable **16**. It is also apparent that with this construction, the length of cable between the exploring unit **15** and the indicating instrument **20** remains constant and there is, therefore, little or no modification of results from change in level of the unit **15** within the borehole.

The exploring unit **15** is shown in greater detail in the middle view of the figure as comprising a container **24** made up of a central tubular section **25** which is closed at its opposite ends by means of caps **26** and **27**, threadably attached thereto. The cable **16** passes through the cap **27** at its upper end and in sealing engagement therewith so as to exclude liquids within the borehole from the chamber **30** within the unit. The tubular section **25** may be of steel but is preferably made of material such as brass or an aluminum alloy which is less opaque to secondary radiations which must pass therethrough in order to obtain desired information as to the strata traversed while the unit is moved within a well bore.

The lower portion of the chamber **30** carries a plug **29** which has a central chamber **31** therein. The plug **29** is desirably of lead but may be of any other suitable material that is sufficiently opaque to radioactive emanations so as to shield the detecting means, hereafter described, from the radioactive substance used. A body of radium or other suitable radioactive substance **33** of any desired size may be deposited in the chamber **31** so that direct radiant energy will pass outwardly through openings **32** to the formations surrounding the unit. It will be noted that the passages **32** are inclined to direct the radiant energy and it seems obvious that such inclination will be such as to obtain the best resultant pickup of radiations. This plug also serves as a shield to prevent direct radiation of the radioactive emanations from the radioactive substance within the chamber **31** to the detector or pickup device mounted within the chamber **30** in the container **24** above the plug.

A detector or pickup device for the secondary radiations from the strata is disposed in the chamber **30** of the container **24** and is here illustrated as including an ionization chamber **35**, shown as a conventional so-called Locher tube which comprises an outer envelope **36** having an axial central terminal **37** and a cylindrical terminal **38**. These terminals are connected by means of conductors **40** to an amplifier unit **41** which is in turn connected through the cable **16** to the indicating instrument **20**. It is thus apparent that the intensive direct radiation to the earth formations from the radium or other radioactive substance will be at once detected as secondary radiation from such formations by the ionization chamber **35**.

Another gamma-gamma logging device has been described by *R.L. Caldwell, G.L. Hoehn, Jr. and T.W. Bonner; U.S. Patent 2,934,652; April 26, 1960; assigned to Socony Mobil Oil Company, Inc.*

NEUTRON-GAMMA LOGGING

While an improvement, gamma-gamma logging is not entirely satisfactory since some non-oil-bearing formations will give a density reading similar to an oil-bearing formation.

This led to the development of a more direct method of identifying an oil-bearing formation, neutron-gamma logging. When fast neutrons are directed into a substance containing hydrogen, the resulting collision or reaction between the neutrons and the hydrogen atoms produces secondary gamma rays. This type of well logging uses an instrument similar to the gamma-gamma type except that a neutron source, typically radium-beryllium, replaced the gamma source. In this type of source the alpha particles emitted by the radium, react with the beryllium target material producing fast neutrons.

One such device has been developed by *R.E. Fearon; U.S. Patent 2,349,712; May 23, 1944; assigned to Well Surveys, Incorporated* and is shown in Figure 41.

Referring particularly to the figure, a drill hole **9** is shown penetrating the formation to be explored. The drill hole may be provided with a tubular metallic casing such as designated by **10**. The presence of the metallic casing in the drill hole is not an essential feature. The casing is merely shown for the purpose of illustrating the conditions under which the method may be practiced and it is understood that the process herein described may be applied in cased as well as in uncased holes.

The exploratory apparatus proper consists of a housing **11** which is lowered into the borehole by means of a cable **12**, containing insulated conductors. The cable has a length somewhat in excess of the length of the hole to be explored and is normally wound on a drum **13** positioned adjacent to the top of the drill hole. The cable may be unwound from the drum to lower the exploring apparatus into the hole and may be rewound upon the drum to raise the exploring apparatus. Between the drum and the hole there is a measuring reel **14** which is adjusted to roll on the cable in such a manner that the number of revolutions of the reel corresponds to the amount of cable which has passed up or down in the drill hole. The reel is mounted on a shaft **15**, and the motion of the shaft is transmitted through a gear box **16** to another shaft **17** which turns a spool **18** to wind a photographic film **19**, the film being supplied from a feed spool **20**.

The housing of the exploratory apparatus comprises three parts respectively designated by the numerals **21**, **22**, and **23**.

Figure 41: Well Surveys, Inc. Neutron-Gamma Logging Device

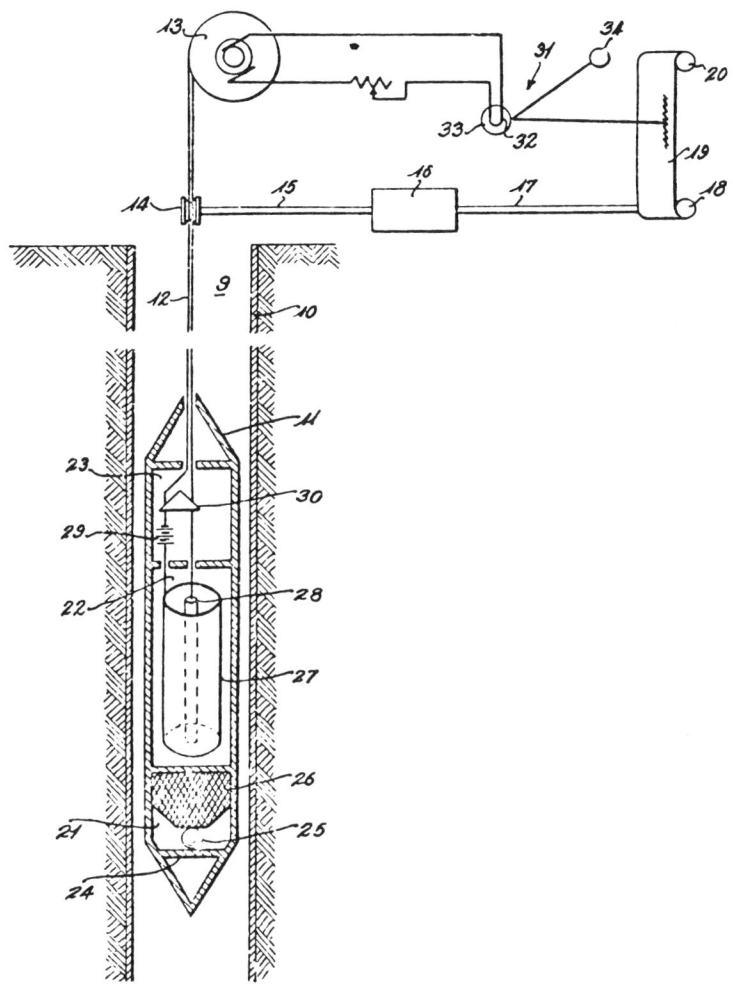

Source: U.S. Patent 2,349,712

In the partition **21** there is provided upon a solid support **24** an appropriate quantity of radioactive material such as, for instance, mesothorium 2, which is designated as **25** and a piece of lead **26** or other material relatively opaque to the penetrating radiations produced by mesothorium 2, which is placed above the mesothorium 2 so as to shield it from a direct communication with the upper partition **22**. Mesothorium 2 is given merely as an example and any other suitable radioactive material, such as for instance a mixture of radium and beryllium may be used, the mixture being characterized by an intense emission of neutrons.

The partition 22 contains an ionization chamber having a cylindrical outer electrode 27 and a central wire electrode 28. The ionization chamber is filled with inert gas such as nitrogen, preferably under pressure of about 300 psi.

The partition 23 contains a battery 29 to apply a voltage to the ionization chamber and an amplifier 30 to amplify the current passing through the ionization chamber. The battery has one of its terminals connected to the cylindrical electrode 27 and the other terminal connected to the input terminal of the amplifier. The central electrode 28 is directly connected to the other input terminal of the amplifier.

The output terminals of the dc amplifier 30 are connected to the cable 12 which conveys the current from the amplifier to a recording galvanometer 31 located at the surface of the earth. The recording galvanometer includes a moving coil 32 connected to the cable and a mirror 33 attached to the moving coil. The mirror is adapted to reflect a beam of light from a lamp 34 onto the sensitive film 19 to produce (after the film has been developed) a record of the well log.

The operation of this device may be explained as follows: The mass 25 is subject to a continuous and progressive disintegration which is well known in the art as a radioactive process and transforms itself from mesothorium 2 into an element known as radiothorium. Various radioactive materials which may be used emit radiations which usually include helium particles known as alpha rays, electrons, known as beta rays, penetrating electromagnetic impulses known as gamma rays, and may include any other radiations or material particles of any nature such as positrons, protons, neutrons and others.

The radiations transmitted from 25 tend to propagate themselves in all directions. There is provided, however, an absorbing block 26 formed of materials, for example, such as lead and paraffin which are relatively opaque to penetrating radiations, the paraffin being relatively opaque to neutrons and the lead being relatively opaque to other radiations. One is, therefore, preventing a direct path between 25 and the ionization chamber. Consequently, the radiations emitted from 25 are directed sideways into the adjacent formations and the amount of radiations going upwards through the absorbing block is negligible.

It is well known by those skilled in the art that when a formation constituting the wall of the borehole is exposed to the radiations which may impinge from a definite direction it becomes itself a source of radiations and these radiations proceed outwards in all directions. These radiations are called scattered radiations and the phenomenon is known as scattering. The radiations coming directly from the mass 25 are called primary radiations to distinguish them from the scattered radiations. It is then apparent that the radiations scattered by the walls of the borehole enter into the partition 22 wherein they are detected by the ionization chamber.

The operation of the ionization chamber is well known in the art. Under normal operating conditions the battery 29 maintains between the central electrode 28 and the cylindrical electrode 27 a voltage of such a magnitude that a discharge will just not pass between them. When, however, a quantum of energy emitted from the adjacent earth stratum and due to radiation scattering enters the ionization chamber and is absorbed in the gas it creates a large number of ions in the gas which permits a current delivered by the battery 29 to pass between the electrodes **27** and **28**. This current becomes amplified in **30** and is transmitted through the cable **12** to the recording apparatus at the top of the drill hole.

It is seen, thus, that the output voltage of the amplifier **30** depends upon the particular formation adjacent to the exploring apparatus and is influenced by the ability of this formation to scatter penetrating radiations produced by **25**. It is also well known in the art that the scattering ability of a given material is directly related to its density, i.e., the larger the amount of scattering produced, the greater is the density of the material. Consequently, the output voltage of the amplifier represents also the density of the material constituting the wall of the formation.

Therefore, when the exploring apparatus is being moved to various depths within the drill hole, the variations in density of various formations cause variations in the output of the amplifier which variations are transmitted to the top of the drill hole through the cable and produce deflections in the galvanometer **31** causing the beam of light reflected from the mirror **33** onto the film **19** to describe an irregular line upon the film.

Another such device has been developed by *R.E. Fearon, J.M. Thayer, and G. Swift; U.S. Patent 2,515,500; July 18, 1950; assigned to Well Surveys, Incorporated.*

Still another neutron-gamma logging technique has been described by *A.H. Youmans; U.S. Patent 3,294,972; December 27, 1966; assigned to Dresser Industries, Inc.*

A technique developed by *D. Duffey; P.F. Wiggins and F.E. Senftle; U.S. Patent 3,638,020; January 25, 1972; assigned to the U.S. Atomic Energy Commission* employing neutron activation analysis is particularly applicable to in situ exploration of the ocean floor where it is expected that mineral values are disseminated as relatively low-grade stratiform deposits. It may also have value in the general field of geochemical mapping.

A marine mineral-detection apparatus is described including a ^{252}Cf source of neutrons and a lithium-drifted germanium radiation detector shielded from the neutron source. The neutron source is mounted on an extensible member supported within a mass of neutron-shielding material. The source is extended outside the neutron shield to bombard surrounding mineral values with neutrons. Elements capturing neutrons give off prompt gamma radiation with discrete energies. The radiation detector resolves the radiation into distinct energy peaks for identifying the elements present in the mineral values.

A method developed by *C.H. Neuman; U.S. Patent 3,817,328; June 18, 1974; assigned to Chevron Research Company* is a method for accurately determining changes in the fractional-volume oil and gas content of a reservoir from a cased well bore. The first step of the method is to record the response of both a thermal-neutron-decay-time log and a neutron activated oxygen log to a reservoir formation traversed by the well bore. A purposeful change is then made in the oil saturation in a given region of the formation surrounding the well bore by injecting fluid under sufficient pressure to displace the connate fluids. This change can constitute the removal of substantially all the oil or the removal of as much oil as can be displaced by a proposed flooding technique.

The combination of the thermal-neutron-decay-time log and the oxygen log is then run again to record the response of the same given region. The difference in the oil content around the well bore is then determined from the differences between the two sets of logs in a manner whose details are described hereinafter. The method can be modified to measure changes in fluid content over long periods by running the combination of the thermal-neutron-decay-time log and the oxygen log when knowledge about current saturations is desired. The oil content between the first and second logging runs can be changed only by production of fluids from the reservoir.

A technique developed by *A.J. Kermabon; U.S. Patent 4,103,158; July 25, 1978; assigned to Syminex, France* involves geophysical prospecting by radioactive diagraphy of a drill hole, using a probe containing a radioactive source and one or two detectors of delayed particles emitted by the terrain bombarded by the source. The source is moved periodically, within the probe, vertically away from the detector, and the detector is moved, in synchronism with the source, in order to locate it opposite the zone which has just been bombarded.

NEUTRON-NEUTRON LOGGING

It was also known that fast neutrons would react with hydrogen containing material and would be partially absorbed or slowed down. This reaction produces slow or thermal neutrons. If the gamma detector were replaced by a neutron detector, these thermal neutrons indicative of hydrogen could be detected. This type of logging is called neutron-neutron logging.

One early device utilizing this technique was that of *B. Pontecorvo; U.S. Patent 2,508,772; May 23, 1950; assigned to Well Surveys, Incorporated.* The construction of this device involved suspending a series of regularly spaced ionization chambers and a source of radiation, from a cable, connecting these ionization chambers to a commutator so that the measurements made by the ionization chambers will be sent one after the other in rapid succession, over the supporting cable to the surface of the earth.

At the surface of the earth the signals indicative of the measurements made by the ionization chambers may be impressed upon one pair of electrodes of a cathode ray tube and the other pair of electrodes may be controlled by an oscillator, matched in frequency to the speed of the commutator, so that the indications on the face of the cathode ray tube will directly form the decay curve mentioned above. In other words, the cathode ray tube will be controlled by the oscillator so as to scan in one direction across its face at the same speed that the commutator scans the ionization chamber connections in the well and the position of the ray in the other direction (at right angles to the direction of scanning) will be controlled by the signals indicative of the measurements. The curve so formed may be photographed by a moving picture camera.

The camera may also be caused to photograph simultaneously an indicator of the depth at which the device is operating, or alternatively the motion picture camera may be driven by a device which measures the depth at which the device is operating, so that the strip of film produced may be calibrated directly in depth.

Another scheme for neutron-neutron logging has been described by *R.G. Norelius; U.S. Patent 2,933,609; April 19, 1960; assigned to Dresser Industries, Inc.*

Still another scheme for neutron-neutron operations has been described by *A.H. Youmans, U.S. Patent 3,200,251; August 10, 1965; assigned to Well Surveys, Incorporated.*

In the borehole, the neutron-neutron log can be used to determine the fluid content of a formation and, therefore, its porosity.

In principle, by drawing a neutron-emitting tool through a borehole to irradiate the surrounding earth formations, radiation patterns ought to be established that uniquely identify the porosities of the formations under consideration. In the field, however, the borehole environment tends to degrade the accuracy of these tools. For example, strings of steel pipe and irregular annuli of cement casing are often used to prevent caving and fluid communication between the different formations traversed by the borehole. Through scattering and absorption, these casing materials sometimes exert such a dominant influence on the neutron distribution that the apparent formation porosity observed in these circumstances is erroneous or misleading. Thus, a need exists for a neutron porosity tool that is not sensitive to changes in the borehole environment.

More particularly, it has been observed that the neutron detector count rate usually decreases in the borehole with increasing separation between the neutron source and the neutron detector. A graph of the logarithm of the neutron detector count rate in a given mineral composition of constant porosity as a function of the separation between the neutron source and the detector may be drawn for a particular borehole condition. Different curves will be obtained, depending on the specific borehole environment as, for example, a particular cement thickness or borehole diameter. These semilogarithmic curves at distances sufficiently far from the source may be approximated by straight lines, each line exhibiting a distinct slope in accordance with the particular borehole variable under consideration.

Changes in the porosity of the formation also produce variations in the observed neutron detector count rate. Consequently, in the field, it is often difficult to determine if a specific change in the neutron detector count rate indicates a change in the formation porosity, or, for example, a change in the cement thickness.

A multiple detector neutron logging apparatus has been described by *N.A. Schuster; U.S. Patent 3,546,454; December 8, 1970; assigned to Schlumberger Limited.* In this apparatus, the influence of variations in the borehole environment on neutron porosity logging tool signals is overcome by combining a signal that reflects these variations with the signal from a neutron detector, or its physical equivalent, that also reflects these same variations, but in a generally different or distinguishable manner.

More specifically, a two-neutron detector porosity logging tool is provided with a primary radiation source that continuously emits neutrons. A short-spaced neutron detector is placed close to this source. A long-spaced detector is separated from the source by a distance that is at least equal to or greater than that which is occupied by the short-spaced detector. The ratio of the neutrons, or counts, registered with the long-spaced and the short-spaced detector accurately identifies the porosity of the formation under study, subject to the influence of the aforementioned variations in the borehole environment.

However, in this device, a third neutron detector (or its physical equivalent) is joined to a two-neutron detector formation porosity logging tool. The third detector equivalent is positioned within the tool at a distance relative to the neutron source and the other detectors that enables changes in the signal from the third detector caused by variations in the borehole environment to be generally distinguishable from changes in the signal from at least one of the other detectors. The signals from the third detector and one of the two other detectors are combined in order to compensate for the influence of the borehole environment.

A device developed by *L.S. Allen; U.S. Patent 4,005,290; January 25, 1977; assigned to Mobil Oil Corporation* is a borehole logging tool which includes a steady-state source of fast neutrons, two epithermal neutron detectors, and two thermal neutron detectors. A count rate meter is connected to each neutron detector. A first ratio detector provides an indication of the porosity of the formation surrounding the borehole by determining the ratio of the outputs of the two count rate meters connected to the two epithermal neutron detectors.

A second ratio detector provides an indication of both porosity and macroscopic absorption cross section of the formation surrounding the borehole by determining the ratio of the ouputs of the two count rate meters connected to the two thermal neutron detectors. By comparing the signals of the two ratio detectors, oil-bearing zones and salt water-bearing zones within the formation being logged can be distinguished and the amount of oil saturation can be determined. Such a device is shown schematically in Figure 42.

The borehole logging tool **10** has a steady-state neutron source **11** for irradiating the formations, two spaced-apart thermal neutron detectors **12** and **13**, and two spaced apart epithermal neutron detectors **14** and **15**. The neutron source preferably is a steady-state Am-Be fast neutron source with an average energy of about 4 million electron volts. The thermal neutron detectors may be proportional counters of the type disclosed in U.S. Patent 3,102,198 and filled with six atmospheres of helium-3 gas.

Detectors of this type are very sensitive to thermal neutrons. A shield **43** protects the thermal neutron detectors from directed neutron radiation from the neutron source. The epithermal neutron detectors may be of the same type of proportional counters as detectors **12** and **13** with cadmium shielding to prevent thermal neutrons from reaching the active volume of the detectors. A power supply **16** is located within the borehole tool for supplying power to the thermal neutron detectors **12-15** by way of conductor **17**. Current is applied to the power supply from the surface by way of conductors **18**.

The outputs of the thermal neutron detectors **12-15** are applied to amplifiers **19-22** which in turn are coupled to conductors **23-26** included in the cable **47**. At the surface, the outputs from conductors **23-26** are applied by way of the slip rings **27** and brushes **28** to conductors **29-32** which extend to amplifiers **33-36**. The outputs of amplifiers **33** and **34** are applied to the thermal neutron count rate meters **37** and **38**, while the outputs of amplifiers **35** and **36** are applied to the epithermal neutron count rate meters **39** and **40**.

Figure 42: Mobil Oil Corp. Neutron-Neutron Logging Scheme

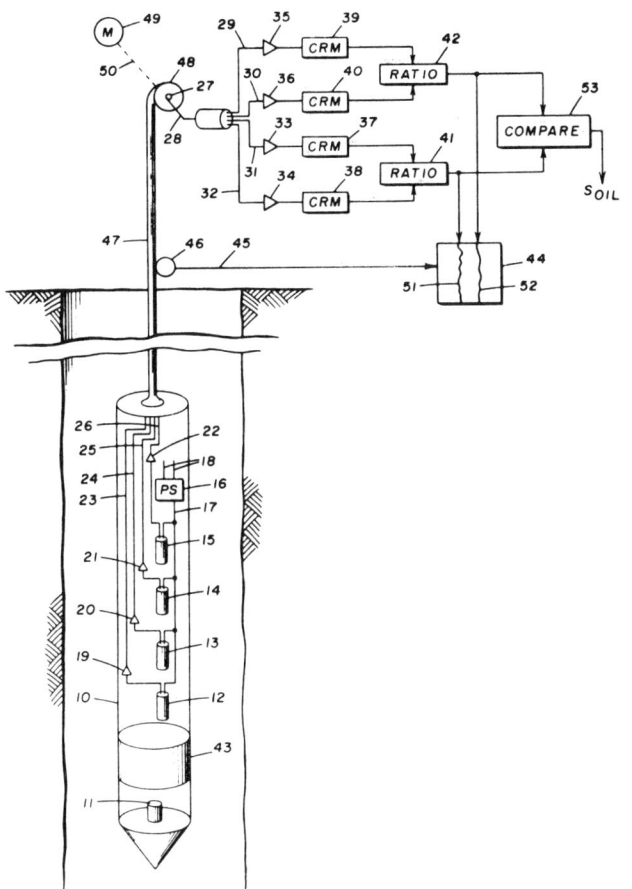

Source: U.S. Patent 4,005,290

The outputs of the thermal neutron count rate meters **37** and **38** are applied to the ratio detector **41**, while the outputs of the epithermal neutron count rate meters **39** and **40** are applied to the ratio detector **42**. Ratio detectors **41** and **42** may be of a conventional type.

By taking the ratio of the counting rates from the two epithermal neutron detectors, a signal is obtained that responds predominantly to changes in the porosity of the given formation. Further, by taking the ratio of the counting rates from the two thermal neutron detectors, a signal is obtained that responds predominantly to changes in both the porosity and the macroscopic absorption cross section of the given formation. It has been found that the salinity of the fluid in the formation affects the macroscopic absorption cross-section of the formation but has no effect on the porosity of the formation. Consequently, by comparing the signals of the two ratio detectors **41** and **42**, an indication is obtained each time there is a change in the salinity of the fluid in the formation. More particularly, when the pores of the formation contain salt water rather than oil, the signal from ratio detector **41** increases while the signal from ratio detector **42** remains unchanged for a given formation porosity.

Figure 43 illustrates in graphical form characteristics of example subsurface formations as might be encountered when logging with the borehole system shown.

Figure 43: Formation Characteristics Shown by Neutron-Neutron Logging Scheme

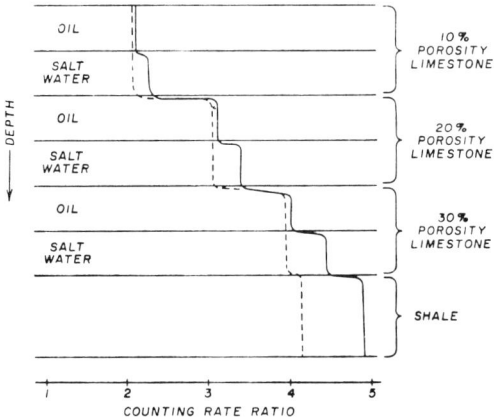

Source: U.S. Patent 4,005,290

The epithermal neutron counting ratio is shown as a dashed line and the thermal neutron counting rate ratio is shown as a solid line. It can be seen from this figure that the thermal neutron counting rate ratio increases when the formation fluid changes from oil to salt water, but the epithermal neutron counting rate ratio remains unchanged. The magnitude of the differential between the two ratios can be further seen to increase as the formation porosity increases.

Referring back to Figure 42, coupled to the outputs of ratio detectors **41** and **42** is the recorder **44**. The ratio signal from ratio detector **41** is recorded as trace **51** and the ratio signal from ratio detector **42** is recorded as trace **52** as the logging tool is continuously moved through the borehole.

By recording the outputs of the ratio detectors as continuous traces, one can readily observe changes in the differential between the magnitudes of the signals from such detectors and thereby distinguish oil-bearing zones from salt water-bearing zones since the magnitude of such differential is much larger for a salt water-bearing zone than for an oil-bearing zone.

The signals from both ratio detectors are applied to comparator **53** which is calibrated to provide an output signal representative of oil saturation, that is, the volume fraction of the fluid in the subsurface formation that is occupied by oil.

Normally in nuclear geophysical prospecting, a plurality of logs, often natural gamma, gamma-gamma, neutron-neutron and neutron-gamma, are made simultaneously.

In a scheme described by *A.S. McKay; U.S. Patent 2,761,977; September 4, 1956; assigned to The Texas Company* an induced gamma ray log or a scattered gamma ray log can be made simultaneously if desired, with another log showing variations in the diameter of the borehole. The latter log can, if desired, be a more or less conventional caliper log in which a device having extending feelers or fingers is passed through the hole, these fingers contacting the walls of the formations and through suitable electrical circuitry providing a record of the changes in the diameter of the hole. It has been found that in a scattered gamma ray log a great deal of the detector response, perhaps a major portion thereof, is due to the presence of liquid in the hole and since the amount of liquid surrounding the logging instrument varies with the hole diameter a conventional scattered gamma ray log can be used as a caliper log.

A technique developed by *E.C. Hopkinson and A.H. Youmans; U.S. Patent 3,691,378; September 12, 1972; assigned to Dresser Industries, Inc.* produces simultaneously a well log of the macroscopic thermal neutron cross section (Neutron Lifetime Log) of formations adjacent a well bore and logs of the capture gamma rays, epithermal neutrons and thermal neutrons returning to a well bore as a result of irradiating the formations adjacent the well bore with pulses of neutrons. The thermal and epithermal neutron logs are obtained by separating the signal from a single detector into two time-dependent groups.

Methods and means are also disclosed for combining the capture gamma ray log with the thermal neutron log to obtain a log indicating the salinity of the fluids contained within the formations. Methods and means are also disclosed for combining the epithermal neutron log with the thermal neutron log to obtain a log related to the macroscopic thermal neutron cross section of the formations. Either or both of these derived logs may be obtained simultaneously with the first suite of logs.

The preferred apparatus includes a pulsed source of 14-MeV neutrons, a gamma ray detector, and a neutron detector sensitive to both thermal and epithermal neutrons in the subsurface instrument. Surface apparatus includes the appropriate gating circuits and ancillary circuits whereby the gamma rays detected while the neutron source is quiescent are used to form three signals corresponding to the gamma rays detected in three time periods. Similarly, the surface apparatus includes gating and ancillary circuits to separate the detected neutrons into two time groups.

A device developed by *J. Tittman; U.S. Patent 3,823,319; July 9, 1974 assigned to Schlumberger Technology Corporation* provides an accurate mudcake correction for sidewall neutron borehole logging tools. A sidewall porosity sonde is equipped with a vertically separated neutron source and a neutron detector that are both eccentered in the tool in order to engage one side of the borehole wall. This basic porosity measuring array is supplemented by a source of lower energy neutrons having appreciably shorter slowing down length and another detector for appropriately sensing the influence of the mudcake. A circuit combines signals from the detectors to produce an indication of the formation porosity.

IMPROVED RADIATION DETECTORS

The limitation of both neutron-gamma and neutron-neutron logging, in the early stages, was that some of the hydrogen containing formations identified did not contain oil. Water for example, gave essentially the same response as oil.

The solution to this problem was the development of better radiation detectors. The early radiation detector, used in the nuclear logging sondes or probes approximately 30 to 40 years ago, was a gas-filled, ionization chamber, often called a Geiger-Mueller detector. This type of detector will give a reading as to intensity or frequency of radiation present but can not accurately define the magnitude or amplitude of the signal current generated. With the development of the scintillation detector, pulse height definition of the output current of the detector was made possible. A scintillation detector is composed of a phosphor material or crystal that produces a pulse or flash of light when a gamma ray or a particle, such as, alpha, beta or neutron, strikes and reacts with it.

This phosphor is placed adjacent the photocathode portion of a photomultiplier tube so that light generated in the phosphor is picked up by the photomultiplier. The output current of the photomultiplier is indicative of the amount and amplitude of the light generated in the phosphor material.

An early disclosure of the use of scintillation detectors in well logging was that by *R.E. Fearon, U.S. Patent Reissue 24,226; October 9, 1956; assigned to Well Surveys, Inc.*

Another use of this technique has been described by *S.A. Scherbatskoy; U.S. Patent Reissue 24,797; March 15, 1960; assigned to PGAC Development Co.*

With the use in neutron-gamma and neutron-neutron logging of the scintillation detector which distinguished energy levels of induced secondary radiation, it was found that each element of the Periodic Table emitted induced radiation at a single distinct energy level or at a plurality of distinct energy levels. This has been discussed for example, by *C.W. Tittle; U.S. Patent 3,184,598; May 18, 1965; assigned to Schlumberger, Limited.*

It has also been discussed by *H.E. Hall, Jr.; U.S. Patent 3,452,210; June 24, 1969; assigned to Texaco, Inc.* Another disclosure was that by *H.D. Smith, Jr., D.M. Arnold and W.E. Schultz; U.S. Patent 3,780,301; December 18, 1973; assigned to Texaco, Inc.*

With a scintillation detector, as depicted in Figure 44 it was possible to identify specific elements present in the various borehole strata.

Fine distinctions were possible. For example, the presence of salt water instead of fresh water could be determined by means of the induced gamma rays produced when neutrons react with chlorine.

This has been discussed by *S.A. Scherbatskoy; U.S. Patent 2,862,106; November 25, 1958.* This is important because salt water is always present in or near oil pools or formations. Also distinctive energy peaks are produced by neutrons reacting with carbon, a major component of oil as noted in U.S. Patent 3,184,598 referenced above.

Figure 44: Nuclear Scintillation Wellhole Logging

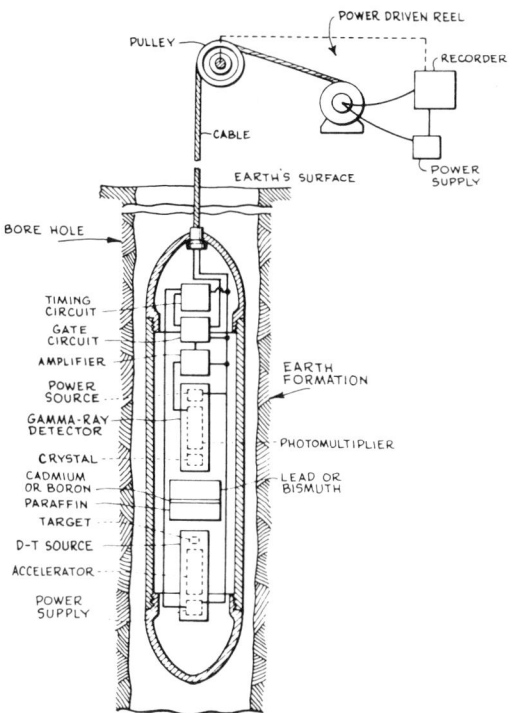

Source: U.S. Patent 3,184,598

A technique developed by *A.S. McKay and H.E. Hall, Jr., U.S. Patent 3,016,961; January 16, 1962; assigned to Texaco, Inc.* utilizes improved radioactivity geophysical exploration apparatus involving the recording of a signal on photographic film in response to detected penetrative radiation.

The use of photographic film to record a signal indicative of penetrative radiation within a well bore provides a useful method of obtaining a permanent record of the detected radiation. However, the conditions found in a borehole, especially relatively deep boreholes, pose serious problems. For example, the temperature within a well bore may range upwards of 300°F, and, in fact, may reach temperatures of 350 to 400°F in deep wells. Especially in the case of a logging-while-drilling apparatus where the film must be kept in such a high temperature environment for relatively long periods of time, such temperature conditions may be of relatively serious consequence.

In this process penetrative radiation is detected and converted to a light signal which is employed to fog a strip of photographic film in accordance with the quantity of radiation detected. The photographic film is maintained in a moisture-free oil bath. The oil bath serves as an optical link to couple the light signal to the photographic film.

Preferably, the radiation is detected by means of a luminophor and the photographic strip is moved past the luminophor at a predetermined rate. The luminophor is optically coupled to the photographic strip by means of an oil medium. Advantageously, the entire quantity of film is maintained within the oil bath. The luminophor may also be maintained in the oil bath for temperature stabilization purposes.

In this technique, preferably, a monitor luminophor together with a source of radioactivity of predetermined intensity is also mounted next to a separate portion of the photographic film in the vicinity of the radiation-detecting luminophor in order to fog a portion of the film as a reference.

A preferred embodiment of the technique involves coupling a luminophor to a photographic film which is maintained in an oil bath while mounted within a drill string within the vicinity of a drill bit during the course of drilling a well.

IMPROVED NEUTRON SOURCES

Further improvements have also been made in the neutron source. Neutron sources first used were of what is usually called the chemical type, such as, radium-beryllium or polonium-beryllium. The high-energy alpha particles emitted by the radium or polonium react with the beryllium atoms, producing neutrons having a plurality of energy levels.

The use of polonium or plutonium sources has been discussed by *T.W. Bonner and R.L. Caldwell; U.S. Patent 2,948,811; August 9, 1960; assigned to Socony Mobil Oil Co., Inc.* in connection with a process for the bombardment of earth formations surrounding a drill hole by activation of boron 10 by alpha-ray bombardment for the production of neutrons and detecting prompt gamma radiation in a specific energy band for determination of the concentration of carbon in such formations.

A suitable form of apparatus for the conduct of such a process is shown in Figure 45. A well logging tool **10** supported by a cable **11** in a borehole **12** includes a capsuled neutron source **13** in which the source is comprised of a target of ^{10}B and an alpha rayer preferably of the class comprising polonium or plutonium. A shield **14** separates source **13** from a detecting crystal **15** and eliminates direct transmission of low intensity gamma radiation which may attend the production of neutrons from bombardment of ^{10}B. Crystal **15** is positioned adjacent a photomultiplier tube **16** whose output is connected by way of amplifier **17** and cable **11** to the earth's surface.

A measuring circuit **20** is connected between the cable on reel **21** and a recorder **22**. A drive connection **23** is provided between a cable measuring element **24** and recorder **22** to drive a log chart such that its length is proportional to the depth of the unit **10**. The measuring circuit **20** preferably is designed to discriminate pulse heights so that the energy of individual gamma rays impinging crystal **15** may be recorded.

Figure 45: Socony-Mobil Process for Logging Based on Neutron Production by Alpha Disintegration of ^{10}B

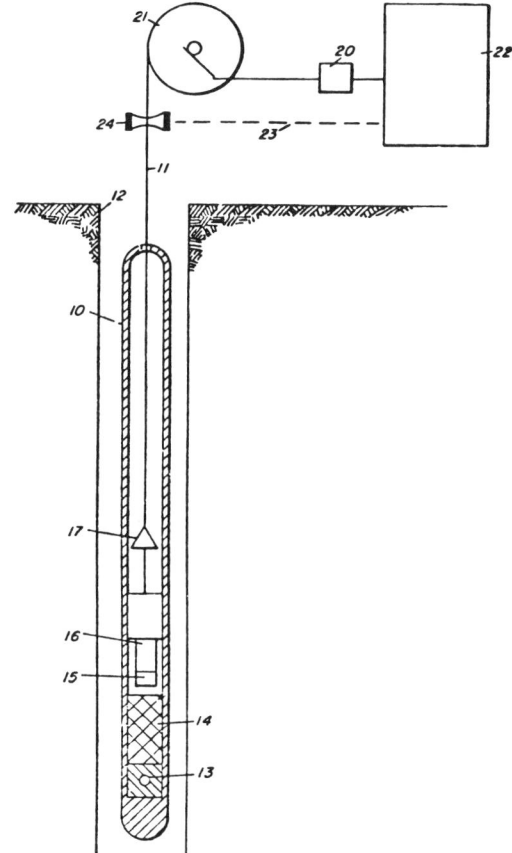

Source U.S. Patent 2,948,811

Further, it is desired that unit **20** eliminate from the measurement all gamma radiation pulses except those within a predetermined band or in the alternative above a predetermined minimum level. It has been found that carbon in formations when bombarded with neutrons emits gamma radiation having an energy of 4.4 MeV. A log of the carbon content of formations may be secured by bombarding with the neutrons produced by alpha particles on boron 10 and measuring gamma radiation above about 4 MeV. This 4.4 level of excitation of carbon 12 results from bombardment by neutrons having energies in excess of 4.8 MeV. Since oxygen would not be excited by 6 MeV neutrons, an integral count of gamma radiation above about 4.0 MeV would depend substantially entirely upon gamma radiation from carbon. If greater definition is desired, a differential count may be made by measuring a relatively narrow band of gamma radiations having energies centered at 4.4 MeV. To this end, the counting circuit **20** may be of the type well known in the art for determining energy levels of radiation detected by scintillating crystals and the like.

A monoenergetic source of neutrons of higher energy level was often desirable in borehole logging. While it was known that monoenergetic neutrons of 14 MeV could be produced when deuterium ions are accelerated at high velocities and directed at a target containing tritium ions, it took a number of years to develop such a neutron source that could be placed in a well logging sonde, the maximum diameter of which is usually about 5 inches. Such a neutron source has been described by *R.E. Fearon, and J.M. Thayer; U.S. Patent 2,884,534; April 28, 1959; assigned to Well Surveys, Inc.*

Another such neutron source has been described by *C. Goodman; U.S. Patent 3,461,291; August 12, 1969; assigned to Schlumberger Technology Corp.* This neutron source is an accelerator tube in which charged particles, e.g., deuterium ions, are accelerated under relatively low voltages toward a target element including tritium. The resulting reaction of the deuterium ions with the tritium target produces monoenergetic neutrons at an energy level of 14 MeV. These neutrons bombard the formations surrounding the well bore and induce a radioactive response, in the form of neutrons and gamma radiation, which is detected by a suitable instrument in the logging tool.

Operation of the accelerator and the detecting circuitry is controlled completely from the surface and the neutron source may be kept inactive until it reaches the logging depth in the well bore and deactivated prior to its return to the surface, thereby minimizing radiation hazards. At the same time, the surface circuitry for providing the output indications may be controlled so that pulsed neutron bombardment and detection during preselected intervals may be effected. The tool may be used to obtain neutron-neutron or neutron-gamma ray logs separately, or both on a single run with appropriately gated detecting circuits.

Most of these chemical or accelerator type neutron sources operated continuously which at times precluded accurate detector results because of source and borehole effects. Hence source pulsing was developed which permitted the source to be turned on and off for specified periods of time. In the chemical type of neutron source this is usually accomplished by a shutter, normally a rotary type, that passes between the alpha emitter and the target material.

One such device has been described by *R.E. Fearon; U.S. Patent 2,275,748; March 10, 1942; assigned to Well Surveys, Inc.* Such a device is shown in Figure 46. The device consists of a tubular shaped capsule or housing **10** adapted to be lowered into a drill hole or other opening in the ground at the end of a cable **11** which serves to support the housing and carry currents indicative of the results of the measurements to the surface of the earth where they are recorded by a recorder which is not shown in the drawing.

Within the housing **10** which is preferably cylindrical in shape and of small enough diameter so that it can be lowered into a drill hole, is a measuring instrument for measuring the intensity of the gamma rays that reach it, and a source of neutrons so arranged as to bombard the adjacent strata from which will come the gamma radiations to be detected by the measuring instrument.

Figure 46: Well Surveys, Inc. Design for Pulsed Emission Device

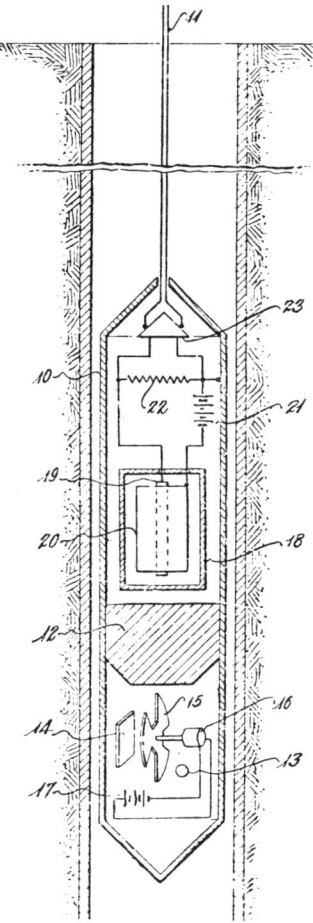

Source U.S. Patent 2,275,748

Separating the two parts of the device is a lead block or shield **12** which lies in the housing **10** between the two parts of the device and is of sufficient thickness so that gamma rays or streams of neutrons will not penetrate through it and affect the measuring instrument.

The source of bombarding neutrons consists of a source of alpha rays **13** and a plate **14**, of or containing boron, beryllium or some other material that will emit neutrons upon exposure to alpha radiations. Among suitable sources of alpha rays are masses of radium or polonium, or entrapped radon.

Positioned between the source of alpha particles **13** and the plate **14** is a revolving shutter **15**, formed of material opaque with respect to alpha rays, driven by a motor **16** which can be powered by any suitable source such as a battery **17**. As the shutter **15** revolves the alpha rays from the source **13** are periodically permitted to fall on the neutron generating plate **14** and when this happens neutrons are projected into the surrounding strata and induce radiations or are scattered and in part returned to operate the measuring instrument.

The measuring instrument consists of an ionization chamber **18** preferably containing argon under around 1,500 to 2,000 pounds per square inch pressure and also containing a pair of electrodes **19** and **20**. The inner electrode **19** is preferably an iron rod about six inches long and the outer electrode **20** is preferably a sheet iron sleeve about two inches in diameter surrounding and insulated from the rod.

Insulated and sealed connections lead through the wall of the ionization chamber, one from the outer electrode **20** to the negative side of a battery **21** and the other through a resistor **22** to the positive side of the same battery. The resistor **22** preferably has a resistance of around 10^{12} ohms. The positive side of the battery is preferably grounded to the case **10**.

An amplifier **23** has its inputs connected across the resistor **22** where it will be affected by the potential drop across the resistor. The potential drop itself is dependent upon the flow of current through the ionization chamber and hence upon the ionization which the gamma rays impinging upon it produce. The amplifier **23** is of the alternating current type so that any constant potential drop across the resistor which results from the relatively steady action of the natural radioactivity of the surrounding earth is ignored. The resultant alternating current from the amplifier **23** passes up through the cable **11** and is recorded on the surface of the earth by a suitable recorder.

A similar device has been described by *R.L. Caldwell, and W.W. Givens; U.S. Patent 3,389,257; June 18, 1968; assigned to Mobil Oil Corp.* It is a mechanically actuated pulsed neutron source formed by a plurality of alpha sources spaced from a plurality of associated targets and having a rotatable shutter therebetween for controlling the passage of alpha particles to the targets for producing bursts of fast neutrons. The sources and targets are geometrically positioned and the shutter is constructed to allow all of the targets simultaneously to be irradiated periodically with alpha particles to increase the neutron yield. The sources and targets are maintained in a chamber which is filled with helium at about atmospheric pressure to increase the neutron yield further.

In the accelerator type of neutron source the pulsing is accomplished by a trigger or switching circuit, which simply turns the accelerator on and off at a very rapid rate. For chemical sources that are pulsed, neutron pulse lengths as short as 1 millisecond down to several hundred microseconds are possible according to *M.E. Dillingham; U.S. Patent 3,885,160; May 20, 1975; assigned to The Western Company of North America.* A pulsed neutron source has a chamber containing a plurality of alpha emitting strips and beryllium targets coaxially mounted. A pulsed source is provided by rotation of the target to on-off positions along with electromagnetic and magnetic devices for positive locking and rotation.

For the accelerator type of neutron source, neutron pulse lengths shorter than one microsecond can be produced as described by *H.C. Pollock; U.S. Patent 2,867,728; January 6, 1959; assigned to General Electric Co.*

Another advancement was the use of a pulsed detector in combination with a pulsed source. This technique substantially eliminates source and borehole fluid effects in the detected results by leaving the detector off during the period when the source is on and for a predetermined period after the source is off as described by *A.H. Youmans; U.S. Patent 3,379,884; April 23, 1968; assigned to Dresser Industries, Inc.* The use of a plurality of detection periods between the pulses was also developed as described by *C. Goodman; U.S. Patent 2,991,364; July 4, 1961; assigned to Schlumberger Well Surveying Corp.*

THE USE OF RADIOTRACERS

Radioactive tracers have also been applied to borehole logging. Porosity or density can be determined by shooting a radioactive bullet into a formation and detecting the degree of penetration by the strength of radiation detected.

One such method has been described by *J. Neufeld; U.S. Patent 2,320,643; June 1, 1943; assigned to Well Surveys, Inc.* Another such method has been described by *R.C. Hoss; U.S. Patent 2,685,038; July 27, 1954.*

Also, radioactive fluid can be injected, subsequently removed and residual radiation detected. Such a radiological logging procedure has been described by *M.M. Albertson; U.S. Patent 2,352,993; July 4, 1944; assigned to Shell Development Co.*

An auto-radiographic technique developed by *S.J. Pirson; U.S. Patent 2,733,353; January 31, 1956; assigned to Stanolind Oil & Gas Co.* provides for (1) accurately and simply making radiographs of the well formations at any time desired during the operation of a well, either before or after well treatments; (2) obtaining detailed photographic information about the porosity and permeability of well formations and the variations thereof in a well producing zone; (3) making radiographic exposures in wells by an instrument completely self-contained and not requiring electrical communication to the ground surface.

The movement of fluids in a borehole can be determined by injecting a radioactive fluid and a nonradioactive fluid simultaneously at substantially the same or varying pressures, with a detector following the movement of the interface between the two. Such a technique has been described by *E.F. Egan and G. Herzog; U.S. Patent 2,700,734; January 25, 1955; assigned to The Texas Co.* and also by *K.C. ten Brink and R.H. Widmyer; U.S. Patent 3,100,258; August 6, 1963; assigned to Texaco Inc.*

A technique developed by *W.T. Cardwell, Jr. and S.B. Jones; U.S. Patent 2,856,536; October 14, 1958; assigned to California Research Corp.* involves exploring for a probable accumulation of petroleum by injecting a radioactive fluid into a favorable geological stratum traversed by an unsuccessful well bore and then measuring, within the same well bore, the movement of the radioactive fluid by detecting its distribution in the stratum, immediately after injection thereinto and at a time sufficiently later to permit formation fluid movement to transport the radioactive fluid in the direction of the entrapped petroleum deposit; the direction being determined by the detection of only those radiations from the fluid, which have the full energy of the emitting source within the tracer, and which are measured independently of radiations, either naturally present, or due to the radiation from the tracer being multiply-scattered, so that they are nonrepresentative of the direction of movement of the formation fluids.

TRENDS IN NUCLEAR GEOPHYSICAL PROSPECTING

As discussed in the preceding text, the major advances in nuclear borehole prospecting have been the development of the scintillation detector; the accelerator type, monoenergetic neutron source; and the pulsing of source and detector for specific periods.

The thrust of the technology at present appears to be in further refinement of systems utilizing pulsed sources and detectors, and the correlation of a plurality of simultaneous logs from these systems by data processing type circuitry.

In this last category is a system developed by *R.J. Schwartz; U.S. Patent 3,609,366; September 28, 1971; assigned to Schlumberger Technology Corp.* It is a well-logging exploring device which includes a pulsed neutron source which is repetitively energized to irradiate a formation with neutrons and means for generating a sync pulse representative of the time of termination of such irradiation. In response to this irradiation, a nearby scintillation counter generates signal pulses at a rate representative of the neutron population in the media surrounding the exploring device.

The sync and signal pulses are then transmitted to the surface of the earth. At the surface of the earth, signal-processing circuits operate to accurately distinguish the sync and signal pulses and resolve any signal pulses which may have become merged due to the relatively low frequency response of the cable. These resolved signal pulses are then applied to computing circuits which operate to determine the rate of decay of the neutron population in the formation by determining the rate of change of the signal pulse repetition rate. In this connection, the signal pulses are weighted and then gated to a binary counter during a cycle of adjustable time intervals.

By appropriately selecting the relationship of the time intervals to one another and the factors by which the signal pulses are weighted during each time interval, the accumulated count in the counter after each cycle of time intervals will be representative of the rate of decay of the neutron population and is used to determine the time intervals. The sync pulses synchronize the surface electronics with the repetitive irradiation of the formation with neutrons.

Another system described by *C.W. Johnstone; U.S. Patent 3,890,501; June 17, 1975; assigned to Schlumberger Technology Corp.* concerns the provision of information as to the reliability, or validity, of logged values of formation and fluid properties which characterize the reactions between the materials composing the formation matrix and fluids and incident irradiations of neutrons. Based on this information, appropriate corrective steps may then be taken, when indicated, to assure that accurate values are obtained.

A technique developed by *H.J. Paap, D.M. Arnold and M.P. Smith; U.S. Patent 3,979,300; September 7, 1976; assigned to Texaco, Inc.* involves measuring simultaneously the thermal neutron lifetime of the borehole fluid and earth formations in the vicinity of a well borehole, together with the formation porosity.

A harmonically intensity modulated source of fast neutrons is used to irradiate the earth formations with fast neutrons at three different intensity modulation frequencies. Intensity modulated clouds of thermal neutrons at each of the three modulation frequencies are detected by dual spaced detectors and the relative phase shift of the thermal neutrons with respect to the fast neutrons is determined at each of the three modulation frequencies at each detector. These measurements are then combined to determine simultaneously the thermal neutron decay time of the borehole fluid, the thermal neutron decay time of surrounding earth formation media and the porosity of the formation media.

SEISMIC GEOPHYSICAL PROSPECTING

To locate and extract deposits of oil and gas in the outer layers of the earth's crust, a profile of the subsurface strata is necessary. To provide this profile the subsurface can be irradiated with acoustic waves, a portion of which, after reflection and/or refraction, can be detected at the earth's surface. The processing of the detected waves in order to facilitate interpretation of the signal information is defined as seismic signal processing.

The basic theory of seismic reflection prospecting and signal processing is depicted in Figure 47. A seismic source (e.g. dynamite, air gun, vibrator, or weight drop) creates elastic disturbances which propagate through the earth.

Figure 47: Seismic Reflection Prospecting

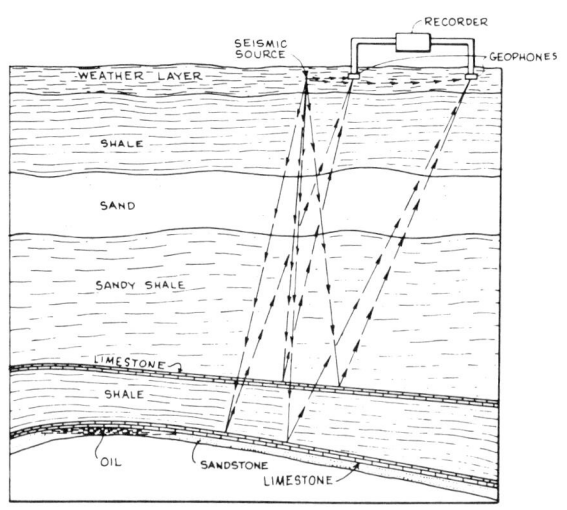

Source: Reference (3)

Seismic Geophysical Prospecting

A very small portion of the downwardly traveling elastic waves is reflected up from each interface between layers of differing hardness and density. The deeper the layer is in the earth, the longer it takes for a signal reflected from that layer to arrive at the surface and be detected by an array of geophones (surface receivers).

These arrays typically contain up to 100 geophones laid out in patterns and are connected by a seismic master cable which ranges from 1 to 3 miles in length (3).

The arrival times of primary reflections (single bounce echoes) at the geophone array permit the determination of depth and inclination angles of the subsurface reflectors when subsurface velocities are known. The lithological characteristics can be determined from the reflection amplitudes and attenuation characteristics contained in the reflected signals. This method makes it possible to detect and map structural features which are associated with accumulations of gas and oil, such as anticlines (Figure 48a; convex upward structures), faults (Figure 48b; linear displacement), salt domes (Figure 48c), reefs (Figure 48d), and stratigraphic traps (Figure 48e; lateral changes in composition).

Where seismic waves are primarily refracted by the subsurface strata, another mode of geophysical prospecting can be pursued. This method depends on the subsurface layers having relative acoustic velocities greater than that of the contiguous overlaying strata in order to diffract the seismic waves. Massive limestones, hard sandstones, salt formations, igneous rocks or basement are the most common subterranean refractors (3).

Figure 48: Various Structural Features Associated With Oil and Gas Accumulations

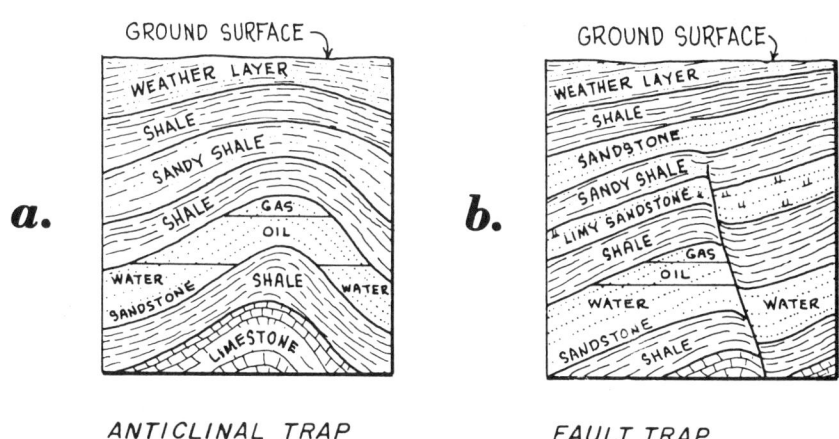

(continued)

Figure 48: (continued)

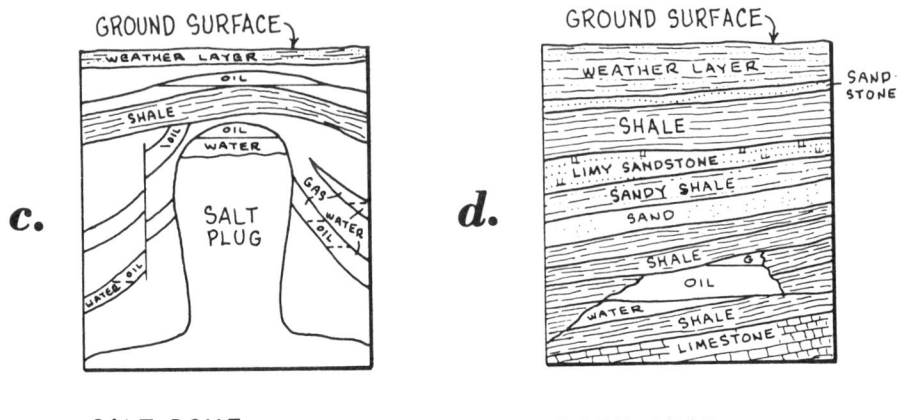

c. SALT DOME

d. REEF TRAP

e. STRATIGRAPHIC TRAP

Source: Reference (3)

The depth of the refracting layers and the speed of sound in the layers are determined from the times required for the refracted wave to travel between the source and the receivers. The refraction method of seismic prospecting is used primarily to determine, in newly explored areas, the existence of high acoustic velocity layers, their depths, and their slopes with respect to the earth's surface.

Seismic signal processing facilitates the interpretation of the received signals and mitigates the effects of signal noise. Figure 49 depicts a cross section of the earth's crust and illustrates the various types of noise which are induced by strata being explored.

Seismic prospecting techniques have been applied to coal as well as to oil as detailed in the proceedings of recent symposia (5)(15).

Figure 49: Strata Induced Noise in Seismic Prospecting

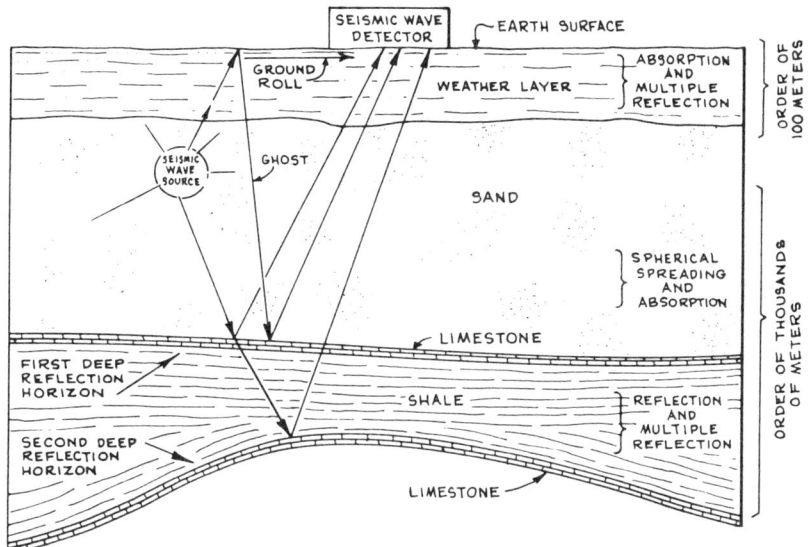

Source: Reference (3)

THE BASIC PATENTS

A basic process for seismic exploration was described by *L. Mintrop; U.S. Patent 1,599,538; September 14, 1926.* Mintrop's method consists principally of detonating a charge of explosive at a point near the surface of the earth and receiving the resultant seismic waves refracted from the subsurface at plural detector points spaced from the position of the explosion. By observing the travel time of the seismic waves, subsurface strata structures along which the seismic waves traveled with greater velocity, could be detected (3).

Following Mintrop's patent, the first of a series of basic patents for the reflection method of seismic surveying was issued to *B. McCollum; U.S. Patent 1,672,495; June 5, 1928.*

McCollum determined the travel velocity of the seismic waves by measurements taken in boreholes in the survey area, and used the velocity data to determine the depth and dip of buried structures in a reissue of U.S. Patent 1,672,495, *B. McCollum; U.S. Patent Reissue 17,242; March 19, 1929; assigned to McCollum Geological Explorations, Inc.*

In 1928 he received a patent for the method of implementing an electrically driven vibrator for producing wave beating between the direct and subsurface reflected seismic waves in order to determine the contour of subsurface structures. This was described by *B. McCollum; U.S. Patent 1,675,121; June 26, 1928.*

B. McCollum; U.S. Patent 1,724, 495; August 13, 1929; and U.S. Patent 1,724,720; August 13, 1929; both assigned to McCollum Geological Explorations, Inc; describes a process for the determination of the slope of subterranean strata by either using plural geophones or successively moving one geophone, detecting reflected seismic waves, and calculating the length differences of the vertical components of the paths of the seismic waves.

A process of common depth point stacking (CDP), developed by *W.H. Mayne; U.S. Patent 2,732,906; January 31, 1956;* is purported to be one of the most significant advances to date in the seismic method and the single most important invention to date in seismic signal processing (3).

In the CDP method, geophone signal traces, which are recorded from shot points at different horizontal locations on the ground surface, but which reflect from the same subsurface point, are stacked (added). The summing of CDP traces improves data signals because primary common reflection point (CRP) reflections are in phase and add constructively, whereas ambient noise and other seismic signals which are not in phase, cancel. This compositing increases reflection signal-to-noise ratios by factors approaching $\sqrt{N}$, where N is the number (fold) of CDP traces summed.

Figure 50 shows the elements of this technique. In the interest of clarity, the effect of firing only two shots is represented and only one interface or subsurface bed is shown. In the first step of the method, the explosive charge detonated in shot hole **10** propagates seismic energy in all directions, and a part of this energy is reflected upwardly from subsurface bed **12** to seismometers **15**, the latter being located at **A** in horizontally spaced relation to the shot hole in the vicinity of the surface **16**. The approximate path traversed by the energy which is reflected and received at each of the several seismometers is illustrated.

The seismometers are then removed to **B**, and shot hole **11** is used for wave propagation, it being noted that the seismometers and the shot hole are still substantially equidistant from a point intermediate the initial location of the shot point and seismometers. Consequently the energy transmitted to each of the several seismometers impinges on the subsurface bed at approximately the same point as in the firing of the first shot, and except for the greater lapse of time between the instant of detonation and the instant of reception of the reflected energy, the received signals should correspond.

Thus when the received energy derived from the two shots is properly combined, the reflected energy will be reinforced whereas random energy, being noncoincident, will be relatively deemphasized, and the clarity of the resulting record accordingly improved.

Figure 50: Representation of Common Depth Point Stacking (CDP) Technique

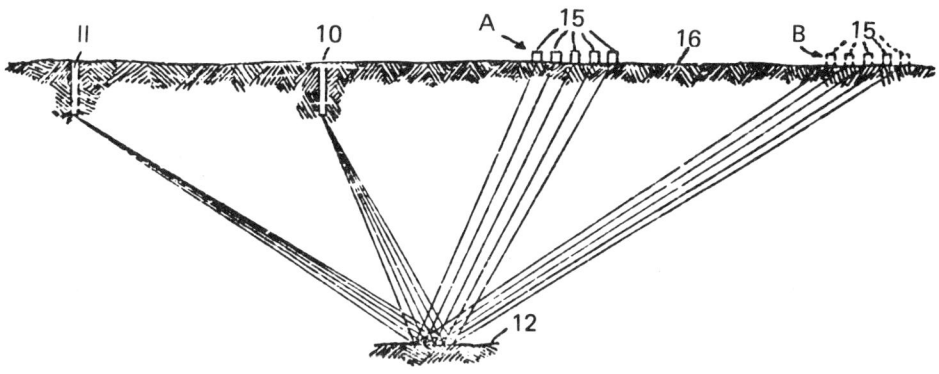

Source: U.S. Patent 2,732,906

PRODUCTION OF SEISMIC WAVES

Seismic waves for geophysical exploration may be generated by one of a number of devices which are described in the pages which follow. These devices may be listed as follows:

- (1) Wave generators using solid explosives
- (2) Implosive seismic wave generators
- (3) Gas exploders
- (4) Air guns
- (5) Seismic vibrators
- (6) Miscellaneous land seismic wave generators
- (7) Miscellaneous marine seismic wave generators

Solid Explosives

In one method of generating seismic waves for geophysical exploration, solid materials are processed to react exothermically and create shock waves. Typically, in offshore prospecting the explosive is cartridge contained and percussion initiated; while in land prospecting the explosive is packaged as a cord or in elongated cylinders and is electrically ignited.

The classical seismic energy source is high explosives. Because of high costs, ecological factors, hazards of operation and time consumed in their use, high explosives have been replaced to a large extent by other energy generators.

Much recent patent activity has been directed to systems using solid explosives in offshore prospecting. The systems provide for handling the explosives in a cartridge mode and projecting the cartridge to the desired explosion area by fluid pressure.

A technique developed by *B.D. Lee; U.S. Patent 3,050,148; August 21,1962; assigned to Texaco Inc.;* is one in which at least 3 explosive charges spaced apart vertically in a borehole are detonated to generate seismic waves. There are means operatively associated with the explosive charges for detonating the charges in time sequence at a velocity substantially equal to the seismic velocity of the medium surrounding the borehole, in order to produce a reinforced seismic wave.

A very similar technique has been described by *O.A. Itria; U.S. Patent 3,048,235; August 7, 1962; assigned to Texaco Inc.*

A device developed by *R.R. Larson; U.S. Patent 3,971,319; July 27, 1976; assigned to Hercules Incorporated;* utilizes an explosive cartridge which is thermally actuated and specifically designed for land seismic use. A thermostatic percussion initiator consisting of a pair of opposed convex and concave curved metal strips (copper/brass and iron/nickel) which have different coefficients of thermal expansion is used to drive a pin, by snap action, into the percussion sensitive area of the explosive cartridge in response to a temperature increase caused by hot fluid directed at the cartridge.

Figure 51 shows the placement of such a device in the earth. The figure shows a cartridge assembly **10** emplaced in a borehole **53** in earth formation **54**, by securing the cartridge to a hot fluid-delivery conduit **20** for connection with a hot fluid supply not shown, and then lowering the cartridge-delivery conduit into the borehole with subsequent stemming by suitable stemming material **57**.

Figure 51: Hercules Inc. Thermally Actuated Percussion Initiatable Explosive Cartridge Assembly

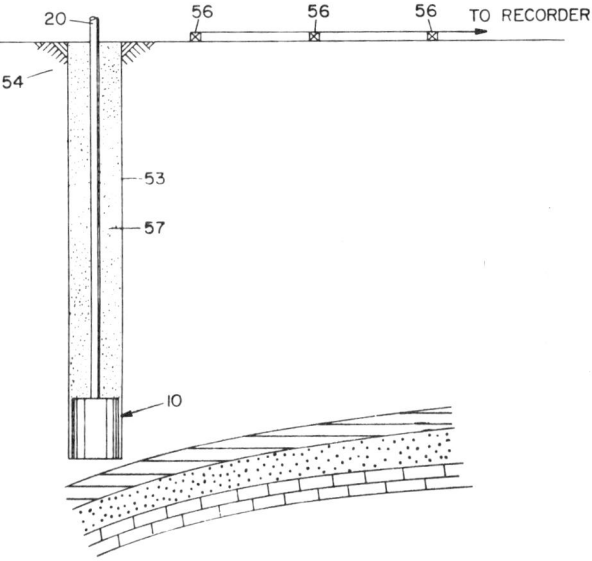

Source: U.S. Patent 3,971,319

Geophones **56** are positioned on the ground at predetermined distances from the explosive assembly for receiving shock waves from refraction and reflection in the underground structure following the shot, and for transmission to the seismic energy measuring system for the seismic record.

The explosive charge for the seismic prospecting operation conventionally is loaded into shot holes or is arranged near the surface in a linear or other appropriate array. In the instance where a linear array is used, lengths of detonating cord are either simply stretched along the surface of the ground or deposited within shallow covered ditches, formed by suitable plows or the like.

However, in many areas, such as those designated as "wilderness areas," tractors and certain wheeled vehicles and plows are not permitted other than on established roads or trails, thus preventing the use of truck mounted drills or ditch forming plows. In such areas, portable backpack drills have been used for drilling shallow holes, known as miniholes. This has proven expensive and the quality of the resultant seismic operation has been less than adequate.

When detonating cord is simply hand deposited on the earth's surface without using a plow, it has been effective in barren or snow-covered areas where the snow provides a tamp to lessen the air blast and quench the fireball generated upon firing the cord. However, where the linear cord is deposited on surfaces covered with dry brush and other vegetation, it creates a safety hazard, starting brush fires that not only destroy the vegetation of the wilderness areas but also cause loss of some of the prospector's equipment such as the geophones and line cables.

As a result, geophysical activity in such areas has declined drastically and data gathering has been limited to roads and trails where wheeled vehicles can be used and where the chances of incendiary ignition of underbrush is minimal.

A system developed by *S.R. Kelly, J.P. O'Brien, and W.C. Morrow; U.S. Patent 4,102,428; July 25, 1978; assigned to Ensign-Bickford Company;* is a system with the seismic energy source exposed on the surface of the ground, which system produces seismic energy with less unwanted background noise than a conventional cord of equal core load, i.e., an improved signal to noise ratio, and results in a seismogram of higher quality than is obtained with miniholes. Included is the provision for not only high record quality with a surface seismic source but also freedom from the incendiary problems previously associated with the surface use of seismic detonating cord.

More specifically, the high core load cord is comprised of an elongated core of detonating explosive having a core load of about 100-400 grains per foot, an encircling layer consisting essentially of flame quenching particulate material adjacent the core along its entire length, a textile covering circumscribably confining both the core and the encircling layer and a fire retardant outer sheath protectively enclosing the textile covering.

Such a cord is shown in Figure 52. As shown, cord **10** includes a central core **12** of detonating explosive material enclosed within one or more layers, such as the plastic sleeve **14** typically used in conventional detonating cord. The core may be of any high velocity explosive conventionally used for detonating cord, such as PETN, PDX, HMX or the like but is of a higher core load than most conventional detonating cords. Accordingly, it exhibits a core load well in excess of the 40 grains per foot used in most conventional detonating cords. Preferably, a core load of about 100-400 grains per foot is used in commercial seismic operations. The preferred explosive material is PETN and when that explosive is used the core load falls primarily within the center of the indicated range with excellent results being obtained at a core load of about 200-250 grains per foot.

Figure 52: Ensign-Bickford Co. No-Flash Seismic Cord Design

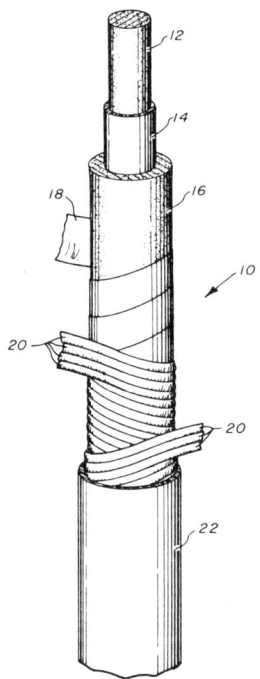

Source: U.S. Patent 4,102,428

If desired, the core may consist solely of the high velocity explosive material. However, as mentioned, the core assembly preferably takes a form similar to a conventional detonating fuse in that the granular explosive material within the core **12** is confined within a suitable confining sleeve or wrapping **14** of one or more layers such as layers of textile, plastic and the like. Due to the preferred high core load of the explosive core, this initial covering will not confine the fireball generated upon detonation of the cord. Accordingly, the present design provides for encirclement of the core by a layer **16** which consists essentially of a flame quenching material. This material is positioned in intimate, adjacent relationship to the core along its entire length and consists of flame quenching and cooling material that absorbs both the flash and flame of the detonating explosive within the core. The material is preferably in particulate or granular form for ease of manufacture and effectiveness of operation.

The materials most effective from both a manufacturing and performance standpoint are inorganic salts, such as diammonium phosphate, sodium bicarbonate, potassium chloride, sodium chloride and the like. As will be appreciated, other known flame quenching material may be used with comparable results.

The flame quenching material is usually present in amounts approximately equal to or slightly more or less than the weight of the explosive within the core. Consequently, the cooling salts are present in amounts of about 100-450 grains per foot and preferably about 250 grains per foot when the cord has a core load of about 200 grains per foot of PETN.

The flame quenching layer **16** is covered by a plastic tape **18** or similar confining layer which in turn is confined by countering strands **20** of textile overbraid or overwrap. The plastic tape is used in the preferred embodiment to confine the particulate material in layer **16**, but is an optional portion of the detonating cord construction. The layer **16** may be formed without the need for confinement or the confining function could be accomplished by the textile overwrap which circumscribably confines both the core and the encircling layers to fully enclose the high explosive core **12** and its flame-suppressing outer layer.

The textile covering provides strength and toughness to the linear detonating cord, without adversely affecting the flexibility thereof. It is of the type conventionally used in the detonating cord and consists of yarn such as rayon, cotton, fibrillated polypropylene or the like. The textile strands are shown as wrapped and counterwrapped to provide a flexible, yet tough covering for the operative layers of the detonating cord. However, variations on the manner in which the strands cover and confine the layer may also be used.

Finally, an outer plastic sheath **22** is applied over the textile wrapping to protectively enclose the assembly and reduce the possiblity of rupture of the wrapping during use. The sheath is a flame retardant plastic material such as polyethylene, vinyl copolymers and the like having flame retardant materials incorporated therein. The flame retardant plastic provides an added barrier to the flash and flame of the core and may be of any suitable commercially available grade, e.g., PE Concentrate 101. It is believed that such material contains a flame retardant of the aliphatic bromine type, such as Citex BN-451.

A scheme developed by *D. Steele; U.S. Patent 3,939,941; February 24, 1976; assigned to Imperial Chemical Industries, Ltd., England* is one in which a detonating fuse-cord is fed in a continuous manner through an advancing tubular conduit to lie in a position substantially parallel to the upper ground layer of a prospect area, the fuse-cord is detonated and the resulting seismic waves are recorded after reflecting or refraction of subterranean rock layer interfaces.

A particular cord construction having a ribbed outer sheath is superior to cords with plain sheaths in respect to flexibility, abrasion resistance and drag and is superior to the plain sheathed cord for burying in seismic prospecting operations and in other operations where there is a risk of damage to the cord by abrasion and snagging.

Such a cord is shown in Figure 53. The core **1** consisted of crystalline PETN loaded at a charge rate of 10 grams per meter. The tube **2** was of glazed kraft paper 13.5 millimeters wide and 0.18 millimeters thick and the central yarn **3** was a single yarn of 380 denier polypropylene. The spun layer **4** consisted of 10 strands of 2,000 denier polypropylene tape (width 2.5 millimeters, thickness 0.01 millimeters) spun at 26 turns per meter and the countering layer **5** consisted of 8 strands of the same polypropylene tape spun at 39 turns per meter. The outer sheath **6** was polyvinyl chloride or polyethylene.

Figure 53: I.C.I. Detonating Cord Construction

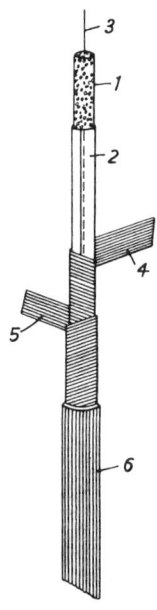

Source: U.S. Patent 3,939,941

A scheme developed by *O.A. Itria; U.S. Patent 3,908,789; September 30, 1975; assigned to Texaco Inc.;* involves seismic delineation of sedimentary section below the surface of a seismic energy propagation medium.

It utilizes a vertical, straight, explosive material, suspended under water by a float, to generate and shape seismic energy pulses. A first downwardly propagating compressional energy pulse is generated by initiating detonation of the upper end of the explosive while a second upwardly propagating rarefactional seismic pulse is generated by initiating detonation of explosive material at a position intermediate the ends of the explosive. Variations in the length of the charge and the position of the ignition points shape the energy pulses. It is asserted that the amplitude of the secondary bubbles and time duration are attenuated by this process.

A device developed by *J.C. Mollere; U.S. Patent 3,968,855; July 13, 1976; assigned to Hercules Incorporated;* is a towed underwater gun assembly which includes an acceleration barrel, which is resiliently coupled to and separable from the firing gun. Percussion initiatable explosive cartridges are accelerated through the barrel by a stream of water with sufficient kinetic energy to initiate the firing pin and ignite the explosive charge. It is asserted that the resilient couple absorbs the shock waves from the ambient charge explosions and, minimizes damage to the expensive accelerator barrel. If the gun is damaged it can be easily removed and replaced.

An apparatus developed by *T.L. Slaven and J.C. Mollere; U.S. Patent 3,912,041; October 14, 1975; assigned to Hercules Incorporated;* is a water operated explosive cartridge launcher for use in offshore seismic exploration having an injection system which prevents air from mixing with the cartridge propelling stream. The cartridge injection system has a cartridge launching plug that is rotatably mounted within a cylindrical bore which carries the water. A barrel, extending out from the plug, which is sealed off from the atmosphere and from the water flow prior to reaching either its loading or unloading position is used for loading the cartridges. A bypass pipe is connected in parallel with the water passageway to maintain uninterrupted water flow in the system.

Such an apparatus is shown in Figure 54. It shows a seismic vessel **14** equipped on its deck **16** with a system for generating seismic waves in a body of water by rapidly and consecutively firing relatively small, explosively operated cartridges **12** from a towed underwater gun **10**.

To operate the gun there is provided a water propulsion system **20** including in its circuit the launcher **19** and a water pump **24** which continuously pumps seawater through a conduit **22**. The portion of the conduit leading from the deck into the body of water is flexible. The explosively operated seismic cartridges are stored in a container **18** near the launcher.

Figure 54: Hercules Inc. Water-Operated Cartridge Launcher

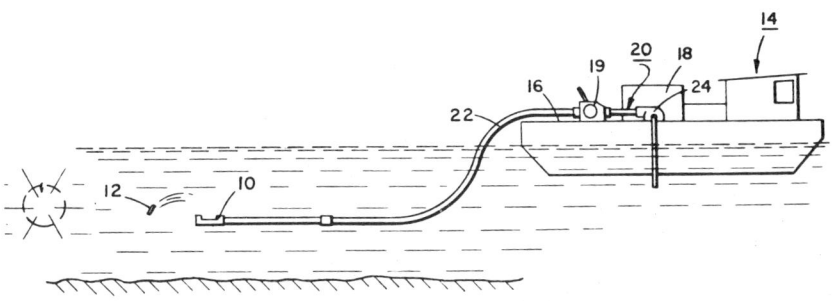

Source: U.S. Patent 3,912,041

A scheme developed by *J.C. Mollere; U.S. Patent 4,132,974; January 2, 1979; assigned to Hercules Incorporated;* comprises trailing a gun adapted to initiate explosively operated charges at a sufficient depth below the water's surface to allow the creation in the body of water of high pressure waves, and detecting the pressure waves with a detector assembly including a housing having a flexible, sound-transmitting wall defining a chamber, an incompressible liquid completely filling the chamber, and a detector probe completely immersed in the incompressible liquid.

A device developed by *H.L. Fitch; U.S. Patent 3,948,177; April 6, 1976; assigned to Hercules Incorporated* is a self-disarming explosive cartridge for use in offshore seismic exploration to eliminate the hazard of unfired live cartridges being washed ashore and exploding. When the shell is fired by striking the primer, a conduit which acts as a primer well, is driven into the cartridge shell so as to break its watertight seal. The ingress and egress of ambient water desensitizes the main charge and renders it harmless. It is asserted that, for an explosive charge of up to 10 pounds under a hydrostatic head of about 50 feet, the period of desensitization is 2 to 16 hours.

Implosive Seismic Wave Generators

These devices are shock wave generators which operate by causing the ambient liquid which surrounds the generator to implode. The implosions are induced either by mechanical movement of the generator or by collapse of a gaseous bubble. These sources are primarily used in offshore seismic prospecting.

Recent research and patent activity in this area has focused on the use of steam or rapid piston movement to produce low pressure areas which will collapse under hydrostatic pressure. The systems are designed for high peak pressure pulse amplitudes and pulses having little or no secondary oscillatory pulses.

A device developed by *J. Cholet; U.S. Patent 3,833,087; September 3, 1974; assigned to Institut Francais du Petrole, des Carburants et Lubrifiants, France;* is an imploder having two stationary cylinders and two slidable interconnected pistons, each mounted in a separate cylinder. When a piston is disengaged from a closed position in a cylinder, the surrounding water abruptly fills up the free space of the cylinder, thereby producing an implosion which generates acoustic waves.

Another device developed by *J. Cholet; U.S. Patent 3,997,022; December 14, 1976; assigned to Institut Francais du Petrole, des Carburants et Lubrifiants et Entreprise de Recherches et d'Activites Petrolieres Elf, France;* is a marine cylindrical imploder which utilizes two hydraulically driven in-line pistons. The rapid movement of the pistons produces a volume variation in the imploder. Acoustic waves are generated by an implosion which is produced due to the rapid total displaced volume variation of the cylinder and concomitantly, of the surrounding liquid medium.

A device developed by *D.H. Reed and A.A. Franklin; U.S. Patent 3,952,833; April 27, 1976; assigned to Atlantic Richfield Company;* is a gas exploder which produces imploding cavitation vapor bubbles by rapid movement of a base plate. The exploder comprises a piston chamber within a closed upright cylindrical housing which is supported beneath a water surface by a floating platform.

Explosions within the chamber lift the piston to compress an air spring within the housing and thereby transmit a lifting force to the housing. As the housing is lifted a cavitation vapor bubble is created along the undersurface of the base plate.

An apparatus developed by *J. Pauletich; U.S. Patent 3,944,019; March 16, 1976;* is an underwater sound imploder consisting of three main units, a boiler, a heat shield and a bell-shaped sonic section. Successive sonic impulses are produced by bubbles of steam released into water which is contained within the imploder. The steam bubbles expand until the pressure of the ambient water causes them to burst, thereby causing the imploder to vibrate and produce sonic waves.

Another device developed by *J. Pauletich; U.S. Patent 3,912,042; October 14, 1975;* is a shallow water imploder which uses either remotely generated steam or gases (oxygen and acetylene), which on mixing burn with a heat output high enough to vaporize nearby water. The steam from an outside steam generator expands within the imploder's outer shield until hydrostatic pressure causes an implosion.

Still another device developed by *J. Pauletich; U.S. Patent 3,858,171; Dec. 31, 1974;* is an imploder which produces steam bubbles in the surrounding water environment by passing high amperage current through triangular-shaped conductors. The conductors become red hot producing steam bubbles in the water which expand between baffle plates until they abruptly collapse, thus causing an implosion.

A device developed by *D.H. Reed; U.S. Patent 3,919,684; November 11, 1975; assigned to Atlantic Richfield Company;* is a torpedo-shaped underwater seismic implosive wave source which consists of a main body section and a separable forward nose section. An axial connecting rod joins the nose section to a piston which is slidable along part of the main body section. Ignition of an explosive gas mixture in front of the piston rapidly drives the piston and the main body section in opposite directions, thereby creating a low pressure bubble of condensable water vapor.

It is asserted that the implosion of this bubble generates a singular seismic pulse having a peak pressure on the order of 10^4 atmospheres.

Figure 55 shows such a device in place on the surface before firing (upper view) and after firing (lower view). The hydrodynamically streamlined seismic gun **10** is supported from a floating vessel **12** beneath the surface of the water at a predetermined depth by means of a pair of flexible cables **14**.

The gun, preferably of elongated hollow tubular shape, consists generally of an aft main body section **15** and a forward nose section **16** interfitted therewith in end-to-end relation. External contact between the body section and the nose section is effected at the interface between the transverse surfaces **17** and **18**, which are separable by relative longitudinal movement between the two sections.

The body section is provided with a cylindrical bore **19** extending over a part of its length. A piston **20** provided with O ring or other suitable peripheral seal means **21** and slidable within the bore is interconnected with the nose section at the surface **18** by means of a connecting rod **22**.

Figure 55: Atlantic Richfield Co. Underwater Imploder Device

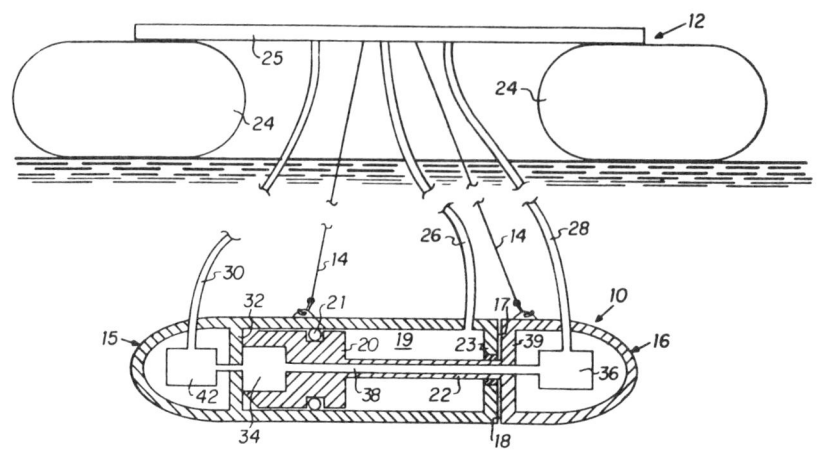

Before Firing

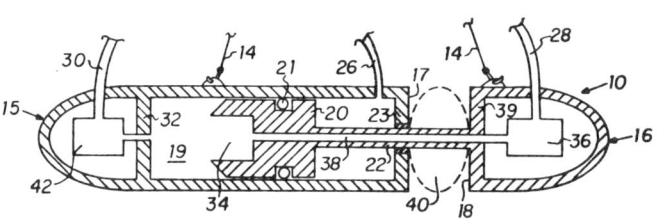

After Firing

Source: U.S. Patent 3,919,684

The rod **22** extends axially along the bore **19** and slidably penetrates an end wall **23** of the body **15** which includes the external surface **17**. Suitable sleeve means make this slidable junction substantially hermetic.

The floating vessel **12** may conveniently consist of a pair of parallel pontoons **24** between which an equipment carrying platform **25** may be supported. Appropriate electrical connections and gas lines to be described may be brought to the gun **10** from the platform by means of downwardly extending flexible umbilical links **26**, **28** and **30**. It should be understood, however, that there are substantially no external constraints on the relative longitudinal movement of the sections **15** and **16** other than the resistance of the water.

Preparatory to the firing of the gun, the volume between the piston **20** and the end wall is pressurized with air introduced through the umbilical **26** at a suitable low pressure, for example, about 10 pounds per square inch gauge pressure. In consequence, the piston 20 is urged away from the end wall 23 and toward the opposite end wall 32, carrying the nose section 16 with it until the surfaces 17 and 18 are in contiguous contact. The bore 19 is of a length such that in this prefiring position, the piston makes contact with the wall 32. The face of the piston may be provided with a cavity or depression 34 of a depth and contour sufficient to create in conjunction with the end wall 32 an explosion chamber of selected volume.

Separate gas lines carrying explosive gaseous fuel such as propane and oxygen may be introduced into a central cavity in the nose section through the umbilical 28 and there fed into a combination gas mixture and spark control unit 36. A longitudinal gas conduit 38, suitably interconnected with the supply of mixed gases, may be drilled through the aft wall 39 of the nose section, through the connecting rod 22 and through the piston to the face thereof. This channel 38 enables introduction into the explosion chamber formed by the volume in front of the piston of an explosive mixture of propane and oxygen in desired proportions.

The pressure of this explosive mixture is maintained at a value below that of the air cushion existing behind the piston in order to insure positive initial seating of the surfaces 17 and 18 which also insures that the volume of pressurized explosive mixture remains constant.

In operation, the explosive gas mixture formed in the control unit is ignited by a suitable electrical signal delivered from the float 12 and initiates a firing train along the conduit 38. When the burning gases reach the face of the piston, an explosion occurs within the adjacent chamber within the cylindrical bore which reaches a pressure peak in a very short period of time, on the order of milliseconds. The force of this explosion drives the piston towards the wall 23, compressing the intermediate air cushion.

By suitable valve means (not shown) the air behind the piston may be exhausted to the surface through the umbilical 26 to lessen the resistance to the travel of the piston. At the same time an equal and opposite force against the wall 32 urges the main body section 15 in the opposite direction. The result is that sections 15 and 16 separate rapidly at the interface between the transverse surfaces 17 and 18. The speed of this separation is aided by providing the extremities of the sections 15 and 16 with a hydrodynamically streamlined shape such as shown.

The travel of the piston is controlled by various factors such as the quantity of explosive gas mixture, the frictional forces present, and the reaction due to pressure build-up within the air spring. In any event, sufficient damping is provided to prevent the piston from contacting the end wall 23.

A low pressure bubble 40 of condensable water vapor as best seen in the lower view in the figure and shown in dotted outline now forms the space between the surfaces 17 and 18, having a growth pattern determined by the speed of separation and whose ultimate size is directly related to the cross sectional surface area of the surfaces and to the distance of travel of the sections 15 and 16 relative to each other.

After approximately 30 to 100 milliseconds, the bubble 40 implodes or collapses responsive to the surrounding hydrostatic pressure to create a seismic pulse at the conclusion of such implosion. Although shown for simplicity as spheroidal, it is theorized that the bubble **40** is probably depressed in the middle into an hourglass shape because implosion has begun there while the remainder of the bubble is still forming. This pulse has an energy content and peak value substantially independent of the depth of seismic gun **10** below the surface of the water. Typically, peak pressures on the order of 10^4 atmospheres may be expected. Since there is little resistance to the initial separating motion of the sections **15** and **16**, the initial positive pulse due to the chemical explosion will have no appreciable effect upon the total pressure spectrum.

After a suitable time interval, the gun may be prepared for the next firing by repressurizing the air cushion from the surface as previously described and by introducing a fresh explosive gas mixture in the chamber in front of the piston **20**. Exhaust gases may be scavenged by the replenished supply of explosive mixture and expelled to the surface through the umbilical **30**. This may be accomplished, for example, by suitable valve means (not shown) within a control unit **42** operated electrically from the surface.

Gas Exploders

These devices are seismic wave generators in which gaseous fuel-oxidant mixtures are rapidly burned to provide compressional waves for geophysical exploration. Typically, the explosions occur either outside the exploder apparatus and provide direct contact between the shock waves caused by the explosion and the ground; or, inside the exploder, where the shock waves are transmitted outwardly by pistons driven by the explosion.

Recent research and patent activity has been directed towards minimizing rebound created noise by guiding land gas exploders in both rise and descent and, in addition, cushioning the descent. Activity has also been directed to providing better acoustic wave coupling between the gas exploders and the ground strata to be explored by sealing the explosive section of the exploder in the ground.

A device developed by *H.S. Field and D.K. Mitchell; U.S. Patent 4,026,382; May 31, 1977* is a gas exploder for use in land prospecting, which uses a long (5 to 8 feet), slender (0.25 inch), heavy-walled (o.d. of 0.54 inch) pipe which is open at the bottom end and is driven into the earth. Oxygen and a combustible gas (preferably acetylene) are passed through the pipe into the voids and fissures in the shallow sediments of the earth. A spark plug at the top of the pipe detonates the gas mixture providing an explosion and concomitantly, seismic waves in the ground.

As the explosion opens up ground fissures and provides a subsequent larger void space, the pipe is pressed further into the ground and the process is repeated until the desired seismic wave intensity is reached.

Such an apparatus is shown in Figure 56. The device which is used as the source of the seismic waves is indicated generally by the numeral **10**. As shown, this comprises a long, slender, thick-walled pipe **16** which has, at its bottom end, a pointed closure **18** and a plurality of holes drilled through the wall **20** near the lower end of the pipe. The upper end of the pipe is closed.

Figure 56: Repetitive Detonation Seismic Surveying Apparatus

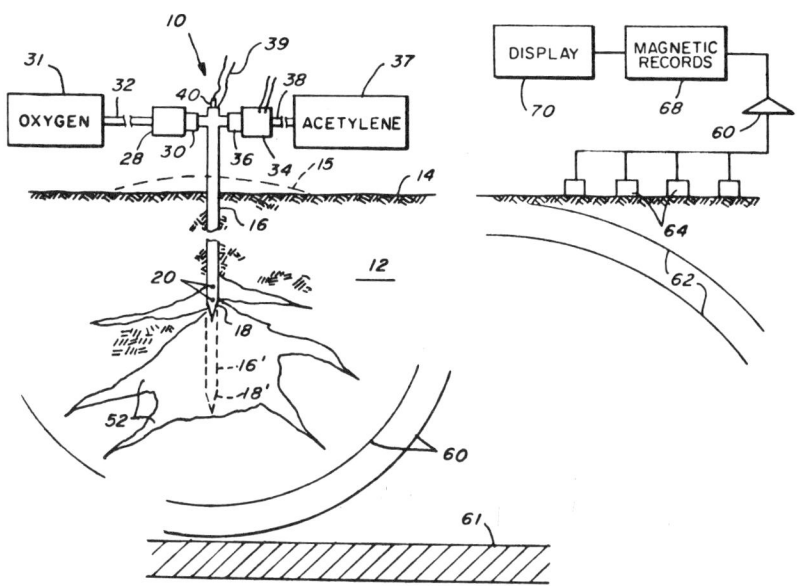

Source: U.S. Patent 4,026,382

There are three points of attachment of auxiliary apparatus. These may be attached in any way desired, but are preferably joined by threaded fittings in the form of a cross at the upper end of the pipe. One arm carries a valve which is preferably an electrically operated remote controlled solenoid valve **28** which contains also a check valve **30**. Oxygen, air, or other oxygen-containing gas is provided by means of a tube **32** from a source **31** of such oxygen-containing gas.

Symmetrically placed to the oxygen arm is a second arm which also contains a remote controlled electrically operated solenoid valve or other type of valve **34** and including also a check valve **36**. This valve is supplied by means of tube **38** with a combustible gas which may be hydrogen, certain light hydrocarbons, acetylene, etc. The tube connects, as shown, to a container **37** of such combustible gas.

A third connection of the pipe includes a spark plug or similar device **40** which on the application of a high potential applied to wires **39** will cause a spark to jump between the electrodes inside the pipe. If there should be an explodable mixture of gas inside the pipe, the mixture will detonate and create a downward moving detonation wave.

The figure also shows a general cross section of the earth **12** with the surface indicated as **14**, the pipe **16** is shown as being driven into the shallow layers of the earth. The point **18** is at a nominal depth, such as approximately a foot below the surface. There will be voids and fissures **52** in the shallow subsurface layer of the earth, which are nominally filled with air. When the valves **28** and **34** are opened, oxygen-containing gas will flow from the container **31** through the tube **32** through valve **28** and check valve **30** into the pipe. Correspondingly, the combustible gas from reservoir **37** will flow through tube **38** through valve **34** and check valve **36** into the interior of the pipe **16**. Here the two gases will mix, and by proper proportioning of the flow, the mixture will be detonatable.

The control for the valves can be any conventional controls. The valves stay open for a selected period of time, during which the gases will flow into the pipe and will mix and travel down the pipe and out through the openings **20**, and will flow into the fissures and porous areas **52** in the earth.

Some of these fissures will be very thin, others will be of greater volume. But, in general, at the start, the volume of porous space will be quite small so that the volume of gas flow into the earth during the flow interval will be small. At the end of this interval, the flow of gas is stopped, the gas mixture in the pipe will be detonated, and the detonation wave will detonate the gas in the pores. The explosion of gas in the earth causes the lifting of the overburden, which will form a new surface in the region of the pipe which is indicated by the numeral **15**.

This explosion will send out a compression wave **60** which will pass downward into the earth, and will reach a stratum **61**, where part of the energy will be reflected and will move up to the surface as spherical waves **62**. Here they will contact the transducers **64**, which may be conventional, and will need no further explanation. The outputs of the transducer are amplified with a conventional amplifier and recorded at **68** in a conventional manner and displayed at **70**.

The reflected waves resulting from the first explosion may be recorded. If this record is satisfactory, no further operations are needed. In general, however, because of the small volume of pore space, the volume of gas which will be available for the explosion will be too small. So a second operation will be required and mixed gas will be introduced again through the valves, which will now fill the enlarged pore space formed by the previous explosion. This larger volume of gas is then detonated and a larger explosion and higher intensity reflected waves will be received and recorded, and so on.

As the volume of void space increases, the pipe can be pressed deeper into the earth (the pipe and closure now assuming positions **16'** and **18'**) and therefore there is obtained a greater mass of overburden to tamp the explosion, and therefore a greater seismic effect. When the volume of gas is sufficiently large so that the detonation provides a seismic wave of sufficient energy, and the record that is received is satisfactory, the operation is complete. The pipe is withdrawn, and the apparatus is moved to a new shot point.

It is clear that more than one of the guns, or pipes, can be used in the form of a multiple point operation as is well known in the seismic art. Thus, for example, four of the pipes may be driven into the earth at the corners of a square and simultaneously filled with explosive gas and detonated. In this way a multiplication of the elastic wave intensity can be provided which makes for a more satisfactory seismic record. In normal operation the number of shots may vary from 5 to 10.

An apparatus developed by *D.H. Reed and J.E. Hardison; U.S. Patent 4,016,952; April 12, 1977; assigned to Atlantic Richfield Company;* is a gas exploder device containing a pair of dish-shaped plates connected to combustion chamber pistons for use in offshore geophysical prospecting, which produces seismic waves by cavitation. Combustion within the piston chambers accelerates the plates through ambient water, producing cavitation bubbles which collapse and generate seismic waves.

A device developed by *T.P. Airhart; U.S. Patent 4,007,803; February 15, 1977; assigned to Atlantic Richfield Company;* is a truck transportable gas-exploding seismic wave generator for land prospecting containing an elongated cylinder with multiple, sequential, vertically spaced gas inlets mounted on a base, and containing a massive free piston. An explosive gas mixture is initially detonated at the bottom of the cylinder to drive the piston upward, sequentially exposing the gas inlets through which the supply of gas is replenished causing additional detonations to develop.

After rising to the top of the cylinder the piston falls, cushioned by the burned gases, to its lowest limiting position. Each detonation sends shock waves to the base, which generate the acoustical pulses.

A somewhat similar device is described by *T.P. Airhart; U.S. Patent 4,100,991; July 18, 1978; assigned to Atlantic Richfield Company.*

An apparatus developed by *C.D. Dransfield; U.S. Patent 4,102,429; July 25, 1978; assigned to Atlantic Richfield Company;* is a seismic wave generator assembly with self-contained guidance and shock-absorbing means which is confined within an elongated, upstanding cylindrical housing which may be lowered into position at the end of a lateral arm or a cable supported from a transport vehicle.

An apparatus developed by *L. Sayous; U.S. Patent 3,981,379; September 21, 1976; assigned to Societe Nationale des Petroles d'Aquitaine, France;* is a gas exploder particularly adapted for seismic exploration use in shallow water regions (lagoons, swamps, ponds, etc.). A differential piston is mounted in a cylinder which serves as both the mobile wall of a pneumatic chamber and the mobile wall of the gas combustion chamber. The pneumatic chamber provides compressed air to push the piston onto a valve and closes the combustion chamber. The fuel-oxidant gases in the combustion chamber are then ignited to drive the piston upward and create seismic waves.

A technique developed by *W.H. Carman, Jr.; U.S. Patent 3,976,161; August 24, 1976; assigned to Amoco Production Company;* is one in which seismic waves are generated by detonating an explosive gas mixture in an earth cavity formed by driving a short earth auger to a desired depth, leaving most of the augered earth in the hole surrounding the auger shaft, and then pulling the auger and shaft upwardly to lift and compress the augered earth and pack it around the auger shaft, thereby forming and sealing the cavity below the auger. The gas mixture is then introduced through the hollow auger shaft and detonated, preferably by ignition propagation through the connecting hose and the auger shaft.

Upon completion of as many gas fills and detonations as may be desired, the auger is reversed and withdrawn, leaving the ground surface virtually undisturbed.

Air Guns

These devices are seismic wave generators which release compressed gases. The gas expands creating a compressional wave front which is transmitted, by acoustic coupling, to the subsurface strata which is to be explored and mapped. Although primarily designed for offshore use, air guns specifically designed for land use have been developed.

Recent research and patent activity has been directed towards the use of mechanical shields and air release controls in order to minimize the noise produced by air bubble oscillation. Multiunit air gun systems have been developed to permit shaping of the transmitted seismic signal by controlling the time of air release, the amount of air released, and the order in which the units release the air.

A system developed by *C.W. Dick, O.A. Johnston, J.L. Paitson, and C.H. Savit; U.S. Patent 4,016,951; April 12, 1977; assigned to Western Geophysical Company of America;* is a high pressure (4,000 to 7,000 psi) air gun surveying system to be used in ice covered waters and at low temperatures (−35° to −60° F). A small enclosed, helicopter transportable, insulated shelter (6 feet high, 6 feet long, and 5 feet wide) is provided to house the air gun, auxiliary equipment, and operating personnel and provide protection from adverse environmental conditions. The heat generated by the auxiliary equipment is used to prevent freezing of the air gun and to provide the operating personnel with a relatively warm environment.

A device developed by *M.G. Bays; U.S. Patent 4,008,784; February 22, 1977; assigned to Seiscom Delta Inc.;* is a vehicle-mounted air gun system for use in seismic exploration over a land surface. Air guns are mounted on the rims of the vehicle's fluid-filled tires to provide compressed air which is directed to impinge upon diaphragms in the tires. The compressed air causes the diaphragm to expand and accelerate the fluid in the tire, which exerts a force on the land surface through the tire, thereby imparting a seismic signal to the land surface.

A device developed by *O.A. Itria; U.S. Patent 4,006,794; February 8, 1977; assigned to Texaco Inc.;* is an air gun which flattens the bubble of released air and prevents the bubble from contracting back into the body of the air gun. Air deflectors are mounted at the exit ports of the gun and are connected to cylinders which straighten and hold the exhausting air in a designated direction. A spring operating at 200 psi is provided to close the air release valves when the air pressure becomes neutralized, just prior to the bubble reaching its maximum size and collapsing back into the cylinder. It is asserted that the elongated and flattened air bubble will collapse under hydrostatic pressure with reduced impact and, as a result, the bubble pulse period and amplitude are considerably reduced.

An apparatus developed by *D. Silverman and J.R. Bailey; U.S. Patent 3,979,140; September 7, 1976; assigned to Senturion Sciences, Inc.;* includes a long, continuous small diameter pipe adapted to be reeled up on a drum mounted on a vehicle, and means to insert the end of the pipe into the mouth of a borehole and to lower it to any desired depth in the earth. Compressed liquids or gases are injected from the surface of the earth into the pipe which extends into the borehole.

Fast action electrically controlled valves permit the compressed fluid to explosively expand into the liquid in the annulus of the borehole and cause seismic waves to be generated. Such explosive injection of the fluid from the chamber into the well will cause a seismic wave to be generated which will expand until it reaches the surface and is detected and recorded by the geophones.

A plurality of geophones are positioned on the surface of the earth in the vicinity of the borehole with conventional amplifying and recording means connected thereto. Electrical control wires may be carried in the pipe means by which the valve can be timed and controlled, so that the travel times of the seismic wave from the chamber to the geophones can be determined, from which the position of the borehole at the depth of the chamber can be determined.

A device developed by *R.A. Kirby and J.F. Bayhi; U.S. Patent 3,896,898; July 29, 1975; assigned to Exxon Production Research Company;* is an air gun which has compressed air stored in a reservoir and which is mounted on a platform for use beneath the surface of a liquid medium. The compressed air is simultaneously released into a plurality of conduits of equal lengths, but unequal diameter, which are arranged in an areal pattern on the platform. The tubes are 0.5 to 5 meters in length and .5 to 5 centimeters in diameter. It is asserted that the simultaneous release of compressed air through the individual conduits will produce seismic waves which add algebraically to produce a single, high energy, high frequency pulse.

A device developed by *L.M. Mott-Smith; U.S. Patent 3,893,539; July 8, 1975; assigned to Petty-Ray Geophysical, Inc.;* is an underwater seismic source for generating an improved seismic signal employing a spaced array of air guns of different sizes to generate an initial bubble impulse and to suppress the cumulative effect of secondary bubble impulses. The combination includes means for providing further secondary bubble suppression integral with one or more of the guns in the array.

The sizes, minimum spacings, and secondary bubble suppression techniques are intimately combined to provide an underwater seismic source having an acoustic pressure signature wherein the amplitude of the initial impulse, including the signal reflected from the water surface, is from 8 to 12 times the amplitude of any subsequent bubble impulses, including their surface reflections.

An elevation of such an array is shown in Figure 57. There is shown an array **10** of air guns which is secured to a ship **12** by conventional towing cables **14**. The ship includes the usual air compressors, plumbing, valves, etc., (not shown), supplying high pressure air of the order of 2,000 psi to the series of guns, through an air line **16**. The guns, herein numbered **18** through **30**, are secured to respective buoys or floats **32** through **44** which support the guns at a predetermined level under the water surface, e.g., 30 feet.

The guns in this exemplary array have sizes, or total chamber volumes, of 70, 90, 120, 150, 200, 300, and 450 cubic inches respectively, and are spaced apart at (increasingly greater) distances of 7, 8, 9, 10, 12, and 14 feet, respectively. As may be seen, the distance between guns generally increases in proportion to the maximum diameter of the initial bubble generated. The distances are chosen sufficie...tly large to prevent interference between guns, i.e., to prevent bubble coalescense.

Figure 57: Petty-Ray Geophysical, Inc. Multiple Air Gun Array

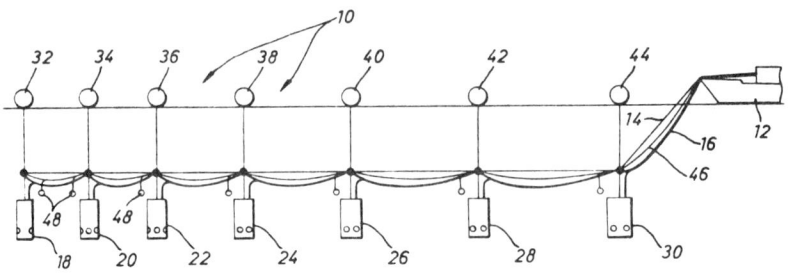

Source: U.S. Patent 3,893,539

The relationship between maximum bubble diameter, air gun chamber volume, the pressure to which the air guns are charged, and the depth of the gun beneath the water surface is generally known in the marine seismic exploration field, and the minimum gun spacings may readily be determined by those of ordinary skill in this art.

The usual electrical wire complex **46** extends from the ship **12** to each of the guns **18** through **30**, and supplies conductors for simultaneously firing the plurality of guns. Because there are differences in the actual firing times of the guns, even though energized with exactly the same firing signal, a hydrophone **48** is placed near each gun and connected to timing equipment on the ship through the wire complex.

Each gun can be fired individually to determine electronically, by a signal generated by that gun's corresponding hydrophone, the exact delay between when the gun was energized and when it actually fired. An adjustable electronic time delay (not shown) may then be employed with each gun to compensate for its particular firing delay characteristic, whereby all guns actually fire simultaneously when energized by a common firing signal.

The ship is depicted here as being attached to the end of the array **10** having the largest gun **30**, since the weight of the larger gun tends to stabilize and hold the array at the selected depth in the water during towing. It will be appreciated, however, that the array may be towed from either end and that the guns do not need to be arranged in increasing or decreasing order of size. Any random sequence or arrangement of guns may be employed, as long as the spacings between the guns are sufficient to prevent coalescence of any adjacent bubbles upon their reaching maximum diameters.

It is asserted that this system generates a seismic pulse which defines a signature having a signal-to-bubble ratio of from 8:1 through 12:1. The air guns' interspacing is set to avoid coalescence of adjacent air bubbles and 25% to 33% of the total air gun volume is used to provide a secondary bubble within the first bubble in order to lengthen the total bubble collapse time.

An apparatus developed by *J.L. Paitson and M.L. Parker; U.S. Patent 4,114,723; September 19, 1978; assigned to Western Geophysical Co. of America;* is an air actuated seismic signal generator, commonly termed an air gun which includes a cylindrical casing that is divided into a control chamber and a firing chamber by a sliding valve. The end face of the sliding valve that is exposed to the control chamber has a greater area than the end face exposed to the firing chamber. An air exhaust port is provided in the firing chamber; the exhaust port may be opened or closed by sliding the valve away from or towards a seat that is associated with the exhaust port. Air is admitted to the control chamber at a relatively low pressure of about 500 psi, and holds the valve in the exhaust port-closed position because of the differential forces on the valve end faces.

The firing chamber is pressurized to a relatively high pressure of about 5,000 psi by a separate air supply. There is no air communication or equalization between control and firing chambers. When some of the air in the control chamber is dumped, the differential forces are upset and the sliding valve moves away from the exhaust port seat, to impulsively release the air in the firing chamber. As the piston moves away from the seat, it momentarily compresses the air remaining in the control chamber. The compressed air acts like a spring, absorbs the motion of the valve and bounces the valve back towards the exhaust port seat. At the same time a fresh supply of control air is admitted to the control chamber to finally push the valve closed again.

The amount of air remaining in the control chamber and hence the "resiliency" of the air spring can be set by means of calibrated orifices in the air dump line. Figure 58 is a schematic showing the utilization of such a device. As shown, a ship **10** tows a seismic signal generator in the form of an air gun **12** through a body of water **14**.

At desired locations, the air gun generates an acoustic pulse produced by releasing high pressure air jets **16** into water **14**. The acoustic pulse travels along ray path **18**, is reflected from subsurface formation **20**, and returns to the water surface along ray path **22**. The reflected signals are detected by hydrophones **24** attached to a long streamer cable **26**, also towed by ship **10**. The detected reflected signals are transmitted to a seismic signal processor (not shown) aboard the ship through suitable conductors or telemetric link (also not shown) in the streamer cable.

Air compressor **28** in the ship feeds air at two different pressures to the air gun through hoses **30** and **31**. Hose **30** supplies control air under a relatively low pressure of a few hundred pounds per square inch; hose **31** conducts air, having a pressure of several thousand pounds per square inch, to the firing chamber of the gun (to be described below). Electrical cable **32** connects electrical triggering and sensing devices in the air gun to a fire control unit **29** on the ship. The fire control unit, triggers the gun to cause it to periodically emit the air jet into the water through exhaust ports **34**. When not in use, the streamer cable and hoses are stowed aboard ship on cable reel **36**. Cable **32** is stowed on reel **38** when the gun is idle.

Control air (200 to 1,500 psi) is fed from the air compressor through a separate hose **30** and a three-way, two-position, spring-loaded solenoid valve **86**. The gun is fired by momentarily opening the valve to vent or dump the control air into the environment. The valve used was model SV-431-532P8P43 (Circle-Seal).

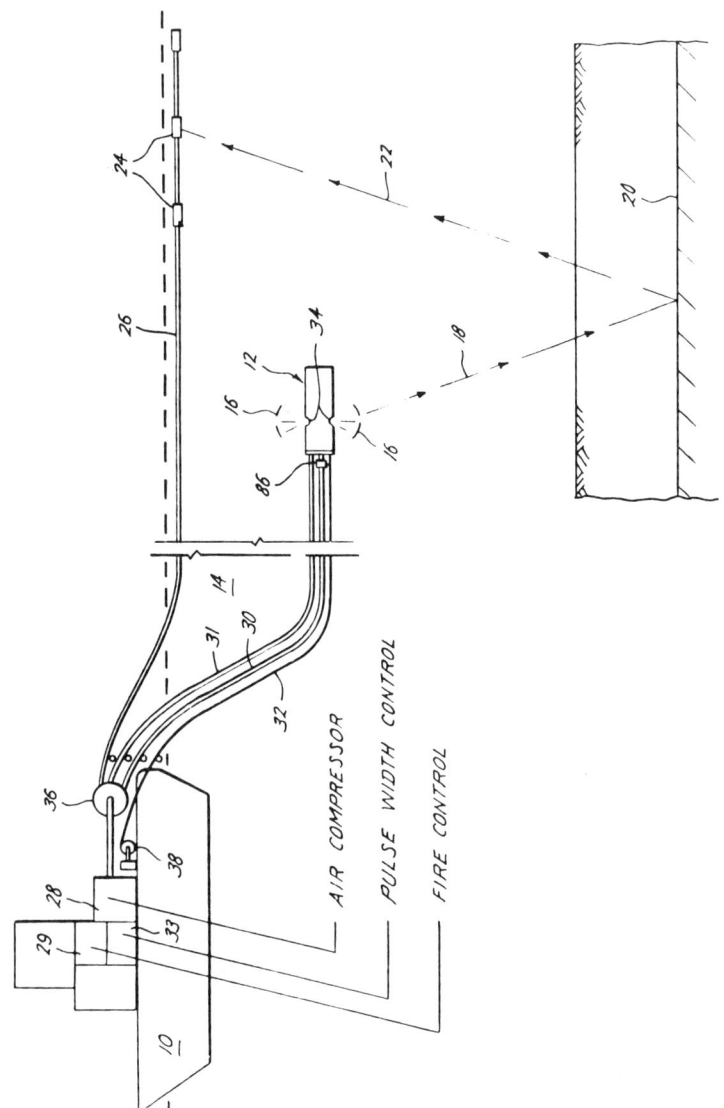

Figure 58: Western Geophysical Co. of America Pneumatic Seismic Signal Generator Installation

Source: U.S. Patent 4,114,723

Seismic Vibrators

These are seismic wave sources which provide an oscillatory vibrational motion in the earth section to which they are contiguous. The vibrators may contain hydraulically driven reaction masses operating at low speeds or electrically driven eccentric masses rotating at high speeds.

One of the most widely used seismic energy sources or signal generators is the shaker or vibrator. A significant advantage of the use of vibrators lies in the ability of such sources to impart large amounts of energy into the earth over a period of time. Techniques well known in the art are applied to the reflected and detected energy to effect time compression of the received signals, thereby resulting in a high signal-to-noise ratio.

It is common practice to use vibratory sources to impart signals having an approximately sinusoidal waveform with a linearly increasing or decreasing frequency. A typical energy transmission of a seismic vibrator would be a frequency sweep extending from 5 to 80 Hz over a period of about 10 sec. A peak force amplitude of 15,000 to 30,000 lb is typically used.

One type of vibrator used has involved counter-rotating eccentric weights. Other vibrators have employed electrodynamic and hydraulic force generators. Eccentric weight vibrators have not been successful because they cannot achieve the desired force level over the required frequency range without changes in rotating weight or changes in eccentricity.

Electrodynamic or electromagnetic vibrators have not been widely used because they cannot economically develop the large forces and displacements necessary in seismic wave generators.

Hydraulic vibrators are widely used as seismic wave generators. A typical system of this type is the Vibroseis vibrator. Hydraulic vibrators have the disadvantage of requiring large power inputs and large hydraulic components with high flow capabilities. These requirements impose high system cost and make it difficult to produce seismic waves with a small amount of distortion.

It is particularly difficult to generate large forces at low frequencies because of the large hydraulic flow required. The hydraulic seismic vibrator becomes more efficient with increasing frequency in the frequency range commonly used.

Recent research patent activity in this area has focused on vibrator phase lock systems and, in particular, phase control systems, which operate in a feedback mode. Recent activity also has been directed to multiunit vibrators or multivibrator systems which modulate the separate units or vibrator outputs in order to approximate an optimum transmitted seismic wave.

A device developed by *J.W. Bedenbender and G.H. Kelly; U.S. Patent 4,026,383; May 31, 1977; assigned to Texas Instruments Incorporated* is a vibrational energy source which has a high inertia gyroscopic mass rotating at high angular velocity about a shaft. A control force is used to impart an oscillatory angular displacement to the shaft in a plane perpendicular to the vector direction of the source. This displacement produces a gyroscopic reactionary moment in the shaft which is mechanically coupled to the earth's surface to produce oscillatory waves.

The truck mounting of such a device is shown in Figure 59. There is shown a gyroscopic vibrator **100** mounted on a vehicle such as a truck **102**. The vibrator is coupled to the truck by a lifting means which includes columns **104** located on either side of the vibrator. The columns are slidably located on brackets **106** attached to the truck, the columns being adapted to move along a generally vertical axis with respect to the truck. The columns are coupled to the vibrator base plate **108** by spring means **110**, such that when the columns are in a lowered position with respect to the truck as shown, at least a portion of the truck weight is applied as a static load to the base plate.

The compliance of the spring means will generally be selected such that the resonant frequency of the truck motion in response to vibrations induced in columns **104** will be substantially outside the range of frequencies induced in the ground plate. Lifting means (not shown) adapted to raise the columns and consequently the vibrator with respect to the truck chassis may comprise one or more hydraulic cylinders associated with each column. Each of the cylinders may be coupled at one end to the truck chassis and at the other end to the columns such that upon actuation of the hydraulic cylinders the columns are caused to move in a generally vertical direction with respect to the truck chassis.

Bumpers or lock devices **112** are provided to limit the travel of the rotor housings **114** and reaction mass **116** when the vibrator is in the raised position. It will be recognized by those skilled in the art that if the vibrator is oriented with respect to the truck as shown, the gyroscopic reaction forces induced by turning and bouncing of the truck during movement will tend to be mutually cancelling.

A conventional electric and/or hydraulic power supply **118** and control system **120** are also located on the truck. These are adapted to provide suitable energization of the lifting means and of the actuating means in the gyroscopic vibrator.

Figure 59: Texas Instruments Incorporated Truck-Mounted Gyroscopic Vibrator Seismic Source

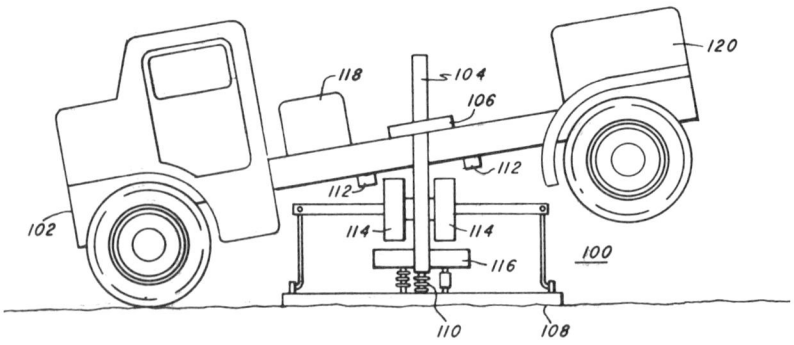

Source: U.S. Patent 4,026,383

A system developed by *D. Silverman; U.S. Patent 3,984,805; October 5, 1976* is a control system for a seismic vibrator, in which the reference, or sweep signal, drives the vibrators directly without the need to control the vibrator so that the seismic signal generated in the earth will be inphase with the reference signal. Instead, the output of one or more sensors on the vibrator is used to generate a transmitted signal which is representative of the seismic signal generated in the earth. The transmitted signal is used to phase-shift the reference signal to provide a counterpart signal which is inphase with the transmitted signal. The counterpart signal is used to correlate with the received seismic signal.

A plurality of vibrators may be used. They may be grouped together at one point, or they may be spaced apart at independent points. The vibrators may use the same or different reference signals, and may start simultaneously or at delayed times.

An improved sensor for mounting on the vibrator is described to provide information as to the absolute displacement of the baseplate during vibrator operation.

A device developed by *D. Silverman; U.S. Patent 3,983,957; October 5, 1976* is a vibrator having a two part chamber; one part in contact with the earth, the other part supporting a seismic reaction mass. The chamber is hydraulically expanded so as to lift the mass above the earth. By modulating the hydraulic pressure in the chamber, seismic waves are induced into the earth.

A device developed by *A.G. Hufstedler, R.R. Weaver and B.R. Slater; U.S. Patent 3,979,715; September 7, 1976; assigned to Texas Instruments Incorporated* is a phase lock control for a chirp signal, i.e., a sinusoidal signal with a linearly increasing or decreasing frequency, driven seismic vibrator. On the upsweep section of the chirp, when the phase error is greater than a preset threshold level, phase lock is achieved by slowing down the drive signal, so as to allow the baseplate signal to fall back into phase with the chirp signal.

On downsweeps, when the phase error is above a preset threshold, phase lock is achieved by speeding up the drive signal so as to allow the baseplate signal to catch up with a reference signal.

An apparatus developed by *R.A. Broding and R.A. Landrum, Jr.; U.S. Patent 3,909,776; September 30, 1975; assigned to Amoco Production Company* is a variable frequency fluidic oscillator which is particularly adapted for use in a well for vertical prospecting. The oscillator operates by the modulation of the pressure and volume of a fluid within a bistable fluidic circuit. It is asserted that the oscillator can be repetitively pulsed at the same frequency-time pattern thereby permitting the summing of geophone responses of a plurality of oscillator cycles.

Figure 60 illustrates generally the application of the device to the determination of the location of a salt dome generally indicated by **11**. This salt dome is in the earth **12** below the surface **13**. Two wells, **14** and **15**, are shown having been drilled adjacent to the salt dome. (Sometimes only one well is necessary for the type of survey being considered.) A string of pipe **28** has been lowered into well **14**. At the lower end of this is located the oscillator, generally indicated by **29**. At the top, this pipe is suspended in a derrick **30**.

Figure 60: Diagram of Application of Amoco Production Company Seismic Survey Technique to Salt Dome Location

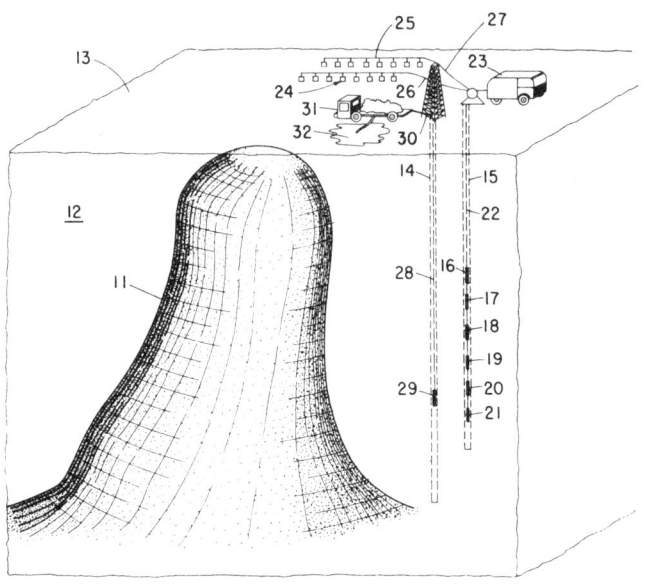

Source: U.S. Patent 3,909,776

The pipe is connected to a pump truck **31**, which, in turn, is supplied with fluid, for example, by pit **32**. Incidentally, it is highly desirable that well **15** not be cased, that is, have no strings of pipe opposite the zone in which the well seismometers **16** through **21** are located. It is also desirable that the oscillator **29** be located either opposite a string of perforations in casing, or in an uncased section of well **14**. It is also assumed that wells **14** and **15** have been surveyed, so that the location of oscillator **29** and of the well-geophones **16** through **21** can be located with the desired precision in three dimensions.

In the other of these wells (well **15**), a plurality of well geophones or seismometers has been lowered on a multiconductor cable **22** from an instrument truck **23**, to which the cable is connected and which contains the usual amplification, mixing and reproducible recording facilities employed in seismic prospecting. Also shown are two additional surface spreads of seismometers or geophones generally indicated by **24** and **25**. These are likewise connected by multiconductor cables **26** and **27** to the recording truck.

Such arrangements of seismometers or geophones are very well known in the seismic exploration art. It is sufficient to state that each seismometer or group of seismometers is connected by a separate insulated line in a cable to the truck recording apparatus. Little need be said about such equipment which may be conventional.

When the equipment has been located as shown, the pumps on truck **31** are employed to send liquid through the oscillator **29**. Discharge from this oscillator returns through the annulus to the surface. This oscillator is a powerful variable frequency fluid oscillator, and it sends out a varying frequency group of seismic waves in a burst or pattern. The response of the vertical spread of geophones **16** through **21** and of horizontal spread on or near the surface, such as spreads **24** and **25**, are recorded as in customary seismic prospecting, in the recording truck **23**. This gives seismic data on the dome for both reflections and refractions.

An apparatus developed by *F.F. Hamilton and J.S. Johnson; U.S. Patent 3,884,324; May 20, 1975; assigned to Hamilton Brothers Oil Company* is a tracked, rough terrain seismic exploration vehicle which is adapted to carry a vibrator on a rocker frame. The vibrator utilizes a reaction mass that can be linearly reciprocated in order to permit a baseplate, which is connected to the reaction mass to be driven into, and out of engagement with the ground.

The vibrator is acoustically isolated from the seismic vehicle by air bags which surround the reaction mass and dampen vertical and horizontal vibrations in a manner that concentrates the application of force in a downward direction.

An apparatus developed by *J.F. Mifsud; U.S. Patent 3,866,709; February 18, 1975; assigned to Exxon Production Research Company* is an electrohydraulically operated vibrator which contains a single power piston connected to a circular baseplate. The baseplate is held against the ground by a reaction mass, e.g., the transporting vehicle. By varying the hydraulic pressure on opposite sides of the power piston, reciprocating movement and concomitant acoustic vibrations are produced.

A device developed by *R.M. Weber and J.W. Bedenbender; U.S. Patent 4,114,722; September 19, 1978; assigned to Texas Instruments Incorporated* is a vibratory seismic energy source capable of generating significant energy over a broad frequency band. The vibrating baseplate and associated structure are designed to have minimum weight while still retaining sufficient structural integrity to permit the use of high actuator forces. This, coupled with a large reaction mass, results in the generation of significant energy levels in the earth at high frequencies.

A system developed by *J.E. Stone; U.S. Patent 4,116,300; September 26, 1978; assigned to Exxon Production Research Company* is a system for controlling the tilt of the vehicle and vibrator when a vibrator is being operated on irregular surfaces. Tilt switches detect excessive vehicle tilt. A bleeder network activated by the tilt switches reduces the pressure in the air bag on the side of the vibrator opposite from the direction of the vehicle tilt. A supply network repressurizes the air bag on the side of the vibrator opposite from the direction of the tilt.

A device developed by *D.W. Fair; U.S. Patent 4,143,736: March 13, 1979; assigned to Continental Oil Company* is a seismic transducer for generating waves in an elastic medium and is particularly applicable to the generation of low frequency shear waves. The transducer is comprised of a mass member having more than one parallel hydraulic cylinder disposed therein, a piston member disposed in each cylinder, each piston member including a piston and a pair of oppositely extending rods, and a frame interconnecting the ends of the piston rods which frame has a base surface thereon for engaging a surface of the elastic medium in energy coupling relationship.

An apparatus developed by *O.G. Erich, Jr.; U.S. Patent 4,143,737; March 13, 1979; assigned to Union Oil Company of California* is a compact, lightweight, rotating eccentric weight seismic source particularly suited for use in relatively inaccessible onshore regions. The source includes an eccentric weight rotatable about an axis and a sensor to detect the instant of peak earthward force developed by the source.

A coded energy signal is transmitted into the earth and the seismic waves returned from within the earth are detected. The polarity of a signal proportional to the seismic waves is periodically sampled to produce a plurality of sign-bit samples. The sign-bit samples are automatically shift-summed in response to a code signal generated by the sensor, thereby forming a plurality of shift-summed samples which are recorded to provide a correlated seismic trace.

A method developed by *D. Silverman; U.S. Patent 4,159,463; June 26, 1979* is a method of seismic prospecting involving at least a first and second spaced apart vibratory sources, cotemporaneously vibrating with the same or different reference signals, and recording into a common geophone at a point distant from both sources.

The method involves recording at least two records, one record includes the first vibrator V1 responsive to a first reference signal R1 and the second vibrator V2 responsive to a second reference signal R2. In the second record the second vibrator repeats the reference signal R2, while the first vibrator is responsive to the first reference signal R1, but in opposite phase, -R1.

When the two received records are added, the components due to the first vibrator are in opposite phase or polarity, and cancel, leaving only the part due to the second vibrator. When the two records are subtracted, the part due to the second vibrator cancels, and all that remains is the part due to the first vibrator.

Miscellaneous Land Seismic Wave Generators

This category of devices includes those seismic wave generators which are used in geophysical exploration on the earth's surface and which do not operate in a vibrator mode and are not powered by solid explosives or by compressed air. Various types of generators exist in this category and no single system is typical of the area. However, electromagnetic generators and the use of heavy weights to impact on the earth comprise the major part of the generators in this category.

Recent research and patent activity in this area has been directed towards the use of seismic generators having modulation control of the wave output. The controls operate on the intervals between the seismic pulses, the output frequency of the pulses and the shaping of the wave fronts of the pulses.

An apparatus developed by *R.A. Broding; U.S. Patent 4,020,919; May 3, 1977; assigned to Standard Oil Company (Indiana)* is a seismic signal generator which operates through the use of a modulated vacuum over the surface of the earth.

A chamber with an air exhaust and a vent is mounted on an inflated compliant vibration isolator (conveniently this could be an inflated tire tube) and degassed to provide an internal low pressure area. By modulating a flow control valve in the air vent both the coupling between the signal generator and the force applied to the earth's surface are varied. This provides for an extremely lightweight apparatus which requires neither a special reaction mass nor a special hold-down mass. Such an apparatus is shown in various views in Figure 61.

Figure 61: Standard Oil Company (Indiana) Lightweight Seismic Source

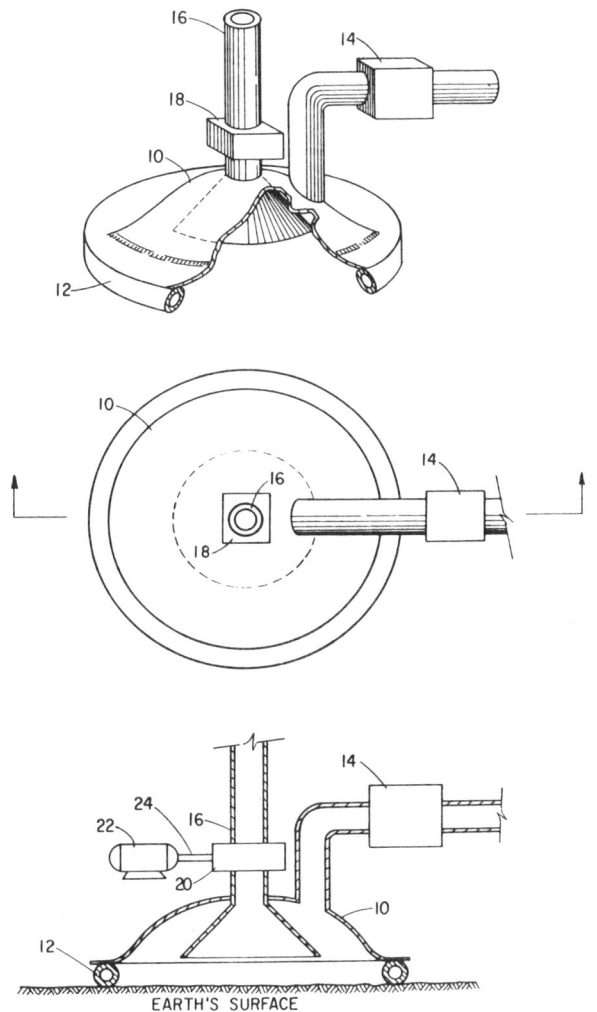

Source: U.S. Patent 4,020,919

Referring to Figure 61, the chamber **10** has a large opening adapted to be positioned adjacent to the earth's surface. A vibration-isolating means **12** is positioned around the large opening in contact with the chamber and the earth's surface. The vibration-isolating means is also compliant to generally form a seal between the chamber and the earth, thus functioning both to seal, as well as substantially reduce vibrations induced into the earth from the chamber's structure. An inflated toroid, such as an automobile tire innertube, conveniently provides a compliant, vibration-isolating means.

The air-exhausting means **14** is connected by piping to the chamber and at least periodically removes air from the chamber to provide a partial vacuum in the chamber. The venting means **16** is connected to the chamber to (at least periodically) allow air to enter the chamber. The flow-controlling means **18** is connected to either the air-exhausting means or the venting means, or both (it must be connected to at least one of them, and in the figure, the flow-controlling means is connected to the venting means), for modulating the partial vacuum in the chamber.

The connections are such as to allow passage of air and are typically, but not necessarily, by means of conventional piping. Preferably, the air-exhausting means, the venting means, and the flow-controlling means are positioned in very close proximity to the chamber to maximize the frequency response and the efficiency of the apparatus.

The figure also shows a plan view of the apparatus. While a circular configuration is shown in the figure, the chamber structure configuration is not critical and other configurations, such as oval or rectangular, could also be used. A vacuum pump can be used for the air-exhausting means, but many other types of exhaust blowers could also be used. A shop-type vacuum sweeper was used successfully in experiments and developed up to about ½ psi of vacuum in the chamber.

The bottom view in the figure shows a rotary valve **20** driven by motor **22** through shaft **24** as the flow-controlling means. The flow-controlling means is connected to control the flow of air through the venting means. The modulated partial vacuum could also be achieved by having a constant vent (in which case the venting means could be a hole in the chamber) and controlling the flow into the air-exhausting means.

In one example of the use of such an apparatus, a 30" diameter bicycle tire innertube was used as a compliant, vibration-isolating means. The chamber was formed by a flat plate on top of the innertube. A rotary valve was used and, with the rotary valve closed, the shop-type vacuum cleaner was capable of generating approximately ½ psi pressure reduction in the chamber. This gave a downward force of about 353 lb from the partial vacuum plus approximately 25 lb due to the weight of the apparatus. By rotating the air valve at various rates, seismic frequencies in the range of 1 to 100 Hz were generated.

Seismic detectors were placed on the surface of the ground and also in a hole approximately 100' deep. The downhole measured seismic energy was relatively high, indicating good coupling of the energy from the apparatus into the earth, and even this relatively unsophisticated apparatus gave efficiencies (signal power/ horsepower of input) as high as commercial mechanical seismic vibrators, and gave about 14 lb of peak-to-peak force/lb of apparatus (compared to less than 1 lb of peak-to-peak force/lb of apparatus for the commercial vibrator).

A device developed by *M. Barbier; U.S. Patent 4,011,924; March 15, 1977; assigned to Societe Nationale des Petroles d'Aquitaine, France* involves the use of a percussion mass for striking the ground with intermittent contact at a frequency of between 0.5 and 100 Hz in order to produce seismic waves for shallow terrestrial seismology. A rotary internal combustion motor drives the percussion mass and the frequency of impact is varied by adjusting the motor's throttle lever and transmission cable.

A scheme developed by *C.W. Dick; U.S. Patent 3,942,606; March 9, 1976; assigned to Western Geophysical Company of America* involves the use of elongated flexible wall tubes which serve a dual purpose in seismic exploration. First, they are connected to a pressure transducer, which monitors seismic waves which are reflected from substrata into the tubes. Second, the tubes are connected to a detonator which creates a linear seismic energy source by igniting a liquid explosive contained within the tube. During seismic prospecting the tubes are consecutively coupled to detectors to detect the explosion of other tubes, and then exploded sequentially.

An apparatus developed by *C.S. Cowles; U.S. Patent 3,974,476; August 10, 1976; assigned to Shell Oil Company* is a borehole acoustic wave generator having an acoustic transducer mounted at one end of a liquid filled tubular housing and an acoustic reflector, which will direct the acoustic waves 90° to the longitudinal axis of the housing at the other end. Two frequencies (4999.5 kHz and 5000.5 kHz) for energizing the transducer are supplied by separate oscillating circuits to produce by parametric generation, a directional acoustic pulse of the difference of the two frequencies (1 kHz).

Similar devices have been used to provide a map of the prominent characteristics of the well bore formations, e.g., the locations of fractures in the rock formations surrounding a borehole. In this type of mapping device the amplitude of the reflected signal is recorded in relation to the position of the tool in the borehole.

All of the above devices require the use of a highly directional sound source for producing acoustic waves. To obtain directionality one must use either an extremely large sound source or extremely high frequencies. The use of large sound source causes problems in the physical size of the device and limits its usefulness. The use of high frequencies reduces the penetration of the acoustic energy and, thus, limits the usefulness of the device.

The device of this process solves the above problems by relying on parametric generation to obtain a highly directional sound source of the desired frequency while using small size sources. Parametric generation depends upon the production in a liquid of two pressure waves having separate frequencies to produce a pressure wave having a frequency equal to the difference of the separate frequencies. For example, if one desires an acoustic wave of 1 kHz, one could use frequencies of 4999.5 kHz and 5000.5 kHz. The use of the parametric phenomena allows one to use smaller transducers while, at the same time, producing highly directional low frequency acoustic waves having good penetration. Such a device is shown in Figure 62.

Figure 62: Shell Oil Company Highly-Directional Acoustic Source for Use in Borehole Surveys

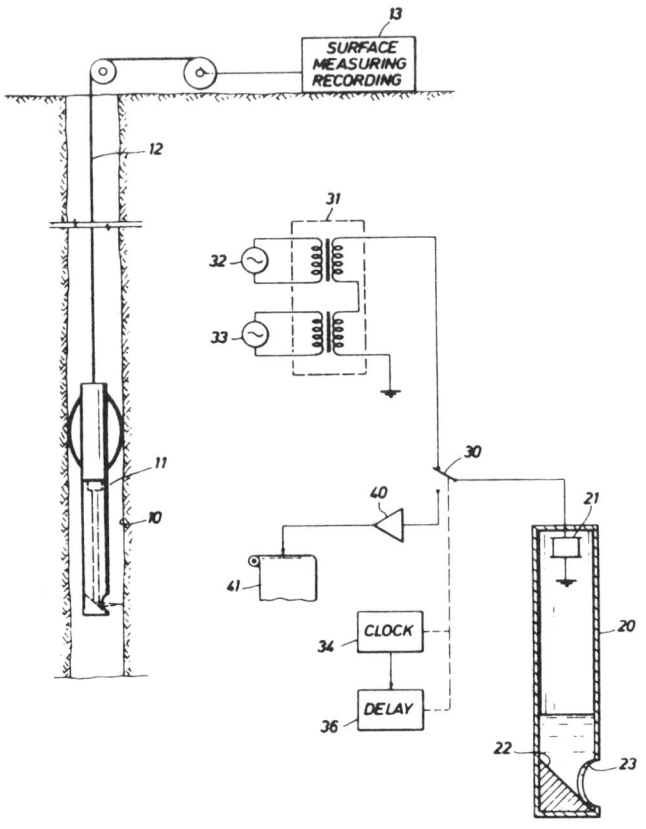

Source: U.S. Patent 3,974,476

There is shown a logging tool **11** suspended in a borehole **10**. The logging tool may be any type of acoustic tool that is designed to inspect the formation surrounding the borehole by means of directional acoustical waves. The logging tool is suspended in the borehole by means of a conventional logging cable **12** which is coupled to the surface recording equipment **13**.

Referring to the view at the lower right of the figure, there is shown the detailed construction of the sound source. The sound source comprises an elongated tubular housing **20** having a transducer **21** mounted on one end and an acoustical reflector **22** mounted in the opposite end. The transducer may be a conventional piezoelectric or magnetostrictive transducer while the acoustical reflector may be a block of some relatively inelastic material such as steel.

In addition, the acoustical reflector is provided with an inclined surface, e.g., a 45° surface to reflect the acoustic waves at right angles to the longitudinal axis of the housing. The reflected acoustic waves will pass out opening 23 in the housing and impinge upon the borehole wall while the wave reflected by the formation can pass through the housing and be reflected to the transducer 21. When the sound source is used in a logging tool the borehole fluid can be used to fill the interior of the elongated housing 20 and provide the next necessary liquid for the generation of the acoustic wave by parametric generation.

While the actual sizes and frequencies used will vary with the desired acoustic frequency, for normal well logging applications one could use a tranducer having a diameter of approximately 7.5 cm. This transducer could be energized with frequencies of 4999.5 kHz and 5000.5 kHz to produce a highly directional acoustic wave of 1 kHz. Also, the elongated tubular housing can have an outside diameter of approximately 8.5 cm and an overall length of 450 cm.

The above conditions and dimensions are applicable to a borehole logging tool where the tubular housing is filled with borehole fluid wherein the speed of sound is approximately 5,000 fps and the desired frequency of the acoustic pulse is 1 kHz. When the tubular housing is filled with a different liquid and one desires a different frequency for the acoustic pulse the dimensions of the housing and the transducer will change. As the designed frequency of the acoustic pulse increases the size of the transducer can be decreased.

Likewise, when the speed of sound in the fluid filling the tubular housing decreases the overall length of the housing can also decrease. Under most conditions, it will not be feasible to reduce the overall length of the housing to less than 2 times the wave length of the desired frequency of the acoustic pulse.

The transducer may be energized through the use of an adding circuit 31 and two oscillators 32 and 33. The adding circuit should be designed to add arithmetically the output of each oscillator and supply the resulting signal to a gate or switch 30. A gate 30 is used for switching the transducer between the adding circuit and a recording circuit. The gate is controlled by a clock 34 and delay circuit 36. Each clock pulse positions the gate to connect the transducer to the adding circuit while the delay circuit positions the gate to connect the transducer to the recording circuit.

The delay between the clock signal and the delay circuit signal should be long enough to produce a difference-frequency pulse and should be shorter than the shortest time between the production of a pulse and the arrival of the first reflection at the transducer. The time period between clock pulses should be long enough to generate an acoustic wave and receive all reflections of interest. The recording circuit may consist of an amplifier 40 and a recorder 41.

A system developed by *C. Baird and W.B. Plum; U.S. Patent 3,938,072; Feb. 10, 1976* is a seismic wave exploration system which uses a collimated energy beam and a seismic wave generator that is also the receiving transducer. A simulated borehole is created through a steel baseplate (16" diameter, 4" wide) which is powered by 20 driving piezoelectric crystals, each rated at 100 W, and arranged in a circle on the upper surface of the plate. When the piezoelectric crystals are energized a collimated acoustical wave is transmitted into the earth until a response is established in the driving transducers. The resonance conditions are indicative of a subsurface strata structure and are monitored by observing and recording the resonance effect on the power supply to the drivers.

An apparatus developed by *D.J. Dowling and J.F. Boyd; U.S. Patent 3,896,413; July 22, 1975; assigned to Texaco Inc.* is an electromagnetic acoustic wave generator for measuring porosity in earth formations adjacent a borehole. The generator consists of a sonde having dual magnetic poles (north and south) between which a current electrode is situated. The fields which are generated by the magnet and electrode create localized increases in pressure in the fluids which are located in the interstices of the formation, thus causing the fluids to flow. This fluid flow creates sound waves by fluid interaction with capillary side walls and surrounding particles which define the capillaries holding the fluid. The noise level and frequency content of the acoustic waves generated are indicative of the formation's permeability. Such a system is illustrated in Figure 63.

Figure 63: Texaco Inc. Permeability Logging Method with Acoustic Listening Devices to Detect Pressure Related Noise

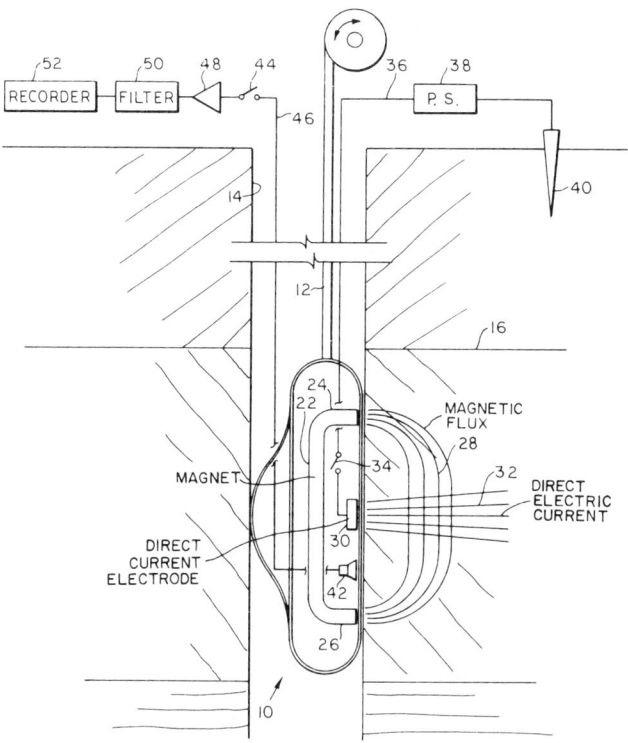

Source: U.S. Patent 3,896,413

In the drawing, a sonde **10** is shown on a cable **12** in a well bore **14** adjacent to a formation of interest **16**. The apparatus provides a measure of the permeability of the formation **16**. The cable **12** extends to a drum at the surface which is rotated to raise or lower the sonde **10** in the well. The cable suspends the sonde in the well. The sonde is pushed to one side by a decentralizing spring. The spring forces the sonde against the formation and contacts certain terminals or electrodes against the formation to obtain better contact.

The sonde incorporates a magnet **22** which includes a south pole **24** and a north pole **26**. The poles are on the wall of the sonde and adapted to be contacted directly against the adjacent formations. The magnet is a powerful magnet which forms magnetic lines of flux in the formation. The numeral **28** identifies a typical flux line which extends approximately parallel to the borehole **14** adjacent to the sonde in the formation. The flux lines pass through the formation of interest, and have a length determined by the spacing of the poles. The magnet is substantially powerful, and may be by way of example an electromagnet.

The apparatus further includes an electrode **30**, which is a current injection electrode. Current flow lines such as the path **32** are formed in the formation. A complete circuit for the current injection electrode is provided. The electrode is connected by means of a switch **34** to a conductor **36** which extends to a power supply **38**. The power supply is also connected to a return electrode **40**. The return electrode may take any form. One suitable form is a ground stake which is driven into the surface adjacent to the formation. The return electrode may also be located on the cable **36** some distance above the sonde.

It is helpful to locate the two electrodes relative to the well **14** such that the current flow, at least in the near vicinity of the sonde, is approximately perpendicular to the magnetic lines of flux. The relative length of the magnetic flux path **28** is measured only in centimeters while the electric current flow path **32** may measure several thousand meters. With these relative scale factors in view, it will be appreciated and understood that the intersection of the proposed field lines **28** and **32** is not precisely perpendicular.

The apparatus further includes a listening device **42** which is connected through a switch **44** by means of a conductor **46** extending to an amplifier **48**. A filter **50** for selecting frequencies of interest is connected in the system. The amplified signal is recorded by means of a recorder **52**.

The apparatus functions in the following manner to obtain a measure of formation permeability. The sonde is lowered to a point adjacent to the formation of interest. The magnet forms flux lines in the formation. They will penetrate only a few centimeters deep into the formation but when the magnetic poles are positioned against the side of the well, the flux lines are formed in the formation. The flux lines at the central portions are approximately parallel to the well bore. The switch is closed, completing an electric circuit between the electrodes **30** and **40**. Electrode **30** establishes a flow path in the formation. The current flow path is approximately perpendicular to the magnetic flux lines in the formation.

The two fields interact with fluids in the formation. The two fields create localized increases in pressure in formation fluids, forcing the fluid to flow in a random direction and with random velocities. The fluid is captive in the interstitial spaces of the formation. The spaces form randomly oriented, randomly shaped serpentine capillaries or conduits. It is not particularly important which way the fluid will flow. The significant thing is that the fluid tends to flow, even microscopically, to interact with the side walls and the surrounding particles which define the capillaries holding the fluid and thereby creates noise with the flow. A substantially large and smooth walled conduit will create little noise.

By contrast, a formation where the capillaries may be described as more serpentine or irregular in shape will create acoustic noise. Thus, a formation with relatively poor permeability will create more noise than a formation which is highly permeable. The noise is heard by the listening device **42** and supplied through an amplification system filtering and recorded. The noise levels and frequency content are related to formation permeability. A standard for a formation of known permeability is established whereupon the logging device and system can then be used for formations of unknown permeability.

A device developed by *J.V. Bouyoucos; U.S. Patent 4,147,228; April 3, 1979; assigned to Hydroacoustics Inc.* is one for generating and transmitting seismic signals which are force pulses shaped to have a spectrum constrained to the range of frequencies which are necessary for penetration to desired depths within the earth and for resolution of the geological reflection surfaces therein.

These pulses are provided in a nonrepetitive or aperiodic train, constructed to produce a transmitted energy spectrum whose mean energy extends smoothly at a substantially constant level over the spectrum frequency range, notwithstanding that the repetition frequency of the pulses may be swept over a frequency band much narrower than the spectrum range. The transmitted spectrum can exhibit an auto-correlation function having a major lobe which is predominant over any side lobes, corresponding to a desired level of resolution of the geological reflection surfaces.

Miscellaneous Marine Seismic Wave Generators

This category of devices includes underwater generators of seismic waves which are used in seismic exploration and which operate by physical principles other than those of vibrators, implosives, solid or gas explosives and air guns. These generators cover a broad spectrum of acoustic wave generation, such as by mechanical impact, electromagnetic, piezoelectric, magnetostrictive, water hammer and laser heating.

Recent research and patent activity in this area has been directed towards directional seismic wave generators which can be used to mitigate the effect of received signal noise. Typical systems utilize plural electromagnetic drivers with lobe design to achieve the desired directivity and passbands. Research and patent activity has also been directed to the use of optical frequency electromagnetic wave sources, particularly laser generated light for seismic wave generation.

A scheme developed by *N.A. Anstey; U.S. Patent 4,006,795; February 8, 1977; assigned to Seiscom Delta, Inc.* utilizes a broad seismic wave source adapted for reconnaissance and use in oil tankers of the 50,000 to 100,000 ton class. A motor driven shaft drives a group of weights against an anvil, acting as a flail, to transmit the shockwaves to the plates of the ship. The bandwidth of the source is modulated by changing either the material of the weight, the material of the anvil, or the mass and rigidity of the anvil plate combination. Such a scheme is shown diagrammatically in Figure 64 which shows the tanker (top view), the detector array (center view) and the seismic source (bottom view).

Figure 64: Seiscom Delta Inc. Scheme for Seismic Prospecting from Bulk Liquid Carriers

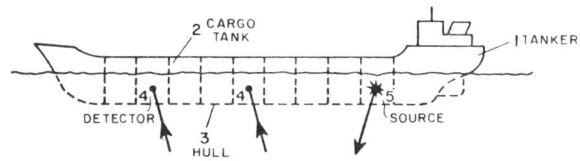

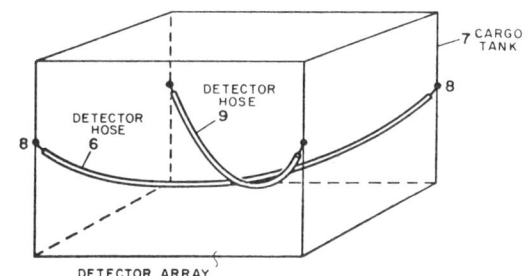

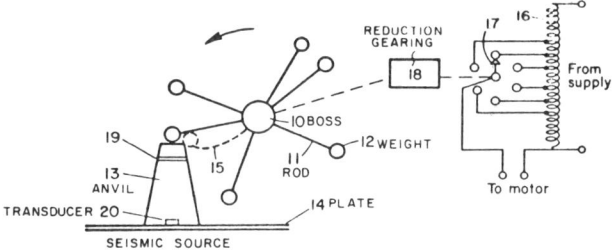

Source: U.S. Patent 4,006,795

As shown, a typical tanker in the 50,000 to 100,000 ton class is illustrated at 1. A typical tanker in the specified class would be 250 m in length and 30 m in the beam, capable of making 17 knots for protracted periods. Of the length of such a ship or vessel, 200 m would be occupied by a plurality of cargo tanks 2 for bulk transports of oil, hydrocarbon, or other liquid. A typical vessel of the specified class has 30 of these tanks, arranged in 3 rows of 10; thus each tank would be 20 m in length, with the center tank of the 3 tanks being 13 m wide and the port and starboard tanks being rather less wide. Each tank **2** is typically 16 m deep, representing 13 m of loaded draught and 3 m of freeboard. The hull of the tanker **1** in the tank is a single thickness **3** of steel, supported by ribs every 1½ to 2 m or so. Of the tanks, several are maintained ordinarily as ballast tanks filled with salt water, while the remainder carry cargo liquid. These salt water ballast tanks, therefore, are acoustically matched to the seawater, but are separated therefrom by 20 to 25 mm of reinforced sheet steel.

The acoustic match between the ballast water and seawater and the other factors set forth above permit installation of seismic detectors in the ballast tanks of the vessel. It should be understood that the seismic detectors may be installed in the oil tanks also, because of similar acoustic properties of oil and seawater. However, electrical connections to transducers within the oil tanks may be unacceptable for safety considerations if the tanks contained oil. Detectors could be used in these tanks, though, during the return journey, when the tanks ordinarily contain saltwater.

It is preferable to use as many tanks as possible due to the desirability of recording with a spread of detectors, and of arranging in-line and transverse arrays of detectors for the improvement of the signal-to-noise ratio of the returned seismic signal. Thus the dimensions of a typical tanker of the specified class allow a spread of 5 detecting stations, each representing an array 50 m long and 30 m wide. In other systems on other vessels, the array employed depends on the number of tanks which are permitted to be used. Two detecting stations **4** are illustrated in the tanks.

In each such tank, an array or subarray of detectors is disposed in a manner to minimize the effect of structure-borne noise in the tanker itself, of flow noise along the hull of the tanker, and of resonant systems established within the tank. A suitable disposition of detectors in a tank **7** is shown in the center view of the figure. The detectors are piezoelectric elements of conventional tube or blender type. A plurality of such detectors, typically 20, are arranged at suitable intervals within an oil-filled hose **6**, to form what is known in the art as an "eel." The density of the eel **6** is adjusted to be only slightly greater than that of the liquid in the tank.

The eel, of approximately 20 m in length, is supported between two suspension points **8** on the diagonally-opposite vertical edges of the tank some 8 m from the bottom of the tank; thus it follows an approximate catenary, of which the lowest point is made some 3 m above the bottom of the tank. The detectors in the eel are thus spaced over that part of the catenary between 3 and 8 m from the bottom of the tank. A second eel **9**, of the same configuration as the eel **6**, is strung similarly across the other diagonal of the tank. Thus a cruciform three-dimensional array of detectors is obtained, having longitudinal and transverse dimensions approaching those of the tank, and a vertical dimension (of 5 m) arranged at a suitable distance below the surface of the fluid in the tank.

It should be understood that other systems of arranging the detectors within the tank may be devised so long as an array well suited to the reception of seismic waves from below is provided, while tending to reduce the contribution from flow noise along the hull, structure-borne noise in the hull, sea-borne noise from the screws, surface waves on the liquid in the tank 7, and standing waves within the tank.

Ordinarily the electrical signals from the eels **6** and **9** are combined to form a single detector output. Similarly the outputs from detector arrays in several tanks may be combined to yield an effectively larger array. If the number of usable tanks allows, a spread may be established by the use of several such arrays along the length of the tank section of the ship **1**.

The electrical outputs from the array or arrays are furnished to suitable recording instruments. These may be of standard type, recording the digitized output from gain-controlled amplifiers onto magnetic tape. The operation of such amplifiers is well known in the art, except for modification of the setting of the frequency bands for the filters in the amplifiers. The lowcut filter is set to attenuate to an acceptable degree the cyclic noise generated by the ship's screw.

For a 5-bladed propeller and a typical shaft speed of 115 to 120 rpm, this means providing significant attenuation at 10 Hz. The setting of the high cut filter may be used to attenuate any tank resonances not sufficiently treated by the array dimensions. With the tank dimensions set forth above, the available frequency spectrum after setting the filter frequency bands remaining for utilization is about 15 to 50 Hz, which is well suited to seismic surveying.

As mentioned previously, the seismic survey end product from the arrays of detectors is a digital tape, which is later processed (into a form representing a cross section of the earth) by computer techniques well known in the art. A rudimentary form of cross-sectional display may also be provided, as is usual, on shipboard to allow immediate monitoring of the quality of the results while the survey is performed.

A first type of source for the detecting system previously described may be a conventional type of source: sparker, boomer, air gun, gas gun, vibrator or any other, towed in the open water at the side of or behind the ship.

An alternative source is available if the ship is propelled by steam turbines. In this case the steam supply may be used to drive any type of steam sound source. Such sources rely on the rapid generation of a bubble of steam in the water; the steam then condenses, and the bubble disappears without oscillation. The bubble may be formed under the ship, or at the end of a trailed hose.

In a preferred embodiment, the location of the source involves its installation within the ship. This may take the form of either of two alternatives. The first is obtained by mounting one or more conventional water-displacement sources within one or more of the tanks (as illustrated generally as **5**). The second is obtained by arranging one or more mechanical impact devices as shown in the bottom view of the figure, to act substantially directly on the inside of the hull **3** of the ship.

A conventional impulse source may be used within a fluid-filled tank of the ship with few problems. As mentioned above, considerations of safety impose a preference for air guns or other sources not involving high voltages or combustible gases. One or more of such sources may then be suspended within a tank, with the supply hoses being conducted through one of the tank covers to the supply equipment mounted conveniently elsewhere on the vessel **1**. Typically the tank or tanks **2** employed for this purpose would be toward the after end of the tank section of the ship (as shown), and the detecting tank or tanks would be toward the forward end to minimize mechanical noise at the detectors, while allowing compressor or other generating equipment associated with the source to be located in or near the ship's engine room.

Although air guns of modest power can be discharged within a few meters of the plates of a ship without adverse effects, there is a power limit which it is undesirable to exceed. If adequate seismic penetration from the source cannot be obtained within this power limit, this scheme makes use of a pulse-compression system. Such systems, as is well known in the art, allow the replacement of the large peak power and short duration of an impulse source by the use of a signal having a smaller peak power and longer duration; thus they provide greater radiated energy (and, therefore, better signal-to-noise ratio after optimum processing) without requiring offensive or dangerous peak power levels. In this scheme, therefore, a pulse-compression system allows good seismic penetration without excessive displacement of the ship's plates.

Another seismic source for use in this scheme involves the installation of a standard marine vibrator in an appropriate tank. In this system, the vibrator is driven by a quasi-sinusoidal control signal whose frequency sweeps from a lower frequency to a higher frequency (or the reverse) while forming the signal; the signal then repeats after a time greater than the greatest reflection time of interest. The control signal is generated by a digital function generator, or from magnetic memory, and the final reflection trace is obtained by cross-correlating the received signal against the control signal. This final reflection trace can offer the same seismic penetration as one obtained from an impulse source necessitating typically ten times as much displacement of the ship's plates.

The vibrator may be suspended from the top or sides of the tank, or from a moored buoy within the tank. The depth of operation of the vibrator is preferably in the range of 6 to 10 m below the surface of the liquid in the tank.

When the source and the detectors are contained within separated tanks, acoustic insulation may be provided in the tanks between them; this is of particular value with a pulse-compressive system, in that it reduces the amplitude of the direct signal (in whose presence the small reflection signal must be detected). This benefit may be obtained without significant loss of the payload or the ballast effect of the tanks, by arranging a curtain of air bubbles within the intervening tanks. One convenient way of doing this is by lining the walls of the tanks by closed-cell expanded rubber or the like, suitably protected to resist the chemical action of the liquid in the tank. The same technique may be used to reduce other unwanted acoustic transmission.

Another suitable seismic source is shown in the bottom view of the figure. To a revolving boss **10** are secured one or more flexible rods **11**, each of which carries at its outer end a weight **12**. The boss is mounted on a shaft and driven by a motor (not shown) through suitable reduction gearing (not shown). In motion, therefore, the apparatus is the general form of a flail. An anvil **13** is placed to receive the impact of the rotating weights, and to transmit the resultant impulses to the plates **14** of the ship. The flexibility of the rods is such that the apparatus may continue to rotate after impact, despite the fact that each ball **12** is momentarily checked; further rotation of the boss **10** allowed by this flexibility pulls the weight off the anvil **13** as shown in phantom at **15**.

The rods **11** are mounted in the boss at irregular intervals, so that the sequence of impacts of the weight **12** with the anvil within one revolution of the boss is random. In a typical case there are a set of six of the weights in one plane, as shown in the figure, and a further set of six weights, also randomly located, in a second plane (not shown) behind and adjacent to the first set. Thus a succession of twelve randomly-timed impulses is produced by each revolution of the boss. The dimensions of the apparatus are typically as follows.

Radius to point of impact	1 m
Diameter of weight	130 mm
Weight of each	10 kg
Separation between flail planes	150 mm
Impact plate of anvil	80 x 250 mm
Height of anvil	700 mm
Mean period of revolution	1 sec
Mean time between impacts	83 msec
Weight velocity on impact	6.3 m/sec

Calculations based on these figures show that the impact energy released by this source is equal to that of a conventional weight-drop truck operating every 30 sec.

In the basic form described above, steady rotation of the boss produces a series of impacts which repeats every 1 sec. Since the energy released is ordinarily sufficient to secure reflections at greater travel-times than this, it is desirable to increase the period of the repetition without reducing the mean energy of the impact. This may be done, for example, by driving the motor from a tapped auto-transformer **16** where the tap in use is selected by a rotating arm **17** driven from the main shaft through reduction gearing **18**. Thus, if there are ten taps and the reduction gearing is 10:1, the motor is driven at different speeds for ten successive revolutions, and then the cycle repeats. This allows without ambiguity the recording of reflections having a travel-time of 10 sec.

The taps may be connected in regular order, as shown in the figure, or in an irregular manner. The combination of the number and arrangement of the taps, the gearing, the number of weights, the moment of inertia of the system, and the spacing of the arms **11** may be adjusted to produce a suitably random signal; the definition of such signals from auto-correlation functions having an acceptable side-lobe level being well known in the art.

The effective bandwidth of this seismic source is defined by the material of the weights, the material of the anvil, the mass and rigidity of the anvil-plate combination, and the seawater. It is ordinarily beneficial to mount the anvil in a large unsupported area of the ship's plates **3**; the plate resonance may be utilized (by mounting the anvil symmetrically between ribs) or minimized (by mounting it unsymmetrically). Additional compliance of the system to desired signal characteristics may be inserted, if desired, by including a layer of appropriate material in the anvil at **19**.

The use of a continuous seismic source of this nature requires the recording of a facsimile of the outgoing acoustic signal. This is conveniently derived from a conventional transducer or motion detector mounted on the ship's plates at 20, or from a small array of such detectors spaced around anvil 13.

A device developed by *R. Hutchins, T.W. Orton and N. Thai; U.S. Patent 3,993,973; November 23, 1976; assigned to Huntec (70) Limited, Canada* is a transient sound generator for producing uniform pulses from a piston plate outwardly repelled into the water by eddy currents induced in the plate by an electromagnetic coil. The pressure at the rear face of the piston, tracks the pressure at the piston front face through the use of a rear compressible air volume which is connected by a bleeder aperture to a noncompressible air volume at the rear face of the piston.

An apparatus described by *J.E. Martin and G. Zilinskas; U.S. Patent 3,992,693; November 16, 1976; assigned to The Bendix Corporation* is a ring shaped seismic wave generator formed of plural segments of an alternately placed active material, e.g., barium titanate, and an inactive material, alumina. The velocity of sound in the inactive material is different from that of the active material. The generator rings are positioned axially along a central support tube and are separated by neoprene spacers.

A device developed by *F.R. Abbott; U.S. Patent 3,990,034; November 2, 1976; assigned to the U.S. Secretary of the Navy* is a cylindrical, elongated, towed, transverse variable reluctance acoustic energy transducer having an axial tunnel running its length through which water flows for cooling the transducer. The transducer operates in a magnetomotive mode wherein magnetically interacting fields of flux, generated by alternating currents and direct currents, cause radial displacment of projecting surfaces. The magnetic poles consist of continuously extending helical strips.

A schematic of the use of such a device is shown in Figure 65. In the drawing a transducer 10 is being towed through the water by a surface craft 11 via a towing cable 12. The towing cable is joined to a yoke 13 connected at diametrically opposed projections forward of the transducer's centroid. A drogue chute 14 is connected to the trailing end of the transducer and serves to orient the transducer in the direction that the surface craft is moving.

Figure 65: U.S. Navy Design of Towable Sonar Projector for Seismic Surveying

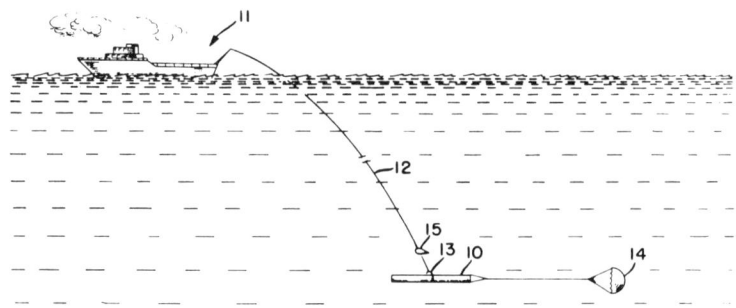

Source: U.S. Patent 3,990,034

While a surface craft having a relatively short length of towing cable is depicted, submersibles or great lengths of towing cables optionally may be used to fit the particular task at hand. Irrespective of what towing mode is chosen, it has been found expedient, however, to couple the towing cable via the towing yoke as shown and connecting a drogue chute to prevent restricting the flow of cooling water through and around the transducer. Additionally, the radiated acoustic energy is not baffled as it emanates from the opposite ends of the transducer.

An ac power supply is carried onboard the surface craft and power lines reached down to the transducer. Because, usually, there are less losses associated with the transmission of high voltage ac power, a transformer and a full wave rectifier circuit may be carried in a faring 15 located near the transducer. From the rectifier circuit dc as well as ac power is fed to the active element in the transducer, a magnetomotor. The magnitude of the power delivered to the motor has approached a 10kW level.

A device developed by *J. Cassand and J.-P. Fail; U.S. Patent 3,949,831; April 13, 1976; assigned to Institut Francaise de Petrole, des Carburants et Lubrifiants, France* is a seismic wave generator which operates by abruptly braking a mass of water in motion. A hollow enclosure is provided with two liquid containing hollow sections interconnected by a narrow passage. The seal between the two sections is intermittently closed so as to create, by hydraulic or electromagnetic drive, a fluid induced pressure differential between the sections. The resultant water hammer effect creates acoustic waves which are used for seismic exploration.

A system developed by *J.V. Bouyoucos; U.S. Patent 3,946,831; March 30, 1976; assigned to Hydroacoustics, Inc.* is a directional acoustic wave transmitter system which contains a seismic wave generator and a wave transmission line. Ports are provided along the length of the line and are surrounded by antiresonant coupler sections which present a high acoustic driving point impedance to their respective ports. It is asserted that the ports evenly partition the seismic waves produced by the generator and concomitantly define an array of equal strength directional radiators.

A scheme developed by *P.J. Westervelt, R.S. Larson and T.G. Muir; U.S. Patent 3,913,060; October 14, 1975; assigned to the U.S. Secretary of the Navy* involves developing coherent acoustic beams in fluid mediums. The beam of a high power laser is directed by a beam splitter, in two separate directions in fluid, and is absorbed by the fluid. The fluid converts the absorbed optical energy to acoustical energy by expansion along the absorption path so that the resultant acoustic beams are perpendicular to the paths of the laser beams.

SEISMIC DATA RECEPTION AND PROCESSING

It is well-known to those skilled in the art that the complex of vibrations received at a given seismic detector array do not consist totally of waves reflected from subsurface boundaries. Rather, the array also detects unwanted random seismic events, as well as various high amplitude modes of spatially-coherent source-generated seismic events whose principal direction of propagation is horizontal, i.e., along and near the free surface. It is essential that the effects of these unwanted horizontally-propagated waves be reduced by utilizing the principles of a directional antenna to attenuate the magnitude of the electrical signal produced by the geophones in response to these waves.

In U.S. Patent 2,698,927 to Parr, there is disclosed a method of reducing the effects of the coherent horizontally-propagated seismic waves. This method comprises assigning relative sensitivity values to geophones in an array. The sensitivity values are selected according to recognized antenna theory so as to reduce the magnitude of the electrical signal produced in response to the unwanted spatially-coherent seismic waves.

Parr refers to his method as a "tapered sensitivity" method, since the sensitivity of the sensing devices in a given array is reduced toward either end of the array from a central point when the transducers are aligned radial to the energy sources used to generate the seismic signal. A combination of the spacing of the individual sensors, the length of the geophone array, and the wavelength bandwidths of the interference to be attenuated comprise the criteria for assigning the relative sensitivity to be employed at each geophone.

A good general review of the weighting of seismometer arrays is given by Parr and Mayne in *Geophysics*, Vol 20, pages 539-564 (1955), and Holzman, in *Geophysics*, Vol 28 (1963), discloses that the optimum attenuation of the effects of coherent horizontally-propagated seismic waves may be achieved by applying Chebychev weighting coefficients to the sensors in an array. The combined teachings of Parr, Parr and Mayne and Holzman are recognized standards for reducing the effects of the unwanted vibrational energy.

There have, however, been other proposals for reducing the amplitudes of the horizontally-propagated energy which is recorded. For example, in U.S. Patent 2,747,172 to Bayhi, two methods are disclosed for obtaining a tapered geophone array that is designed to have a response which attenuates the electrical signals produced in response to the unwanted vibrational energy. The first method involves constructing an array having a plurality of geophones at each location in the array. The number of geophones is maximum at the center point of the array and tapers off in the direction of the ends of the array.

The second method disclosed by Bayhi is to use a single geophone at each location of the array and to install a voltage divider network across each geophone in the array. The voltage divider network at each geophone consists of resistors, and the values of the resistors used are chosen so that the geophone in the physical center of the array has the greatest sensitivity, while the geophones at the end of the array have the least sensitivity. The weighted geophone array of Bayhi is apparently not bidirectional, and it appears that difficulty in maintaining a substantially constant damping factor between all geophones in the array will be encountered with the array of Bayhi.

Later, in U.S. Patent 3,096,846 to Savit, there is disclosed a method of determining the seismometer weights to be applied in array tapering by using a moveout criterion. The results of Savit's method is that the distance between individual seismic detectors in a given array may not be uniform and the sensitivity of the individual seismic detectors will vary according to the moveout criterion.

From practical considerations it has been found expedient to approximate a desired weighting by constructing an array having a plurality of seismic detectors at each location in the array (e.g., as according to Bayhi), with the number of seismic detectors at each location dictating the weighting coefficient of that location. Since it is generally agreed that the Chebychev coefficients are the optimal weights and since these coefficients are not integral numbers, the actual number of individual seismic detectors that would be required to implement (even approximately) these coefficients is very large. Hence, for this practical reason, Chebychev-weighted arrays have not generally been attempted nor realized.

Two recent patents disclose apparatus for applying Chebychev-weighted coefficients to the seismic detectors in an array. In U.S. Patent 3,863,200 to Miller, there is disclosed a built-in seismometer amplifier which permits the sensitivity of the individual seismometer to be adjusted at a given location. It will be noted from the Miller patent that a separate pair of wires is required to convey the signal generated at each seismometer back to a suitable recording point. Consequently, a multipair cable is required between the array of seismometers and the recording point in order to utilize the built-in seismometer amplifier that Miller discloses.

In U.S. Patent 3,863,201 to Briggs there is disclosed a seismometer weighting apparatus to apply weighting coefficients to individual seismometer signals at a recording point. Briggs states that the apparatus may be utilized with a uniformly weighted and uniformly spaced array. It will be noted, however, that in the Briggs scheme, a multipair cable is required between the recording point and each detector in the array.

The ideal detector array is one which (1) provides weighted sensitivity at the individual seismic detectors in the array, (2) maintains essentially constant damping between seismometer units, (3) substantially reduces the number of seismic detectors to achieve weighted sensitivity, (4) has bidirectional capabilities, and (5) supplies data from all seismic detectors to a given end of the array over a single signal-carrying medium.

Seismic Detectors

In the reflection seismic method, seismic waves or impulses are generated at a point at or near the earth's surface, and the compressional mode of these waves is reflected from subsurface acoustic impedance boundaries and detected by arrays of seismic detectors located at the surface. The seismic detectors convert the received waves into electrical signals which are sensed and recorded in a form which permits analysis. Skilled interpreters can discern from such an analysis the shape and depth of subsurface reflection boundaries and the likelihood of finding an accumulation of minerals, such as oil and gas.

It has been found that better and more accurate determinations of the profile of subsurface structures can be obtained using detectors which are directionally sensitive. The directionally sensitive detectors are insensitive to components of the seismic waves, except the component in a particular direction. For example, detectors may be used which are sensitive only to the component of the seismic wave impinging the detector in a line parallel to the longitudinal axis of the detector.

Attempts to orientate the directionally sensitive detector in the proper direction have required complicated servo arrangements for orientating the detector in any desired direction and have also required complicated means for recording at any given moment the direction in which the detector is orientated as it is lowered within a borehole.

A technique has been developed by *R.H. Schmuck; U.S. Patent 2,959,240; November 8, 1960; assigned to Jersey Production Research Company* which provides the art with an improved method for accurately determining the profile of subsurface structures and includes a detector system with at least one directionally sensitive detector whose axis may be orientated as desired. In its broader aspects the method consists of first producing seismic waves. The seismic waves are detected by at least one directionally sensitive detector assembly. The directionally sensitive detector assembly is adapted to automatically orientate the sensitive axis of the detector, or detectors, in the assembly to receive the desired component of the seismic waves.

An apparatus developed by *E.M. Hall, Jr.; U.S. Patent 4,125,823; November 14, 1978; assigned to Western Geophysical Co. of America* is a seismic transducer which comprises a hollow, elongated casing having a bore defining a chamber. An opening extends through the wall of the casing into the chamber. A hydrophone is removably disposed inside the chamber. The hydrophone comprises an elastomer body in a portion of which is embedded a rigid cage marginally supporting a pressure detector assembly. Means are coupled to the casing for exerting a static pressure on the elastomer body. The cage has a geometric configuration adapted to substantially shield the detector from the static pressure which causes the end portions of the elastomer body to expand outwardly, whereby each end portion forms a fluid-tight seal against the wall of the chamber.

A device developed by *D.E. Miller; U.S. Patent 4,126,203; November 21, 1978; assigned to Continental Oil Company* provides for improved detection of reflected seismic energy through detection of air pressure variations in air above an air/earth surface interface. One form of apparatus suitable for detection in air of seismic energy waves emanating from beneath the earth surface consists of supporting a collecting reflecting member over a designated earth surface site and detecting air pressure variations at a focal point within the reflector member; and, thereafter, transmitting or conducting the reflected seismic energy indications for processing in accordance with the seismic energy source requirements to obtain seismic energy reflection data for the particular locale.

Figure 66 shows the overall arrangement of a seismic prospecting system using such a detector. It illustrates a seismic system **10** having a seismic source **12** disposed to impart input seismic energy through the earth's surface **14** into earth **16** for subsequent refraction, reflection and detection thereby to provide distinctive lithological data for geological interpretation.

Earth is shown as including a first substratum **18** and a second or deeper substratum **20**, and it is noted that seismic energy waves emanating along ray path **22** are refracted along substratum **18** via ray path **24** with subsequent return to the surface via ray path **26**, while energy along ray path **22** is also reflected from the interface of substratum **20** for surface return along ray path **26**.

Figure 66: Continental Oil Company System Employing Air-Coupled Seismic Detector

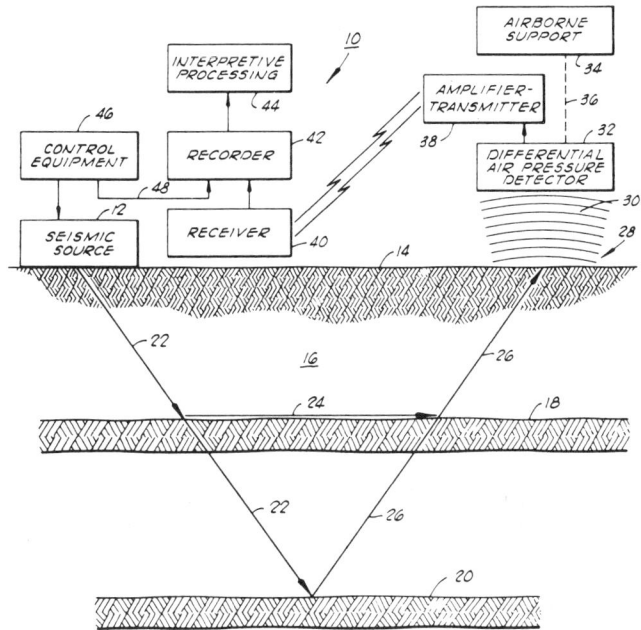

Source: U.S. Patent 4,126,203

While the seismic energy propagation is illustrated in simplified ray form, it is well understood by those skilled in the art that the seismic wave propagation is of a spherical character and omnidirectional as from its initial point source or seismic source **12**.

Seismic energy propagating along ray path **26** intersects earth surface **14** within a predesignated earth aperture **28** whereupon the energy is coupled upward in the form of air pressure waves **30**. The air pressure waves are then detected by an air pressure detector **32** which includes collecting and/or focusing structure as will be further described below. The air pressure detector is suitably supported by an airborne support **34** by means of a supporting link **36**.

The airborne support may be any suitable device such as a tethered balloon, dirigible, hover aircraft or the like. The air pressure detector may be any of various commercially available pressure-measuring elements. For example, while conventional surface energy detectors require measurement of vertical particle velocities as small as 10^{-4} cm/sec, the amplitude of the resulting pressure waves **30** in air are in the range of 10^{-3} to 10^{-2} dynes/cm^2 which can be easily detected by commercially available pressure responsive transducers.

232 Geophysical and Geochemical Techniques for Exploration

The detected pressure variations in the form of electrical output may then be transmitted or conducted to a surface location for further recording, processing and the like. The output electrical signal or signals from air pressure detector **32** are applied to a conventional form of amplifier-transmitter **38** which then transmits utilizing telemetric modes to a ground receiver **40**. Ground receiver **40** then preamplifies the received signal for output to a conventional form of field recorder **42** wherein the seismic data is available for further interpretive processing in stages **44**.

Control equipment **46** provides basic signal generation and control for seismic source **12** and, in the event that vibratory or similar seismic energy input is utilized, a replica control signal may be applied to the control equipment via lead **48** to the recorder. This then enables subsequent received signal processing which includes normalization, correlation and other well-known techniques which provide optimum presentation of the finally processed seismic data.

Figure 67 shows the air-coupled detector **50** in more detail. In this case, airborne support is provided by a balloon **52** maintained in proper position by means of a requisite one or more tethering ropes **54** which may be controllable from earth surface **14**, by truck-mounted winch or the like, to maintain and adjust position of the balloon. A plurality of hanger ropes **56** are then affixed about the horizontal circumference of the balloon and attached at the lower end to suspend seismic detector **58** in desired attitude.

Figure 67: Detail of Continental Oil Company Air-Coupled Seismic Detector

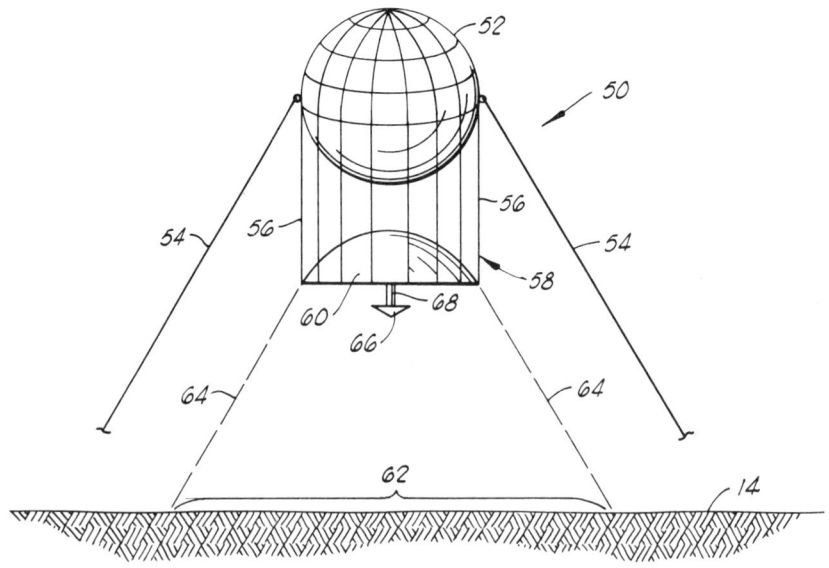

Source: U.S. Patent 4,126,203

Seismic detector **58** is a parabolic acoustic energy reflector **60** as may be constructed of any lightweight, rigid, acoustically reflective material such as aluminum or fiber glass, and is suspended to open downward thereby to define an effective aperture **62** as defined by the paraboloid extension lines **64**. Thus, it may be noted that the effective aperture or area of energy detection may be adjusted as desired, and as dictated by frequency response, in accordance with raising or lowering of balloon **52** and the suspended seismic detector.

A pressure detecting sensor **66** is suspended by means of a support bar **68** to be positioned in the focal point of the reflector in such a manner that it is responsive to all pressure variations emanating from the aperture. The pressure detecting sensor may be any device or cluster of devices which is sufficiently sensitive in the seismic band of frequencies, preferably from 5 to 500 Hz, and is capable of providing a few μV output for the pressure variations in the range of 10^{-3} to 10^{-2} dynes/cm^2.

The diameter across the lower or open end of reflector **60** may vary from a few meters up to tens of meters, depending upon the degree of directionality, i.e., the effective cone of detection, which is desired for a particular band of frequencies. The lower the frequencies, the longer the wavelength and the greater the diameter necessary to obtain a given directionality.

An apparatus described by *C.S. Cowles; U.S. Patent 4,134,097; January 9, 1979; assigned to Shell Oil Company* provides a pressure-sensitive seismic detector, such as hydrophone, in combination with a velocity-sensitive seismic detector, such as a geophone, to measure the seismic waves resulting from seismic disturbances.

This particular device incorporates the geophone-hydrophone combination in a single housing. In particular, the geophone is rigidly mounted in the housing so that the velocity of the housing will be detected by the geophone while the hydrophone is mounted so that the housing may be buried in a solid material. The hydrophone is mounted in a section of the housing having a flexible diaphragm which can be exposed to contact with the material used for burying the housing; thus, the hydrophone responds to pressure in the earth produced by seismic waves.

In addition, the housing includes electronics so that the hydrophone can be made the electrical equivalent of the geophone. By converting the hydrophone to the electrical equivalent of the geophone, the hydrophone will have the same response characteristics to pressure as the geophone has to velocity. Thus, one can easily combine the signals, either by adding them or subtracting them, to detect the presence of up- and down-traveling seismic waves. The electronic circuit consists of a transformer so designed that its equivalent circuit, when combined with that of the hydrophone, produces a circuit which is the electrical equivalent of the geophone.

In typical land seismic exploration operations, the equipment employed uses multiple conductor cables that are extremely long. When they are spread out along the surface of the earth, they will extend for several thousand meters from end to end. The cables which are employed are constructed in sections for portability and the sections are terminated using electrical connectors. It has been found that during use in the field operations, there are discontinuities in the cables which would develop due to such things as rodent and livestock bites, as well as vandalism and vehicle traffic, in addition to old age of the cables.

Since a seismic field operation involves a number of personnel that are skilled and professional workers, the cost of making seismic surveys is quite high. Consequently, it is important to have the operation continue with dispatch, and to minimize the idle time caused by malfunction such as discontinuities in the cable of the type just mentioned. It has been necessary for a worker to walk along the cable and carry a radio for communication with the recording station. Then, an intentional short circuit would be made, while the circuits were monitored from the recording station, in order to locate the position of a discontinuity.

A scheme developed by *F.L. Lankford, Jr.; U.S. Patent 4,134,099; January 9, 1979; assigned to Texaco, Inc.* provides a system for quickly locating discontinuities in land seismic geophone cables, which cables have plural sections with electrical connectors at the ends of each section. One of the connectors of each section has a relay incorporated with it, and includes a circuit for connecting a short circuit between all of the plural geophone circuit conductors. The short circuit connection is made when the relay is actuated, and the relay is actuated upon command from a recording station.

A device developed by *W.O. McNeel; U.S. Patent 4,144,520; March 13, 1979; assigned to Geo Space Corporation* overcomes certain of the disadvantages of the prior art vibration sensing transducers, seismometers, and geophones. One advantage of this device is that the geophone utilizes a coil form which is resiliently supported for a movement in a direction substantially parallel to the axis of the permanent magnet. The movement of the coil form is responsive to a sensed vibration. The coil form has at one end thereof a passageway for supporting a coil in spaced axial alignment to the permanent magnet.

The other end of the coil form has a band of highly conductive metal positioned therearound and in axial alignment with the permanent magnet. The band of metal is responsive to transversing the magnetic lines of force from a permanent magnet to generate an electromotive force and current therein which produces a magnetic field acting in a direction opposite to the direction of transverse of the coil form causing damping of movement of the coil form relative to the permanent magnet.

A system developed by *W.H. Mayne, A.S. Badger and W.S. Hawes; U.S. Patent 4,151,504; April 24, 1979; assigned to Geosource, Inc.* is one in which a seismic array has a plurality of seismic detector connection points, and a seismic detector is located at each seismic detector connection point. The outputs of the seismic detectors are electrically isolated from each other, and weighting may be applied to the output of each seismic detector in the array.

Two signal-carrying media are also provided in the array, and the weighted outputs of the seismic detectors are conveyed to the first end of the array over one signal-carrying medium and to the second end of the array over the second signal-carrying medium. A Chebychev-weighted array is achieved by a proper selection of weighting.

A device developed by *E.M. Hall, Jr.; U.S. Patent 4,159,464; June 26, 1979; assigned to Western Geophysical Co. of America* is a conventional electromagnetic geophone which includes a coil assembly that is spring-suspended in an air gap between concentric pole pieces of a magnet. The coil assembly is wound on the outer wall of a bobbin. The bobbin is typically made from or includes a metallic material to provide a mass for mechanical damping and a conductor for electrical eddy-current damping of the spring-suspended coil assembly.

The improvement embodied in this particular device lies in mounting on the bobbin, a vernier mechanical damping mass made of a coil of insulated wire. The coil is open-ended and the mechanical damping force is adjustable in accordance with the number of turns in the coil.

Such a device is shown in cross section in Figure 68. The geophone is generally designated as **10** and comprises a combined mass-coil assembly **12** concentrically mounted within a stationary magnet assembly **13**.

Figure 68: Western Geophysical Co. of America Geophone with Damping Coil

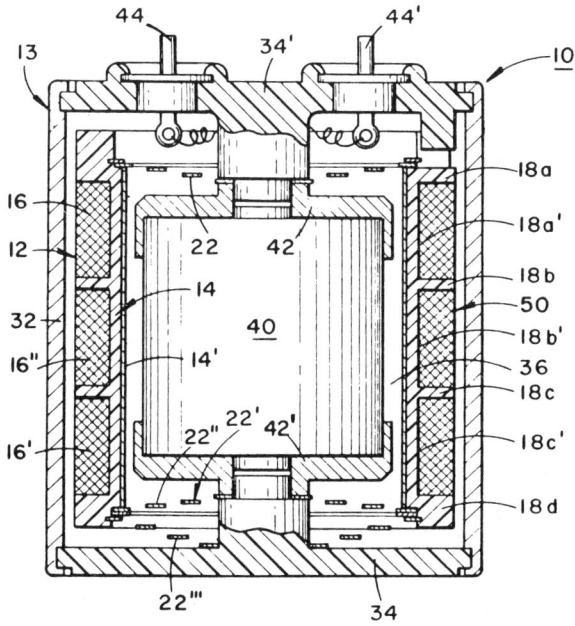

Source: U.S. Patent 4,159,464

The mass-coil assembly includes a unitary, generally-cylindrical bobbin having a thin wall provided with four radially-extending shoulders **18a** through **18d**. Shoulders **18a–b** and **18c–d** define therebetween two outer cradles **18a'** and **18c'**. Shoulders **18b** and **18c** define therebetween a center cradle **18b'**. Cradles **18a'** and **18c'** accommodate two induction coils **16** and **16'** connected in series. Each coil has a plurality of turns of fine insulated wire. The body of bobbin **14** is made from a suitable plastic material. End shoulders **18a** and **18d** support spring spiders **22** and **22'**, respectively. The mass-coil assembly **12** is resiliently suspended from the two springs. Spring **22'** is a compound spring having a flat portion **22"** and a conical portion **22'''**. This compound spring is substantially linear in tension and compression. The construction and arrangement of springs **22** and **22'** is more fully described in U.S. Patent 3,878,504.

The stationary, magnet assembly **13** includes a hollow, outer cylinder **32** which is closed at the top and bottom by insulating cover plates **34** and **34'**. In addition to rendering the inner volume **36** of case **32** fluid tight, cover plates serve as support and centering elements as well.

The magnetic field is created by a permanent magnet **40** provided with upper and lower pole pieces **42** and **42'**, each having a substantially Z-shaped cross section to secure and centrally maintain the magnet within the casing. A pair of outer terminal posts **44** and **44'** extend outwardly from the upper plate **34'** to allow the coils to be electrically connected to an outside utilization device, typically via a geophone cable. An inspection of the drawing will reveal that the various components of the mass-coil and magnet assemblies are press-fitted whenever possible and mounted inside the casing **32**.

In operation, the mass-coil assembly **12** is the inertial element of the geophone and is suspended by the springs in the cylindrical air gap **36** provided by the magnet assembly. The earth's vibration causes relative motion between the induction coils and the magnet assembly, thereby generating a voltage in the coils, proportional to particle velocity, which becomes available at the output terminals.

Inside the plastic bobbin is a hollow cylindrical mandrel **14'** made from a lightweight, conducting metal preferably aluminum. The outer diameter of the mandrel is slightly larger than the inner diameter of the bobbin to obtain a tight fit therebetween. Sufficient clearance is provided by the width of gap **36** so that the concentrically-mounted, mass-coil assembly can move freely in the air gap. Although a small amount of mechanical damping is applied by element **14'**, it primarily reinforces the thin cylindrical wall of plastic bobbin whose wall thickness may be on the order of only 0.005".

In the center cradle **18b'** is provided a vernier mechanical damping mass, generally designated as **50**. In practice, member **50** is a wound coil **16"**, preferably of fine insulated wire such as #40 AWG. The wire is preferably copper but it may be aluminum.

Coil **16"** is open-ended. Accordingly, since the open-ended coil offers a substantially infinite impedance, the current due to the induced emf will be virtually zero. Because the cross-sectional area of #40 AWG wire is very small, about 0.003", eddy-currents will be minimal. The wire is preferably enameled so that electrical contact between turns is not possible. If the adjacent turns of wire were to become short-circuited, the cross-sectional area of damping member **50** would effectively increase, allowing undesired, large eddy-currents to flow.

The mass of mechanical damping member **50** is made continuously adjustable between any desired limits, by simply winding a desired number of turns and a fraction thereof, in cradle **18b'**. For reasons recited above, electrical-damping counterforces are virtually nonexistent regardless of the size of the member. Thus, the mechanical damping member is a vernier mass, used with a main mass composed of bobbin **14**, mandrel **14'**, and the active signal producing coils **16** and **16'**, in order to obtain a fine adjustment of the total mass.

Accordingly, a significant advantage of this device lies in the fact that it allows one to easily increase or decrease the inertial mass of the coil assembly **12** and thus to correspondingly decrease or increase its mechanical damping in order to adjust the peak frequency, without effecting an off-setting change in its electrical damping. Also, by precisely controlling the number of turns of the insulated copper wire in the damping coil **16"**, it is possible to precisely control the damping characteristics of the geophone. With an automatic winding machine it is easily possible to continuously change the geophone's damping as a function of wire length.

Another significant advantage of this device lies in the fact that a single standard bobbin can be constructed for geophones having a wide frequency range. For example, a standard bobbin can now be constructed such that the damping of the geophone would be optimum to tune its natural frequency to 28 Hz without the use of damping coil **16"**. For 7 Hz, this geophone would be overdamped. But merely by winding the required number of turns of insulated copper wire in the center cradle **18b'**, the mechanical geophone damping is reduced to precisely tune the geophone to a desired natural frequency of 7 Hz, without rebuilding the bobbin. Previously, additional mass necessitated the addition of weight rings. A mass reduction in the above example would involve trimming of excess metal from the bobbin itself.

It is true that the coil and bobbin mass could be changed by changing the number of turns of wire in the induction coils themselves but to do so would adversely affect the geophone output signal which would decalibrate the sensor. Hence the need for a third, independent, unterminated coil.

Recorders and Displays

Recorders and displays which operate primarily in conjunction with seismic prospecting equipment, may also be used to record and display nuclear and electrical-magnetic geophysical information. Typical systems record and display seismic output data which has been corrected by a computer and processed so as to facilitate the operation of the desired mode of display.

Recent research and patent activity in this area has been directed towards quantitatively displaying multiple seismic signal functions or parameters by a single display with the use of multiple colors, varying areas, or varying densities. Computer controlled cathode ray oscilloscope display systems are being developed to provide pseudo-real time displays during signal processing.

A system developed by *J.W. Elliott; U.S. Patent 3,986,163; October 12, 1976; assigned to Schlumberger Technology Corporation* is one in which well logging data in analog or digital form is recorded by a cathode ray tube recorder. A representation of the cathode-ray tube (CRT) beam is repetitively swept across a recording medium while being modulated with representations of the well logging signals. This modulation varies as a function of the rate of change of the well logging signals to produce an even density recording.

Coding of the recorded lines or traces and the areas between selected traces can be accomplished. This coding of the lines is also varied in dependence on the well logging signal rate of change to produce a uniform coding presentation. Depth information can be recorded on the recording medium by writing depth numbers and depth lines on the recording medium. Moreover, a section by section visual presentation of the data can be produced while it is being recorded.

An apparatus developed by *N.A. Anstey; U.S. Patent 3,961,306; June 1, 1976; assigned to Seiscom Delta Inc.* utilizes a seismic color graphic display having differing colors quantitatively identifying differing values, or ranges of values, of digitized seismic data. Numerical codes are assigned to specify the visual image densities and the relative density of each component color. The component color displays are superimposed into a composite color graphic display.

A typical seismic survey for hydrocarbons is conducted as a series of intersecting profiles along seismic survey lines on a rectangular grid. Each such profile yields a seismic reflection section, which displays, generally to a first approximation, the configuration of the rock interfaces under the line of profile.

In the traditional practice of seismic interpretation, the first operation is to select, or "pick", as the process is known in the art, the reflections which represent these interfaces. This is a process involving considerable judgment, based on knowledge of the regional and local geology. Often the form of a reflection on one or more sections is obscure, and closure of the pick around a loop of the grid is necessary to resolve it. It is therefore common practice to pick each of several reflectors and to check all loop ties before proceeding further.

The next operation is to "time" each of the picked reflections at points spaced at suitable intervals along the surface, and to "post" these times, for each reflector, on maps of the profiles. Suitable contour intervals are then interpolated between the posted times, and a contour map is then constructed for each picked reflector. Perspective views of the contoured surface may then be generated, and numerical integrations may be made to assess volumes.

In the practice of the art, the timing, posting, contouring and later operations are mechanized using a digitizer, a computer and a computer-controlled plotter. The picking operation, however, remains one which must be done by a skilled person. Decisions must be made on the geological likelihood of one possible interpretation against another, geological faults must be identified by visual character correlation, and a knowledge of the general geological history of the area must be introduced into the picking process.

These matters are often very problematical, and their resolution very tedious; each possible interpretation of a difficult segment on one section requires the closing of a loop involving other sections, and each of the latter sections may raise its own ambiguities of interpretation. Furthermore, each reference to another section represents a cumbersome and time-consuming operation, so that the sheer volume of work tends to limit the number of interpretive iterations which can be made.

These difficulties are intensified by the general problem of visualizing three-dimensional subsurface features of the earth from two-dimensional seismic sections. An additional weakness of the traditional practice is that many subtleties which may be present on the seismic sections themselves, and which subtleties may be important to the search for hydrocarbons, are smoothed out in the picking so that their existence is not apparent on the contour map.

A scheme developed by *N.A. Anstey; U.S. Patent 3,931,609; January 6, 1976; assigned to Seiscom Delta, Inc.* provides a three-dimensional seismic display which preserves substantially all the relevant seismic data, with a minimum of exclusive judgment being exercised during its formation, until the implications of the totality can be observed.

This display also facilitates the picking of seismic horizons around closed loops. It also provides a contour indication as part of the basic display. It also allows geological faults and minor and subtle geological features to be readily included in the interpretation.

Where the seismic sections include the display of a plurality of seismic variables, it also facilitates the areal and three-dimensional correlation of these variables. For example, a seismic variable capable of association with the likelihood of hydrocarbons may be displayed as a color modulation of the individual sections, and the areal extent of a particular color modulation may be taken as indicating the extent and volume of a hydrocarbon accumulation.

The display also facilitates the lateral migration of seismic reflections, to accommodate cross-dip.

In a first embodiment, these advantages are achieved by a method of making a sectionalized model of the earth, comprising the steps of preparing a plurality of seismic reflection sections at a suitable scale, cutting the sections in a manner such that each cut edge is related to the observed configuration of a particular seismic reflection, and disposing and securing the sections in a spatial interrelation representing the lines of profile from which they were derived.

Supplemental benefits of this method are obtained when the seismic sections include a color modulation representative of particular geological conditions (including the presence of hydrocarbons), and when contour indications are added. Further benefits are obtained if it is desired to provide lateral migration of the seismic reflections, or to make simple measurements of geological volumes, or to provide photographs of perspective views of the geology.

In a second embodiment, the above advantages are achieved by a method of making a sectionalized model of the earth, comprising the steps of preparing a plurality of seismic reflection sections at a suitable scale, cutting the sections into slices along lines of equal reflection time or equal reflection depth, disposing and securing the section slices in a spatial interrelation representing the lines of profile from which they were derived, and disposing the mounted section slices one above the other to represent successive layers of the earth.

A system developed by *W.P. Neeley; U.S. Patent 3,916,370; October 28, 1975; assigned to Mobil Oil Corporation* is a seismic data processing system which comprises a multiplexer for applying seismic data from magnetic tape storage to a CRT to modulate its electron beam. A photographic drum plotter provides a recording of the seismic data displayed on the face of the CRT. A timing line generator, a timing number generator, a trace mark generator, and an annotation generator provide for modulation signals which are applied to the CRT for the production of timing lines, timing numbers, centerline trace markings, and alpha-numeric annotation data on the photographic recording.

An interface controller provides for timing and control signals for selecting the rotational speed of the drum plotter and the data rate at which the seismic data is recorded. The interface controller further detects the rotational position of the drum plotter and enables the timing line, timing number, trace mark, and annotation generators to provide the appropriate modulation signals to the CRT at the proper times during the revolutions of the drum plotter.

A scheme developed by *R.G. Quay and C.H. Ray; U.S. Patent 3,899,768; August 12, 1975; assigned to Petty-Ray Geophysical, Inc.* is one in which new insight is gained into geological interpretation of seismograph record sections by use of a new technique for extracting and displaying the lateral variations of seismic properties, such as the peak frequency and energy weighted average frequency of seismic waves, occurring at or between time varying seismic boundaries, associated with the geologic contacts of rock strata, as preoutlined upon such a record section.

In one form, lateral variations in the seismic property, associated with specified intervals or zones of seismic data, are displayed as a pair of envelopes plotted about a reference axis which is the mean time between the upper and lower boundaries of the specified interval or zone of data from which the property was extracted.

An upper envelope displays the local magnitude of the seismic property, associated with the zone; and a lower envelope displays the local ratio of this magnitude to the magnitude of the same property observed at any underlying interval or zone of data. Thus, the upper envelope shows the lateral variance of the seismic property whereas the lower envelope provides a means of estimating whether anomalies in the upper envelope are being caused by changes in the common zone overlying both intervals.

Separate transparent overlay sheets for each extracted property can be used for the plotting of the reference axes and their envelopes. These overlays can be superimposed upon each other or upon the corresponding seismic record section to the same scale. The visual correlation of anomalous variations in local seismic properties relative to the structural interpretation based upon the seismic record section yields new dimensions in the possibilities of interpreting geologic conditions favorable for the accumulation of petroleum.

A scheme developed by *R.W. Brittian, F.L. Malarcher and W.A. Schneider; U.S. Patent 3,882,446; May 6, 1975; assigned to Texas Instruments Inc.* is a programmed computer-human interaction edit method and system for stored seismic horizon data. A two-dimensional graph of such primary horizon data is placed on a data tablet input to the programmed computer and wherein phantom horizon data with reference to coordinates of the graph are generated in response to human contact through the graph to the data tablet for direct input to the computer.

Phantom horizon data is stored in a horizon segment file with primary segment data while preventing entry to the horizon segment file of horizon segment data beyond preselected constraints. Responsive to human contact through the graph to the data tablet at the location of phantom horizons and to stored horizon segment data, a first display of segments of two contiguous phantom horizons is produced with all constraint satisfying segments on the graph within a selectable time gate above and below both of the phantom horizons.

A second display is produced of depth point-rms velocity profiles for all segments on the first display. A third display is produced of depth point-interval velocity data for the earth section between the horizons on the first display. Upon deletion of any segment from the first display, automatically and substantially simultaneously the second display and the third display are modified to reflect the removal of data corresponding to any deleted segment.

A device developed by *C.L. Dennis and J. Zemanek, Jr.; U.S. Patent 3,967,235; June 29, 1976; assigned to Mobil Oil Corporation* is an acoustic velocity logging tool which employs a transmitter and a pair of receivers. Acoustic pulses from the transmitter pass through the formation surrounding the borehole to the receivers. The received signals, along with control signals, are transferred to an uphole recording system. A record unit operates to apply these receiver and control signals to a magnetic tape recorder. A playback unit transfers the recorded receiver signals to the intensity modulation input of a cathode-ray oscilloscope (CRO) and, in response to the recorded control signals, applies a trigger pulse to the sweep input of the CRO. A film recorder makes a continuous film recording of the receiver signals as they appear as variable-density traces on the face of the CRO. Figure 69 shows such a system with the borehole logging tool featured in the upper view and a block diagram of the recording system in the lower view.

In the upper view, there is shown a borehole logging tool **10** suspended within the borehole **11** by means of logging cable **12**. The tool includes a transmitter **T** and two receivers **R1** and **R2**. A pulser circuit **13** energizes the transmitter to transmit high-frequency acoustic pulses into the surrounding earth formation **14**. For each acoustic cycle the pulser sends a transmitter trigger pulse uphole by way of the logging cable. The acoustic pulses are detected by the receivers. During the first acoustic cycle, the receiver gating circuit **15** sends the detected signal from receiver **R1** uphole by way of the cable. During the next acoustic cycle, the receiver gating circuit sends the detected signal from receiver **R2** uphole by way of the cable.

These cycles are repeated, with successive acoustic pulses detected by the two receivers; the receiver outputs are selectively gated for sending detected pulses uphole. The receiver gating circuit also sends a receiver select signal uphole by way of the cable, indicating which receiver output is being gated at any given time.

In the lower view, there is illustrated in block diagram form the recording system. During field operations, the transmitter trigger pulse, the receiver signal, and the receiver select signal from the borehole tool are all supplied to a record unit **30**. These pulses and signals are processed by the record unit for recording on the magnetic tape unit **31** along with the depth pulses from the borehole tool. During recording operations in the field, these units are supplied with 60 Hz power from a portable power supply **32**. After the logging operation is completed, the magnetic tape may be taken to a processing center where the data recorded on the tape may be further processed through a playback unit **33** to an output device (CRO)

34, which may be recorded on film recorder **35**, with the movement of the film across the face of the CRO synchronized with the recorded depth pulses.

Figure 69: Mobil Oil Corporation Acoustic Velocity Logging System

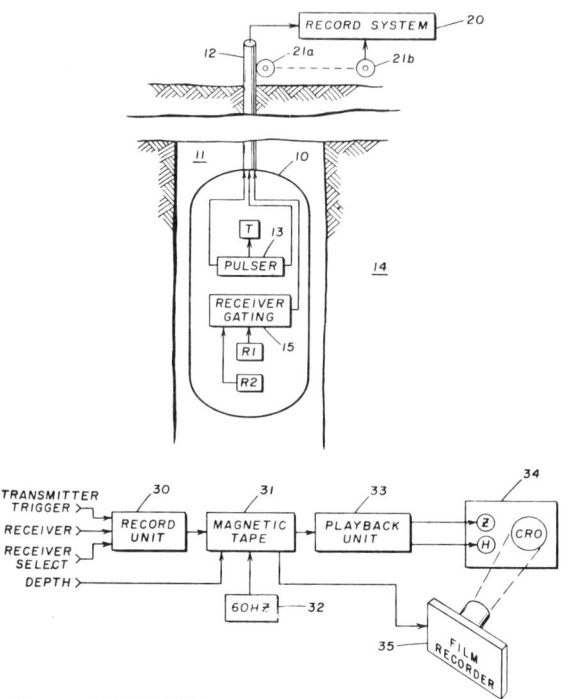

Source: U.S. Patent 3,967,235

In a preferred form, magnetic tape unit **31**, CRO **34**, and film recorder **35** are conventional. A Sony EV-210 Videocorder is used as the magnetic tape unit. It has four recording channels: a video, control, and two audio. While recording in the field, this unit is interfaced with borehole tool **10** through record unit **30**, such that the high-frequency transmitter trigger pulses and receiver signals are recorded on video channel, depth signal on one audio channel, receiver select signal on the other audio, and timing control signals synchronized with 60 Hz power supply **32** on the control channel. In recording signals, the tape speed is 7.8"/sec and the recording head drum speed is 1,800 rpm. Video head-to-tape speed is 590"/sec. High-frequency response of recording tracks of the video channel is 2 MHz. Frequency response of the two audio and one control channels is 50 to 12,000 Hz.

A Tektronix Oscilloscope, Model 461, is used as CRO 34.

A Beattie Magazine Camera (DSV 13302) and periscope (Model 13301) is used as film recorder **35**.

Signal Amplification

During the early 1930s, automatic gain control was developed to suppress large amplitude seismic signals that tend to obscure or confuse subsequent weaker reflections. Seismic traces seldom exceed six seconds in length because hydrocarbon reserves rarely occur below 30,000 feet in geological basins. At these depths acoustic wave velocities average 15,000 ft/sec. In the 1 to 6 second time period that is required for seismic signals to travel from the shot point to subsurface reflectors and back, automatic gain control offsets the tremendous fall-off in signal strength from the early, or shallow, reflectors to the late, or deep, reflectors.

One early patent dealing with automatic gain control was to *W.T. Born; U.S. Patent 2,003,780; June 4, 1935; assigned to Geophysical Research Corporation.* Another early patent in this area was to *J.H. Hammond, Jr.; U.S. Patent 2,008,698; July 23, 1935.*

Multichannel Signal Recording

In the mid 1930s, multichannel seismic recording utilizing multiple receivers was developed. Originally seismic records contained as few as two traces as described, for example, by *H.R. Prescott and F.L. Searcy; U.S. Patent 2,046,843; July 7, 1936; assigned to Continental Oil Company* and shown in Figure 70. As shown there, a seisphone 1 is connected to a suitable amplifying unit 2 by means of conductors 3 and 4. The output of the amplifier is led by conductors 5 and 6 to a recorder 7 in which is positioned a core 8, around which a coil 9 is wound. The output of the amplifier is passed through coil 9. Pivoted on pivot 10 is a magnetic armature 11 carrying a mirror 12. The armature is provided with an arm 14, the end of which is secured to a hair spring 15, which acts as a restoring means. The rays of incandescent light 16 are focused by lens 17 upon mirror 12 and reflected along the path 18 to a moving photographic film 19 which is driven by suitable mechanism 20 operated by motor 21.

At a suitable distance from the seisphone 1, preferably at some distance below the surface of the earth in a suitable hole 22 is positioned an explosive charge 23 which may be dynamite or other suitable material. Within the explosive charge 23 a blasting cap is positioned comprising a shell 24 in which is lodged a booster charge 25 and a detonating charge 26. The detonating charge surrounds bridge wire 27 which bridges electrodes 28 and 29, which are connected to conductors 30 and 31, respectively. Conductor 31 is connected to one pole of battery 38. The other conductor 30 is connected to the other pole of the battery 38 through resistance 32 and firing key 34 which is adapted to complete the circuit by making contact with point 33 which is connected to the battery. Conductor 50 is connected to one terminal of the resistance 32 at one end and to the coil 35 which is housed in recorder 7, wound around core 35 therewithin.

Within the recorder is positioned a second armature 40 having a mirror 41. The armature is pivoted at 42 and is provided with an arm 43 terminating in a restoring hair spring 44. A second incandescent light 45 is housed within housing 46. The rays of light 45 are focused by lens 47 upon mirror 41 and reflected along the path 48 to the photographic film 19. The other end of coil 35 is connected by conductor 37 by an adjustable contact to the resistance 32. It will be observed that, by moving the resistance to the left, the amount of current flowing through coil **35** may be regulated.

Figure 70: Continental Oil Co. Seismic System Showing Data Recorder

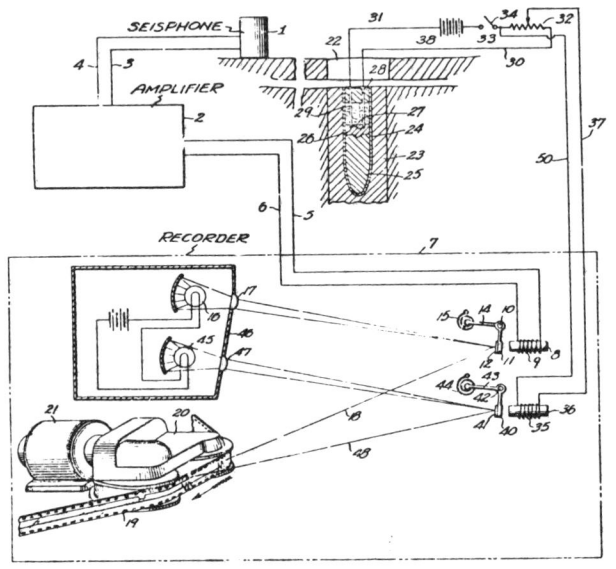

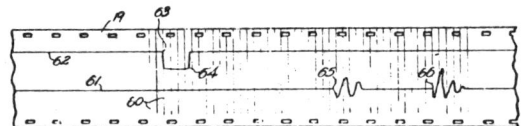

Source: U.S. Patent 2,046,843

In operation, when it is desired to make an exploration, the apparatus is set up and the firing key **34** depressed. It will be observed that current will flow through the circuit comprising the battery **38**, the firing key **34**, the entire resistance, conductor **30**, electrode **28**, bridge wire **27**, electrode **29**, conductor **31**, to the cathode of battery **38**. It will also be observed that current will flow through the circuit comprising battery **38**, key **34**, conductor **50**, coil **35**, conductor **37**, part of resistance **32**, conductor **30**, electrode **28**, bridge wire **27**, electrode **29**, conductor **31**, to the cathode of the battery.

The current flowing through the detonator circuit described above will heat bridge wire **27**. The current flowing through the recorder circuit described above will flow through coil **35**, energizing the electromagnet which will attract armature **40**, depressing the beam of light **48**. As long as the bridge wire **27** is intact, the beam of light **48** will remain depressed. When the explosion takes place, the bridge wire will be ruptured, simultaneously breaking both circuits inasmuch as both circuits have a common return through conductor **31**. At this instant, the electromagnet becoming deenergized will permit the armature **40** to assume its original position.

The velocity of the detonation of fulminate of mercury is 3,920 meters per second. There is only about three-quarters of an inch of fulminate of mercury in the blasting cap. It will be obvious that the detonation of the fulminate and the detonation of the charge are substantially simultaneous as the time lag is so small as to be practically negligible. It can be assumed, therefore, that the instant the bridge wire breaks is the instant of the detonation of the charge without introducing too great an error. It will also be obvious that, if it is to be assumed that the time of the explosion is the time of the application of current, that is, the closing of key **34**, that a large error will be introduced. The magnitude of the error can be appreciated by reference to the detail at the bottom of the figure which shows a reproduction of a photographic record taken with the device. The photographic film **19** has time lines **60** traced thereon by any suitable means, as for example, by an oscillograph controlled by a tuning fork. Usually the interval between the time lines **60** represents one-hundredth of a second.

The trace **61** represents a curve traversed by beam **18** which is controlled by the seisphone. The trace **62** represents the path of the beam **48** which is controlled by the current flowing through the bridge wire. Point **63** upon trace **62** represents the instant contact was made by the firing key **34** in completing the circuit. Point **64** represents the instant the circuit was broken. Point **65** represents the first arrival of an elastic wave generated by the detonation of the charge at the seisphone.

In the making of geophysical exploration, the time lapses from the instant of the blast to the various arrivals are of extreme importance and are necessary in making the necessary computations involved. The accuracy of the work depends upon the accuracy of the instant of the blast. It will be obvious that, in the case shown, over 0.035 of a second will elapse between the closing of the firing circuit and the instant of the blast. This variation in the instant of the blast would seriously affect the accuracy of the computations if it were assumed that the point **63** were the time of origin of the elastic waves.

A technique developed by *H. Salvatori and D. Walling; U.S. Patent 2,087,120; July 13, 1937; assigned to Western Geophysical Company* provides a method of seismic surveying comprising drilling two spaced shot holes in the earth approximately 400 to 3,000 feet apart to depths at least as low as the bottom of the weathered formation and placing a spread of at least two seismometers near the surface of the earth.

The seismometers are placed between and roughly in line with the shot holes, the seismometer spread occupying most of the distance between the shot holes, generating vibrations of the earth in one of the shot holes at a point at least as low as the bottom of the weathered formation. The arrivals of the refracted waves and waves reflected from subsurface formation are recorded at the seismometers, and the operation is repeated using the second of the shot holes without altering the positions of the seismometers, whereby the weathering correction and the dip of the subsurface formations can be determined accurately.

By 1936, the standard number of traces employed was 6; by the early 1940s, 12, as described by *L.W. Blau; U.S. Patent 2,148,422; February 28, 1939; assigned to Standard Oil Development Company* and *L.G. Ellis; U.S. Patent 2,243,729; May 27, 1941; assigned to Sun Oil Company*.

Magnetic Signal Recording

Magnetic recording tape was introduced in the 1950s in order to provide a viable method of recording the entire reflection signal. This eliminated the need for more than one shot at each survey point, and reduced consumption and cost of seismic wave sources. Magnetic tape also provided a mode of replaying the data signals with different filter settings, thereby permitting the determination of the optimum corrections for the recorded data.

In early days, the process of seismic exploration was carried out by making seismic records in the field. These records were then physically transported back to a data processing location where they were analyzed. This was time consuming and final test results were often not available until days after the test was performed.

In a scheme developed by *A.R. Aitken, J.A.F. Gerrard, G.P. Sarrafian and H.J. Jones; U.S. Patent 3,075,607; January 29, 1963; assigned to Texas Instruments Incorporated* the binary data including the sampled amplitudes and the assigned time intervals may be delivered through a connecting lead to and recorded by a magnetic tape recorder to provide a record of field data which is also delivered through a transmitter and then transmitted by radio link to a receiver in the data processing center. The data received is delivered and recorded in binary digital form on a magnetic tape loop. The data is then fed from the magnetic tape loop to a computer where it is processed. In the computer the data is digitally filtered, corrected for time displacement, the reflection zones determined which is called reflection picking, and then the depth and dip information computed.

Another scheme for processing seismic data using magnetic tape has been described by *V.R. Johnson and J.D. Skelton; U.S. Patent 3,047,836; July 31, 1962; assigned to Jersey Production Research Company*.

Three-Dimensional Representation

Holographic systems have been developed to provide a three-dimensional view of the subsurface strata. In this process coherent acoustic energy is transmitted into the earth and detected by an areal array of geophones. A reference signal obtained from the energy source is delayed, amplified and mixed with the signals from the areal detectors to form the acoustic hologram. Reconstruction of the subsurface image from the hologram shows the reflecting surfaces and the source pattern image.

One such process has been described by *E.D. Riggs; U.S. Patent 3,691,517; September 12, 1972; assigned to Atlantic Richfield Company* and is illustrated schematically in Figure 71 shown on the following page.

Another holographic technique has been described by *D. Silverman; U.S. Patent 3,622,968; November 23, 1971; assigned to Amoco Production Company*.

Figure 71: Seismic Holography

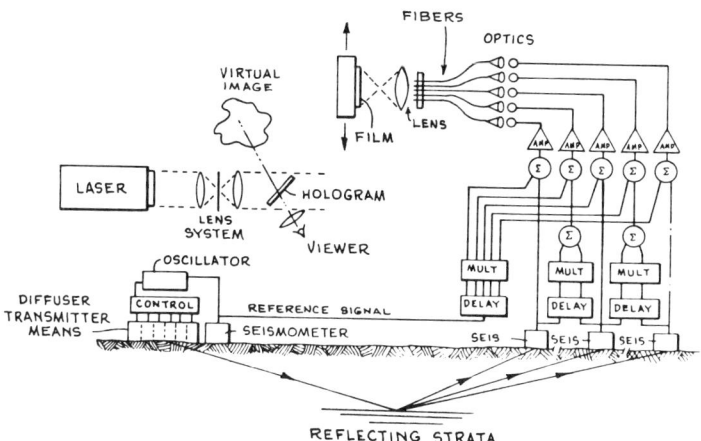

Source: U.S. Patent 3,691,517

Still another application of holography has been described by *J.B. Farr; U.S. Patent 3,729,704; April 24, 1973; assigned to Amoco Production Company*. In this process, a reference wave is produced by generating a low amplitude cyclic vibration essentially at the earth's surface and receiving this under steady-state conditions at the various geophones in the seismic spread. The resultant signal is recorded and its square wave fundamental component used in determining the reference wave employed in making the ultimate hologram.

A beam steering method has been developed to provide three-dimensional visualization of the subsurface strata by emphasizing signal energy received from a particular direction. In this method successive channels are delayed so that events of a certain dip moveout (or apparent velocity) occur at the same time and are then summed. This procedure is repeated for successive different dip moveouts, in effect, to steer the seismic waves for other dips.

A process described by *R.D. Judson, R.J.S. Brown, I.R. Malarky and F.G. Blake; U.S. Patent 3,597,727; August 3, 1971; assigned to Chevron Research Company* is one for collecting and processing seismic data to determine the attitude of strata below the surface of the earth. A spread of seismic detectors is employed in conjunction with an array of obliquely aligned seismic sources (oblique with respect to a survey line at the surface of the earth) such that:

(1) center points between all possible source-detector pairs form a grid of center points having sets of cross-aligned center points perpendicular to the line of survey yet
(2) respective pairs of detector station-shot points associated with any one set of cross-aligned center points have appreciably different horizontal spacings.

Multitrace records of seismic data are produced representing energy reflected from the strata after initiation of each seismic source individually. These traces are rearranged and then processed by the technique known as beam steering to produce traces that not only provide directional information as to the emergence angle of the reflected energy from the strata but also significantly attenuate multiple reflected signals present in the original traces.

Similar ground is covered by *R.J.S. Brown, I.R. Malarky and F.G. Blake; U.S. Patent 3,529,282; September 15, 1970.*

Signal Enhancement

Digital Wiener shaping filters have been developed as described by *D. Silverman and S. Treitel; U.S. Patent 3,680,040; July 25, 1972; assigned to Amoco Production Company* to extract the desired seismic signal from ambient noise. Criteria have been developed for deriving filter operators which, when convolved with recorded traces, yield an output as close as possible to the desired signal. Wiener filters have been designed to remove undesired repetition (ghosts) of primary reflection signals caused by downward reflection of energy from the shot at interfaces above the shot as described by *M.R. Foster, R.L. Sengbush and R.J. Watson; U.S. Patent 3,550,073; December 22, 1970; assigned to Mobil Oil Corporation.*

Weiner filtering has also been applied to the problem of removing coherent noise that has a moveout time or apparent velocity different than the desired reflection events. This filtering operation is based on the time differential, for a particular event, between adjacent seismic signal traces.

Low cut frequency filtering and wave number filtering have been developed to reduce the low velocity, low frequency, surface wave interference (groundroll), which, due to its high amplitude level, masks useful reflection information.

In a technique developed by *C.G. Dahm; U.S. Patent 2,473,469; June 14, 1949; assigned to Socony-Vacuum Oil Company, Incorporated*, there are utilized several pairs of geophones for the production of each record trace; the detectors of each pair are suited, by their horizontal spacing, by the phasing of their individual outputs and by an associated filtering and mixing means, to provide a joint output in which is predominant the reflected wave energy; horizontally traveling energy the same frequency as that of reflected energy having been eliminated as the spacing of the geophones affects them in phase opposition. The spacing of the geophones is different for the different pairs so that collectively their outputs cover the entire band of seismic frequencies. By combining the outputs of these pairs of geophones in a single recorder circuit, there is produced a trace of the reflected waves to the substantial exclusion of much stronger horizontally traveling waves.

In a technique developed by *J.P. Lindsey; U.S. Patent 3,432,807; March 11, 1969; assigned to Phillips Petroleum Company*, a plurality of seismometers are closely spaced to one another on both sides of the shot point. The signals from the seismometers are combined into a single trace by a velocity filtering procedure to reduce ground waves and shallow multiple reflections. In one embodiment, the filtering is accomplished by the use of tapped delay lines and summing amplifiers.

Filters have also been developed to preserve relative amplitudes in seismic signals. A technique developed by *P.L. Goupillaud; U.S. Patent 3,652,980; Mar. 28, 1972; assigned to Continental Oil Company* is capable of unscrambling received information in order that a more meaningful interpretation can be isolated or indicated. Such procedure may be likened to unscrambling techniques as employed in other technologies wherein electrical signal scrambling may be the result of predetermined coding schemes, such intentional scrambling being capable of reversal with designated unscrambling equipment for reconstitution to its original form. However, in the case of seismic information, the scrambling is due to natural causes which tend to defy analysis into the various subcomponents of the overall signal indications. In such case, a discrimination of the subcomponents of returned seismic signal energy must rely upon analysis in terms of probability of occurrence of the particular components for a selected terrain.

This method consists of constructing an ideal filter for each of selected primary, multiple and/or interference components of a seismic tract group, such filter construction being carried out by determining the pseudoinverse or generalized inverse of a polynomial matrix as determined for the particular seismic traces to be examined, with application of the polynomial matrix for preadjustment of a multichannel pattern discriminating filter which, when energized by the input seismic traces under examination, will produce separately the requisite primary, multiple and interference outputs which represent the respective components present. Apparatus for carrying out the method may consist of plural channels of time domain filters connected in series with a weighting device for convolving respective plural seismic traces; each of the convolution filters is adjusted to function at predetermined times and amplitudes in accordance with the matrix function or operator determined for primary, multiple and interference trace functions of a particular group of seismic traces.

Filters have also been invented to preserve relative amplitudes in seismic signals. The amplitude information has been used primarily in amplitude anomaly (bright spot) studies to pinpoint changes in rock composition, strata thickness and stratigraphic conditions. The bright spot technique is based upon the principle that gas saturated sands have a lower acoustic wave velocity than adjacent water or oil saturated sands, so that the velocity contrasts across surfaces bounding the gas zones, above or below, would cause reflections of higher amplitude than would be observed from the same interface on either side of the gas zones. In order to obtain accurate amplitude measurements, corrections are made for spherical spreading, attenuation, and interference (constructive or destructive) between reflections from nearby interfaces, as well as distortions due to processing operations, e.g., convolution and focusing effects due to the curvature of reflecting interfaces.

In a process developed by *K.H. Waters; U.S. Patent 3,921,126; November 18, 1975; assigned to Continental Oil Company* a first seismic survey of an exploration area is conducted thereby giving the reflection characteristics of subsurface layering followed by the introduction of gas into one or more selected formations. The gas is caused to flow into and through the permeable portions thereof, the gas having the characteristic of increasing amplitudes of seismic signals from the formations when it is present therein. A second seismic survey of the exploration area is conducted after the introduction of gas thereinto and the first and second seismic survey results are compared to delineate the locations of gas in the formations and the permeable extent thereof.

In order to preserve relative reflection amplitude variations in seismic signals, inverse gain functions have been developed in two categories, statistical and deterministic. Both approaches attempt to correct traces for average attenuation rates while preserving instantaneous variations caused by changes in subsurface acoustic impedances. Deterministic approaches define general models to describe many of the possible factors affecting amplitudes, such as diverging wavefronts, selective frequency absorption, reflection and transmission losses and source plus receiver array effects. Statistical approaches produce average gain functions based on collections of traces sorted by a common range, source or receiver. Such a technique has been described by *P.L. Goupillaud; U.S. Patent 3,774,146; November 20, 1973; assigned to Continental Oil Company*.

Seismic data is also enhanced by deconvolution filtering as described by *T.J. Hollingsworth and J.D. Morgan; U.S. Patent 3,599,175; August 10, 1971; assigned to Petty Geophysical Engineering Co.* This type of processing is generally applied to reduce the peaking of the frequency response due to the reverberation of the seismic signal between two strata surfaces, and to improve the resolution of seismic signals by building up higher frequencies as the spectrum is widened.

Seismic data processing systems have also been developed to control the emitted seismic signals in order to optimize the characteristics of the signal received for further processing. One such system applies a variable frequency signal at the surface of the earth for a substantial period of time (13 seconds) as described by *W.E.N. Doty and J.M. Crawford; U.S. Patent 2,688,124; August 31, 1954; assigned to Continental Oil Company*. In this system the waves received at the various geophone locations, set up to receive the vibrations reflected from subsurface discontinuities, are cross-correlated with a signal representing the vibrations transmitted into the ground to produce a correlated signal in which the various reflected waves appear as discrete wavelets. Later developments to this process have provided for the use of discrete wave groups of constant frequency in series. These later developments have been described, for example, by *B. McCollum; U.S. Patent 3,182,743; May 11, 1965*.

In a scheme developed by *J.F. Mifsud; U.S. Patent 3,259,878; July 5, 1966; assigned to Esso Production Research Company* the desired effective downtraveling signal is a composite signal and is obtained from a series of elementary signals, each elementary signal having a rather narrow bandwidth. Each elementary signal is an elastic wave and can consist of a pulsed sine wave; that is, a sinusoidal wave of finite duration. The elementary signals are generated at a point near the earth's surface and are reflected from subsurface strata. The center frequencies or carrier frequencies of elementary signals normally are different and are spaced more or less uniformly in the desired frequency spectrum of the effective downtraveling signal. The reflection of each elementary signal from subsurface strata is detected and recorded independently.

Thereafter, the recorded reflected signals of all the elementary signals are added together with the proper relative phase and amplitude to form an effective or composite reflected signal or seismogram. Preferably, the effective or composite downtraveling signal is obtained by adding the elementary signals in proper relative phase and amplitude to give the desired waveform. In most cases it is more practical to record the elementary signals and then decide what relative phase and amplitude changes are necessary to obtain the desired effective downtraveling

Seismic Geophysical Prospecting

signal. The relative phase and amplitude of the received signals are then adjusted in the same relationship.

The synthesized reflection seismograms at several geophone stations (with vibrator location approximately fixed) or at one geophone station (with several vibrator locations) can be displayed in any conventional manner such as in a side-by-side relationship to form a seismic section which approximates a cross-sectional view of the earth through which the signals were propagated. The composite reflection signals can also be operated upon or processed in any conventional manner.

A technique developed by *W.H. Mayne; U.S. Patent 4,004,267; January 18, 1977; assigned to Geosource Inc.* is one in which a vibratory source transmits into the earth a series of discrete wave packets of different center frequencies, at least some of which are not uniformly spaced in frequency. Some of the center frequencies may, for example, be related to each other as a geometric progression. Each wave packet that is transmitted into the earth preferably has a modulation envelope according to a sinusoidal modulating wave that eliminates transient effects due to termination of the wave packet and minimizes the number of discrete wave packets required to define a given frequency bandwidth. The received waves are recorded and later summed with a zero phase relationship relative to a common reference time. The composite record is similar to that obtained from an explosive source.

Another method developed to optimize received seismic signals by minimizing side lobes is the injection of randomly time-spaced signals and cross-correlating the received signals.

A technique developed by *M.T. Taner; U.S. Patent 3,697,938; October 10, 1972; assigned to Seismic Computing Corp.* is one in which an injected signal is used in a reflection seismographic method, which signal includes a sequence of elastic impulses generated by a vibrator or the like as individual pulses rather than a continuous waveform. The time spacing between pulses is substantially random and nonrepetitive, at least within the average travel time for the reflected seismic signals. A large number of impulses are transmitted, over a time period exceeding the travel time, providing the advantages of multiple shots, but yet this may be accomplished in a very short time compared to the total of as many individual runs. Not only is time saved, but also the number of data recordings to be processed is reduced.

More significantly, this technique results in a composite record which exhibits greatly enhanced signal-to-noise ratio, and minimizes the side lobes compared to the main lobe in correlation. The compositing technique used may be simple additive correlation, taking into account the same time intervals used in time spacing the injected signal. The degree of cross-correlation between channels using this technique may be expressed as $N(N-1)/2$, where N is the number of pulses in the sequence; a dramatic enhancement in the number of cross-correlations that will show an increase in the main peak return, compared to noise, is noted.

A technique developed by *M. Barbier; U.S. Patent 3,866,174; February 11, 1975; assigned to Societe Nationale Des Petroles D'Aquitaine, France* is one in which a sequence of waves is produced in the medium in the form of long continuous oscillatory signals emitted by a number of sources. The reflected waves are received by at least one receiver, and the times taken by the waves to travel are determined by cross-correlating the signals received and recorded with the references for the signals emitted.

This method is characterized by the fact that the interval between the times at which two sources or emitters begin to emit consecutively is less than the sum of the duration of emission by the first emitter and the time α taken for the longest distance covered by the waves in the medium being explored, and that the times at which the various sources begin to emit and the polarities of their signals, are such that when the sequence of times at which emission of all sources begins, each combined to the polarities of the emitted signals, are cross-correlated with the sequence of times at which at least an emission of one of these sources begins combined to the polarities of the signals it emits a function is obtained which presents negligible correlation residues for the length of time α.

This method can be applied to seismic exploration in several dimensions, with a marked reduction under certain conditions in organized surface noise reaching the receivers.

Optical filtering systems have been invented to remove noise from seismic signals by spatial filtering, to remove filter reverberation noise by deconvolution, and to perform multidimensional Fourier transforms to conserve phase and amplitude information. The systems generally use coherent light from a laser to produce a two-dimensional Fourier transform diffraction pattern by passing the beam through a transparent reduction of a variable-density or variable-area record section. The seismic signals on the record station act as an optical grating and diffract the laser light. By obstructing portions of the pattern corresponding to particular frequencies or dips on the seismic section, such frequencies or dips can be removed from the reconstructed image.

A technique developed by *E.D. Riggs; U.S. Patent 3,620,591; November 16, 1971; assigned to Atlantic Richfield Company* is one for optically processing seismic or other data by spatial filtering in order to discriminate against optical noise and enhance recoverable information. Optical elements acting on the seismic signals are mounted in special assemblies and rotated at different angular velocities with respect to one another. A series of partial exposures are made of the output information at selected time intervals and added to give a composite exposure. Also disclosed is a preferred process for preparing spatial filters by photographic reduction.

The deconvolution of variable optical density records has been described by *D. Silverman; U.S. Patent 3,524,195; August 11, 1970; assigned to Pan American Petroleum Corporation*. A first strip of variable optical density comprising a multifrequency time function may contain frequency components, the amplitudes of which are excessive or deficient compared to that of other frequency components. A second variable optical density strip can be generated in which the range of amplitude with frequency has been decreased by forming a one-dimensional Fourier transform of the first strip in an optical processing device.

Differences in amplitude with frequency on the first strip are represented on the transform by variation in intensity of illumination at distances corresponding to frequency along a line from the center of this plane. A mask with a narrow opening is moved at varying speed along the frequency axis of the transform plane. This speed is related to the light intensity passing through the narrow opening. Accordingly, the light passing through this opening ultimately generates the second record with more uniform amplitude-frequency relationship than was present on the first strip.

A technique developed by *E.L. Green; U.S. Patent 3,872,293; March 18, 1975; assigned to U.S. Secretary of the Navy* is one in which a coherent beam of light is modulated by signals from an array of hydrophones. Multidimensional optical Fourier transform processing is accomplished in a sequence that includes optical Fourier transform and remapping operations that conserve phase and amplitude. In such optical processing of signals from the array the first or temporal Fourier plane of multichannel frequency analysis is scanned to perform a sequence of one- or two-dimensional spatial transforms. The spatial transforms, each of which corresponds to a discrete acoustic frequency, are performed after remapping the frequency analyzed data into an optical space model of the acoustic array. The means for remapping the data is a set of dielectric wave guides. Alternatively, optical signals are physically measured and then remapped.

A process developed by *R. Peraldi; U.S. Patent 4,122,431; October 24, 1978; assigned to Societe Nationale Elf Aquitaine (Production), France* is one in which seismic surveying is conducted with an emission source and a multiplicity of aligned punctual receivers at the surface of the medium to be surveyed. Adjacent receivers are spaced apart a distance not greater than the maximum space frequency of the waves to be detected. The outputs of the receivers which include surface waves and organized noise are individually recorded for various positions of the receivers to form at least 100 seismic traces. A seismogram is formed from the traces and the seismogram is treated to remove space frequencies. In a variation, a multiplicity of punctual emission sources are used with a receiver.

Digital Data Processing

Since the basic objective of all seismic processing is to convert the information recorded in the field into a form that best facilitates geological interpretation, recent efforts have been primarily directed to digital data processing of the seismic survey data. For a typical survey, to build up a three-dimensional picture of the subsurface geology in a one square mile region, 50 million bits of information are obtained for processing. Processing of such massive amounts of data necessitates the use of digital computers.

In the mid 1950s, digital processing and recording were first applied to seismic signal processing as described by *R.R. Unterberger; U.S. Patent 2,968,022; January 10, 1961; assigned to California Research Corporation* and by *J.F. Bucy, Jr.; U.S. Patent 2,972,733; February 21, 1961; assigned to Texas Instruments Incorporated*, as well as by *R.J. Loofbourrow; U.S. Patent 3,241,100; March 15, 1966; assigned to Texaco Inc.* Digital techniques, such as binary gain ranging as described by *R.S. Foote and G.P. Sarrafian; U.S. Patent 3,134,957; May 26, 1964; assigned to Texas Instruments Incorporated* and floating point amplification were incorporated into seismic recording instruments as described, for example, by *R.A. Harris; U.S. Patent 3,562,504; February 9, 1971; assigned to Texas Instruments Incorporated* and by *R.M. Braham and J.W. Keowski; U.S. Patent 3,924,260; December 2, 1975; assigned to Petty-Ray Geophysical, Inc.*

Digital seismic recording systems presently have dynamic ranges of approximately 80 dB. Reflection amplitudes decay about 100 dB in the first 4 seconds of recording, primarily due to attenuation losses along the travel path. Consequentially, amplifier gain levels change many times during recording in order to preserve signal amplitude for subsequent processing.

Digital processing was also applied to seismic signals in order to provide static and dynamic corrections. Static corrections consist of the application of a time shift, or translation, to an entire signal trace so that a constant time correction term is added or subtracted from all reflection times, regardless of record time or reflector depth. Static techniques have been described by *A.W. Musgrave; U.S. Patent 3,697,939; October 10, 1972; assigned to Mobil Oil Corporation.* Correction for weathering (velocity changes near the earth's surface due to the makeup of the strata) and elevation are static type changes.

Dynamic corrections vary with record time and therefore depend on reflector depth. Dynamic corrections have been discussed by *W.A. Schneider; U.S. Patent 3,629,800; December 21, 1971; assigned to Texas Instruments Incorporated.* Correction of CDP traces for travel time differences, caused by varying ray path distances prior to stacking, is termed "normal moveout" (NMO), which is a dynamic correction as it depends on the depth (record time) to the reflecting horizon according to *E.F. Greene, Jr.; U.S. Patent 3,747,055; July 17, 1973; assigned to Texaco Inc.*

A technique described by *R.F. Carter; U.S. Patent 4,146,872; March 27, 1979; assigned to Chevron Research Company* is one in which connection (and disconnection) of a plurality of geophones associated with a digital seismic data acquisition system is greatly simplified without sacrificing flexibility as to the number of geophone flyers per station or the interval spacing per station and also without the addition of separate subcabling in the internal linkage of the flyers to the field digitizing equipment. The apparatus includes a multiple-phone flyer-jumper connected to each channel of a remote data acquisition and telemetering circuit (RDATC). Each RDATC acts in combination with a group of geophone flyers, say four, to store, amplify, filter, gain control and digitize analog data from the flyers and then telemeter the data to recording circuitry in the field truck.

SEISMIC BOREHOLE LOGGING TECHNIQUES

This category consists of mechanically transmitting compressional waves into and/or receiving compressional waves from an earth section, for the purpose of identifying subsurface structure. The transmitting and/or receiving occur in a shaft or a deep boring in the earth. Seismic data processing systems and mechanical acoustic wave generators particularly adapted for use in boreholes are included.

Recent research and patent activity in this area has been directed towards minimizing road noise by threshold type data processors or mechanical acoustic wave attenuators. Activity has also been directed towards determining well-pipe and well-casing condition by amplitude, frequency, and polarity analysis of reflected acoustic signals.

A technique developed by *R. Gabillard, F. Louage and R. Desbrandes; U.S. Patent 3,690,164; September 12, 1972; assigned to Institut Francais du Petrole, des Carburants et Lubrificants, France* is one in which signals are transmitted from successive depth locations in a borehole and received at a surface location spaced from the location of the borehole. The successively transmitted and received signals are compared with signals indicative of the known characteristics of one of the strata layers to determine the characteristics of the other strata through which the successively transmitted signals have traveled.

A type of construction developed by *P.C. Escaron; U.S. Patent 3,991,850; November 16, 1976; assigned to Schlumberger Technology Corporation* produces acoustic well logging tool bowspring positioners which maintain the tool's centering within a well bore. In order to attenuate the acoustic noise, generated by the rubbing of the outward faces of the springs along the well bore wall, the inward faces of the springs are covered with an elastomer which has interspersed throughout its length (30% of the composite volume) lead balls or pellets, 1.0 to 1.25 mm in diameter.

An apparatus developed by *J.E. Berry and A.J.D. Straus; U.S. Patent 3,982,606; September 28, 1976; assigned to Mobil Oil Corporation* has seismic wave attenuators intermediate the transmitter and receivers. The attenuators consist of an acoustically absorptive material, e.g., rubber sections placed in overlapping relationship and extending a distance sufficient to contact the bore. Two attenuator shapes are disclosed, rubber fingers, arranged in overlapping vertical rows, and discs, having a series of relatively large openings and stacked so that the openings are misaligned. The misalignment permits the requisite displacement of borehole fluid while attenuating impinging acoustic waves.

A technique developed by *D. Silverman and J.R. Bailey; U.S. Patent 3,979,724; September 7, 1976* involves the determination of the position of the bottom end of a long drill pipe in a deep, water filled, borehole relative to the earth's surface. A shock wave is generated in the water, at the surface end of the pipe, and travels down the drill pipe to the bit and to the earth contiguous and adjacent to the bit. A plurality of geophones positioned in a two-dimensional array on the earth's surface detect the arrival of the shock waves emanating from the bit. The position of the bit is determined by the knowledge of the shock wave generation time, the time of travel of the shock wave in the pipe, the length of the pipe, and the arrival time of the shock wave at each geophone.

A system developed by *R. Delignieres; U.S. Patent 3,961,683; June 8, 1976; assigned to Institut Francais du Petrole, des Carburants et Lubrifiants, France* is one particularly adapted for determining the shape and the filling rate of an underground cavity to be used for storing petroleum products. A sequence of acoustic pulses is transmitted along each of successive transmission axes in the direction of the cavity walls from reference locations within the cavity. The average propagation time intervals of selected acoustic wave echoes are measured for each different location and direction of the logging tool in order to provide tool-to-wall distances.

A process developed *E.P. Howell; U.S. Patent 3,962,674; June 8, 1976; assigned to Atlantic Richfield Company* involves the determination of subsurface formation porosity and permeability by ultrasonic (one kHz to one MHz) logging. A uniform sinusoidal acoustic pulse is transmitted into the formation to be logged;

the pulse is converted by the formation into an early arriving primary formation pulse and a later arriving fluid pulse which travels exclusively through interconnected fluid paths in the formation. These two pulses are detected, identified and quantized with cross-correlation techniques to give a direct indication of formation permeability, to identify the formation fluid, and to determine the formation porosity.

A system developed by *C.B. Vogel; U.S. Patent 3,949,352; April 6, 1976; assigned to Shell Oil Company* is an ultrasonic well logging system which detects shear waves reflected from the surrounding formation. Highly directional piezoelectric transducers designed to operate at frequencies greater than 50 kHz are used both as the ultrasonic wave transmitters and wave receivers. The direction of maximum response of the transducers is made to be included at an angle less than 90° (preferably 10° to 30°) from the normal to the borehole wall. The detected shear waves are processed to determine both the transmission time and the attenuation of the shear waves between the transmitter and receiver. Such a system is shown schematically in Figure 72.

Figure 72: Shell Oil Co. Velocity Logger for Logging Intervals

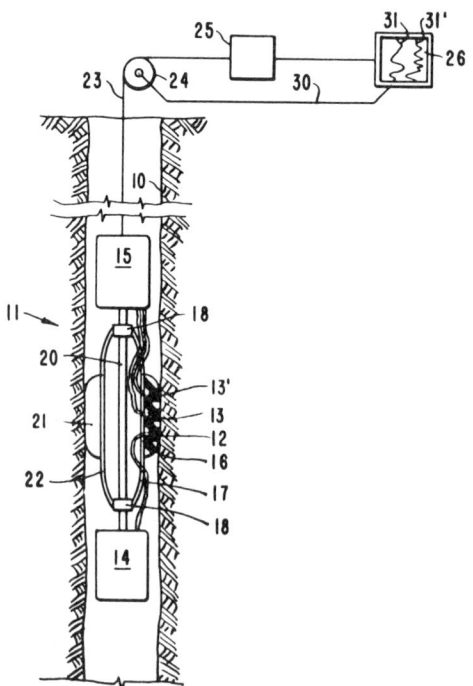

Source: U.S. Patent 3,949,352

There is shown a downhole logging tool **11** which is suspended in a borehole **10** at the end of a multiconductor cable **23**. The downhole logging tool has a transmitting transducer **12** and one or more receiving transducers **13** and **13'**. Preferably four receiving transducers are used, though only two are shown. The transmitting and receiving transducers may have substantially the same construction and may be of the piezoelectric type. In addition, both transducers should operate at 50 kilocycles or more and preferably at a frequency higher than 100 kilocycles in order that the transmitter will generate acoustical impulses having a frequency of at least 50 kilocycles and the receiver will discriminate against impulses below 50 kilocycles. Furthermore, the receiving and transmitting transducers have the required directionality and directions of maximum response. All the transducers 12, 13 and 13' are mounted in a pad or shoe 16 and are spaced from each other a distance of a few inches and in all cases a distance of less than one foot.

The required electrical circuits for the transmitting transducer 12 are carried in a housing 14 at the end of the logging instrument while the required electrical circuits for the receiving transducer 13 are carried in a housing 15 at the upper end of the instrument. The instrument housings 14 and 15 are mounted on the ends of a supporting rod 20 which also serves to maintain them in a spaced relationship. The pad 16 is mounted on a bow spring 17 at the center thereof with the ends of the spring secured to separate collars 18. The two collars 18 are disposed for sliding movement along the rod 20 which couples the two housings 14 and 15 together. A second pad 21 is mounted on a second bow spring 22 whose ends are attached to the two collars 18 diametrically opposite the ends of the bow spring 17.

From the above description it is easily seen that a logging instrument has been provided in which two or more closely spaced transducers are mounted in a pad which is biased into contact with or close proximity to the wall of the borehole 10. This contact is insured by mounting the two pads 16 and 21 and associated springs diametrically opposite each other. The instrument also provides suitable housings for the electrical circuits required for energizing the transmitting transducer 12 and transmitting signals derived from the receiving transducers 13 and 13' to the surface over the conductors contained in the cable 23.

The cable 23 passes over a measuring sheave 24 at the surface which preferably includes a selsyn type unit. Thus, the position of the instrument 11 within the borehole 10 may be accurately determined at all times. The cable 23 is connected to a suitable electrical circuit 25 which converts the signals derived from the receiving transducers 13 and 13' to a related electrical analog signal. The electrical analog signals are functions of the velocity or other transmission properties of the acoustical pulses through the formation between the transducers. The electrical analog signals are recorded on a strip chart recorder, an oscilloscope, or other suitable device 26 and may appear as recordings 31 and 31' thereon. The selsyn unit in the measuring sheave 24 is coupled to the recorder 26 by a conductor cable 30 in order that the chart record of the recorder will be advanced in direct proportion to the position of the instrument within the borehole.

OFFSHORE SEISMIC PROSPECTING

This category consists of signal communication systems and apparatus in which the signal is carried underwater by compressional waves and is utilized for geophysical exploration. Seismic data receiving and processing systems which are particularly adapted for offshore use are in this category also.

Recent research and patent activity in this area has focused on hydrophone receiver array configurations which reduce noise and act as bandwidth filters. Activity has also been directed to determining hydrophone cable position relative to the towing boat and earth reference points by satellite communication or by the forces exerted by the towed cable on sensors connected to the towing boat.

A system developed by *D. Michon and P. Staron; U.S. Patent 4,020,447; Apr. 26, 1977; assigned to Compagnie Generale de Geophysique, France* is an offshore seismic prospecting system utilizing a long cable with a large number of detectors spaced relatively far apart (e.g., 48 detectors, 50 meters apart) and a short cable with a small number of detectors placed relatively close to each other (e.g., 6 detectors, 25 meters apart). The electrical signals produced by the detectors of both cables in response to received seismic waves are successively recorded in common, without commutation, and are distinguishable as a result of relative amplitude and time separation.

A device developed by *J.B. Farr; U.S. Patent 4,011,540; March 8, 1977; assigned to Standard Oil Company (Indiana)* consists of a continuous line hydrophone streamer with multiple coaxial transmission cable segments each containing two concentric conductors and an electret spacer between the conductors. A voltage controlled oscillator is coupled to the first of the cable segments to generate a carrier wave signal. Seismic waves which impinge upon the electret-transducer segments generate a voltage change in the conductors and modulate the carrier wave. The carrier waves from the first transmission segment are transmitted to the next successive segment so that seismic signals from the most instant cable segment are coupled through the entire length of the streamer to a recording system.

A system developed by *J.R. Rogers; U.S. Patent 4,011,539; March 8, 1977; assigned to Texaco Inc.* comprises an array of detector groups towed behind a marine seismic vessel. The detector groups are separately spaced along the length of the array and each detector group has a plurality of hydrophones spaced so as to conform to a Gaussian frequency distribution. The detector spacing operates to truncate the frequency passband at frequency limits according to the degree of noise suppression, or array response level required. The responses of each detector group along the cable are processed and analyzed to determine the likelihood of the existence of hydrocarbon reserves in the submerged formations.

Such a system is shown in Figure 73. As portrayed there, a seismic exploration vessel **V** is shown towing a marine seismic cable **C** through a body of water **10**. A seismic source **S** with the vessel **V** emits seismic or acoustic energy waves which travel through the body of water **10** and a floor **12** of the body of water. As will be understood by those in the art, portions of the energy of the seismic waves from the source **S** traveling downwardly as indicated by a dashed line **14**, are reflected by interfaces or horizons between submerged formations, such as a seismic horizon **16** between formations **18** and **20**. It should be understood that usually several horizons are present. The reflections from the submerged horizons between formations travel upwardly as indicated by a line **22** and are sensed by hydrophones, also known as seismic sensors or detectors, in the cable **C**.

As the vessel **V** travels through the water, boat noise indicated by a line **24** travels rearwardly from the vessel **V** and is sensed along with the desired seismic reflections from submerged formations by the detectors in the cable C. Typical sources of boat noise include acoustic vibrations transmitted into the water from the propellers of the vessel V, noise from the flow of water passing the moving vessel V and noise from vessel equipment which is transmitted into the water through the hull of the vessel V.

The marine seismic cable C is stored on a reel or winch 26 and is deployed therefrom behind the vessel V during seismic explorations. In the cable C, the detectors are arranged into a suitable number of detector groups along the length of the cable C, one of which is schematically indicated at 28. A typical number of detector groups is 24 or 48, although it should be understood that other numbers of detector groups in addition to the foregoing numbers may be used as well. A typical detector group in the cable C contains from 6 to 40 individual hydrophones or detectors, electrically connected together in parallel and schematically indicated at 30. The responses from the hydrophones within each detector group are summed and recorded by conventional seismic signal storage and processing equipment on the vessel V, as will be understood by those in the art.

Each detector group 28 in the cable C thus functions as one common detector having an enhanced signal response characteristic due to the summation of the response of numerous detectors 30, with the effective location of such detector being at the midpoint of the detector group. The responses of each detector group along the cable C are then processed and analyzed in order to determine the likelihood of existence of hydrocarbon reserves in the submerged formations.

Figure 73: Texaco Inc. Offshore Seismic Surveying System

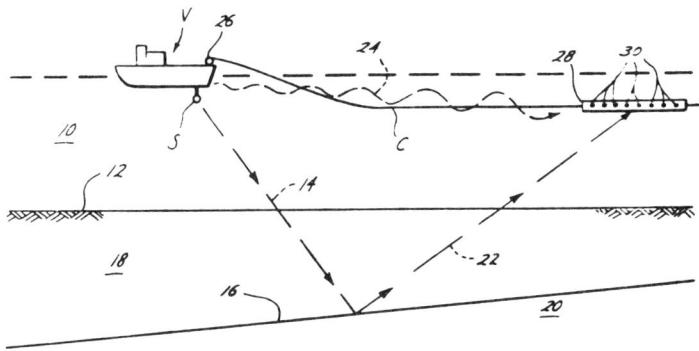

Source: U.S. Patent 4,011,539

A device developed by *G.K. Miller; U.S. Patent 3,978,446; August 31, 1976; assigned to GTE Sylvania Incorporated* comprises a hydrophone array for marine seismic exploration having a longitudinal series of electrically separate sensor sections. The sensor sections have radially spaced coaxial inner and outer conductors and longitudinally extending electret cables in the annular space between the conductors. Each cable has a conductive lead wire insulated by an electret covering, such as tetrafluorethylene, which has been electrostatically charged. An outer jacket of neoprene covers the outer conductor and seals the cable.

A technique developed by *R. Le Moal, J. Cholet and D. Saussier; U.S. Patent 3,953,827; April 27, 1976; assigned to Entreprise de Recherches et d'Activites Petrolieres (E.R.A.P.), France* involves determining the position of a plurality of hydrophones on a towed seismic marine cable by interpolating values obtained by measuring the angle between the tangent to the attachment point of the towed cable and a fixed and known direction. A potentiometric magnetic compass is used for the angle measurement; a hydrophone is used to measure the distance between the seismic wave generator and the compass; and a potentiometric manometer measures the depth of the compass. The hydrophone cable position data is used to indicate which seismic data pertains to a particular subsurface area.

It has been known for some time that in marine seismic operations, there are particular problems that are not encountered in land operations. One aspect stems from the fact that in marine operations the seismic detectors placed below the surface of the water are sensitive to seismic waves in the water regardless of their direction of travel. Furthermore, pressure-type detectors are usually used, whereas in land operations the detectors are ordinarily a displacements or inertia type.

Heretofore, in marine operations, seismic waves were often generated by detonation of an explosive charge which was usually placed at a depth of ten feet or less below the surface. This avoided interference produced by the phenomenon commonly called "bubble bounce" which interference is generated by charges fired at greater depth. The shallow depth of charge also would avoid the problem of ghost reflections which are encountered in land-type shooting where the charge is detonated in competent earth material some distance below a good reflector.

However, the additional problem remained in offshore seismic exploration which was created by the vertically traveling reflection signals that will reflect back down from the surface of the body of water and, consequently, will create an interference pattern.

A technique developed by *A.L. Parrack; U.S. Patent 3,952,281; April 20, 1976; assigned to Texaco Inc.* employs a dual spread of hydrophones which are vertically spaced apart. The signals produced by the uppermost hydrophones, in response to an upgoing reflected seismic wave are delayed so that downgoing acoustic wave signals are in time coincidence with downgoing seismic wave signals from the lower hydrophones. The signals produced by both hydrophone spreads are then algebraically combined so as to cancel the undesired downgoing reflected wave signals.

Many techniques have been employed to improve the signal-to-noise ratio of seismograms produced in marine seismic exploration. Conventional noise suppression, dereverberation and stacking techniques are useful. However, in some cases, the presence of multiple reflections is so pronounced that the dynamic response of the recording system is exceeded. In this case, no amount of post-shooting processing will make the seismograms useful. It is desirable to perform the exploration with an array of sources and detectors which discriminates against these multiple reflections.

A scheme developed by *W.H. Ruehle; U.S. Patent 4,134,098; January 9, 1979; assigned to Mobil Oil Corporation* is one in which the length of the marine source array is changed to obtain rejection of multiple reflections by normal moveout discrimination.

Multiple rejection is achieved by normal moveout discrimination at selected different trace distances for each reflection. This suppression occurs prior to recording the data and is in addition to the suppression produced by the stacking process.

In this scheme, a split source array is used. One section of the array can be extended or retracted so that the total length of the array can be changed without changing the number of sources which are fired. In this manner, the length of the array can be changed in accordance with the geological characteristics of the subsurface formations being explored and the reflection time of reflections from formations of interest. In this manner, the source and detector arrays are used to reject multiple reflections by normal moveout discrimination.

Another technique developed by *W.H. Ruehle; U.S. Patent 4,146,870; March 27, 1979; assigned to Mobil Oil Corporation* is one in which reflected seismic energy is detected by hydrophones to produce seismograms to emphasize a dipping formation which is of interest. The seismograms are produced by directing seismic energy from an array of sources at an angle such that it will reflect vertically from the dipping formation and intersect the hydrophones orthogonally. Directivity of the array is obtained by delaying the seismic pulses produced by each source.

Still another technique developed by *W.H. Ruehle; U.S. Patent 4,146,871; March 27, 1979; assigned to Mobil Oil Corporation* is one in which seismograms recording reflected energy generated from the firing of a seismic source are used to measure the characteristics of the medium through which the energy passes. These measurements are converted into control parameters for an array of seismic sources. When the array is fired under control of these parameters, seismograms having enhanced reflections are produced. As an example, at least one seismic source is fired to produce seismograms which are used in the measurement of water bottom reflectivity and water depth. The measured reflectivity and water depth are converted into a time domain operator representative of the inverse of the reverberation distortion in the water layer. An array of sources is then fired in a sequence which produces an acoustic pressure wave having the inverse time domain characteristics.

A technique developed by *M. Manin; U.S. Patent 4,136,754; January 30, 1979; assigned to Compagnie General de Geophysique, France* is particularly applicable to marine seismic exploration. It consists in simultaneously emitting disturbances from a plurality of sources $S_1 \ldots S_n$ immersed at various depths and positions chosen so as to maximize the power transmitted vertically and to minimize the power transmitted in certain determined oblique directions. The reflected or refracted seismic signals are gathered by hydrophones. By this means certain parasitic refractions are eliminated.

TRENDS IN SEISMIC GEOPHYSICAL PROSPECTING

The major forseeable trend in geophysical processing is the use of the minicomputer in field data acquisition systems in order to provide both automatic control functions and preliminary real time processing of the data. Since data quality often varies widely over relatively short horizontal distances due to surface weathering, layer irregularities, ghosts, etc., high density, separately recorded geophones will be used to determine both static and dynamic corrections more accurately. Higher channel density recording will be utilized to permit more accurate vertical velocity determinations.

Beam steering and seismic holography present great potential for pointing out structure and resolving stratigraphic problems when used in conjunction with two dimensional arrays of closely spaced geophones which are steered, focused, recorded, and processed by an appropriately programmed computerized field system.

Continued emphasis is predicted for the development of improved digital filters, automatic static and dynamic correction algorithms, automatic reflection picking, source wavelet extraction, three-dimensional dip migrations, and computer determinations of lithologic ratios.

EXPLORATORY WELL DRILLING

Just as "the proof of the pudding is in the eating" so the actual existence of an economically attractive body of ore or hydrocarbons can only be ascertained by test drilling. This is, however, a very costly process and follows only after surface indications and geophysical tests have been made.

FOR MINERAL DEPOSITS

The most common method for prospecting for subsurface uranium deposits is gamma-ray logging. This method detects the presence of gamma-ray emitting daughters of uranium, principally bismuth 214. In several cases and particularly in uranium deposits located at depths greater than several hundred feet, gamma-ray logging has proved an inadequate method for detecting uranium deposits. For instance it is not uncommon for uranium in the subsurface to migrate away from its gamma-ray emitting daughters. If uranium migration occurred within the past 500,000 years, uranium may not have emitted sufficient daughter elements and, therefore, its presence will not be detected by gamma-ray detectors.

When such a situation occurs, uranium deposits only can be located if an exploratory drill hole passes through the ore body and the drill cuttings are analyzed for uranium by either of several chemical methods.

The large volume of drill cutting generated by an exploratory drilling program and the fact that uranium accumulations commonly occupy only a fraction of the vertical extent of the host unit combine to make this latter geochemical prospecting method cumbersome and impractical.

Several other methods also detect the presence of subsurface uranium deposits by analyzing for regional concentrations of radon 222, a gaseous daughter element of uranium. However, secular variations in background levels of radon 222 and the short half life of radon 222 (3.8 days) combine to complicate the interpretations of the data.

To overcome these disadvantages, it has been proposed to analyze earth samples taken in depth of about 2 inches to 5 feet from the surface for their polonium 210 (^{210}Po) content. This ^{210}Po content is indicative that radon 222 has been at the respective location and correspondingly that uranium is presently or was at some time in that general area.

This process of analyzing earth samples taken from the surface for ^{210}Po content is not accurate for prospecting for uranium deposits significantly below the surface, e.g., several hundred feet below the surface, because the radon gas migrates slowly through the formation and due to its relatively short half life time of 3.8 days, only a small quantity may reach the surface. Furthermore, the radon gas may not migrate vertically so that a correlation of the ^{210}Po surface analysis data and an expected uranium body location in deeper formations becomes difficult. Thus exploration and development of uranium deposits, and particularly deep deposits, requires a different tool than the surface prospecting.

A technique developed by *G.L. Bartz; U.S. Patent 4,081,675; March 28, 1978; assigned to Phillips Petroleum Company* is one in which the analysis of formation samples from drill cuttings for ^{210}Po allows the tracing and detection of possible uranium deposits from a larger distance. Furthermore, in connection with chemical uranium analysis and equivalent uranium analysis from gamma-ray countings, anomalies such as multiple bodies and the existence of bodies in disequilibrium can be detected.

FOR HYDROCARBON DEPOSITS

Exploratory well drilling is both the last step and the first step in establishing the existence of a new reservoir of oil or gas. It is the last step in the sense that topographic, magnetic, gravitational and seismic studies may have been conducted to indicate the likelihood of an underground producing structure. It is the first step in that it will provide the first real proof of the existence of underground hydrocarbons, their location and extent.

Determining Formation Characteristics

Oil well drilling is extremely expensive and thus it is desirable to choose drilling locations which have a relatively good possibility of providing sufficient yields to justify the drilling costs. In order to select optimum drilling locations it is necessary to know the properties of the subsurface geological structure. These properties are determined by measuring pressure at various depth plots. The plots are then hydrodynamically analyzed to determine the continuity or discontinuity, both laterally and vertically, of pressure systems within the geologic column. Pressure is normally measured during a stabilized shut-in pressure buildup.

In this technique, the pressure and temperature sensors are lowered through the drill stem and a packer is placed above the transducer to seal the drill stem. Pressure below the packer then rises to a stabilized level and the measurements are taken. Measurements are then repeated at a different depth with the packers moved to seal off different geologic zones. After samples are recovered from the drill stem at various depths, the hydrocarbon recoveries can be related to their respective hydrodynamic systems in order to approximate the location of the gas/water or oil/water contact.

Various types of geographical representations of pressure data can then be generated. A potentiometric surface map shows the potential of a given horizon to support a column of free-standing water of known density at a given point expressed in feet of water. A potentiometric surface map generally defines areas of continuous permeability and indicates the presence of possible barriers between these areas which may constitute stratigraphic traps. A barrier to fluid migration is indicated by a rapid change in potentiometric values.

Another type of geographic representation of pressure data is the pressure deflection map. This is a map of pressure values at various points with respect to a key hydrodynamic system. Barriers to fluid migration may be inferred by sudden changes in the pressure deflection values.

The contour interval selected for either the potentiometric surface map or the pressure deflection map is limited largely by the measurement error inherent in the pressure measuring device. With conventional pressure measuring devices it is not possible to measure pressures with sufficient accuracy to allow relatively small contour intervals. This limitation may reduce the reliability and usefulness of such geographical representations of pressure data.

Drill stem pressure tests are also performed in order to measure the properties of an oil well and the surrounding structure in order to calculate the total production of the well as well as the optimum production rate. According to this technique, a packer is utilized to seal a drill stem and the subsequent pressure increase below the packer is measured. The rate of pressure increase provides an indication of the porosity of the structure the oil is in as well as the production rate. Also, flow from the well can be increased until pressure starts to drop thereby providing a good indication of the rate at which flow can be sustained.

In an interference test, pressurized water is injected into a first well and the pressure response in a different well is measured. In a pulse testing mode, the water is injected into a stimulus well at periodic intervals and the pressure is recorded in an observation well. Although pulse testing theory is well developed, the lack of an extremely sensitive pressure gauge has always limited practical applications because the effects of pressure at the observation well are usually small.

The most important advantage of pulse testing is that transients observed as a result of the pulse stimulus are easily distinguished even in the presence of unrelated dynamic reservoir pressure behavior. The results of the pulse tests allow the calculation of in situ permeability and formation thickness between wells.

The most common device currently used for oil well pressure tests are analog pressure transducers connected to conventional strip-chart recorders. These devices are not sufficiently accurate to be useful in many applications and it is difficult to accurately correlate the position of the markings on the strip-chart with time.

Another commonly used device is a self-contained pressure transducer and recorder which is lowered into the drill stem. The primary disadvantage of this device is that the pressure measurements cannot be read until the recording medium is processed at a distant location.

Although pressure is the most important measured property, temperature is also measured in order to normalize or calibrate the pressure measurements and to measure properties of fluids in the drill stem. For example, the pressure increase in the drill stem depends not only on the production rate of the well but also on the viscosity of the oil.

In order to determine the true porosity of the structure surrounding the well and the oil well production rate, it is necessary to know the viscosity of the oil which can be inferred from knowing the temperature of the oil. Also, a knowledge of the characteristics of temperature variations can indicate the presence of a gas rather than a fluid.

In drilling earth formations, it is desirable to log or graphically record the resistivity, radio-activity, gas content and hardness of the earth formations traversed during the drilling operations in such manner that all characteristics can be simultaneously compared to determine the type of formation found at any level of the earth formations traversed. It is also desirable to obtain identified samples of the earth formations traversed which can be studied and analyzed in laboratories remote from the drilling site.

The recorder and sampling device should operate in synchronism in order that the samples from a given level of the well may be compared with the recorded characteristics of the earth formation at the given level. The recorder and sampling device should be automatic in operation in order to obviate the need of the constant presence of an attendant to operate them.

A device developed by *J.B. Brown; U.S. Patent 2,740,291; April 3, 1956; assigned to Addeco, Inc.* is a device for obtaining such samples of formation cuttings entrained in drilling fluids.

An apparatus developed by *B.W. Sewell; U.S. Patent 2,740,292; April 3, 1956; assigned to Esso Research and Engineering Company* is an apparatus with which a continuous log of the nature of the cuttings carried by a returning stream of drilling mud can be made. The apparatus relies upon the difference in hardness of various types of minerals such as sandstone, limestone and shale, as the nature of the vibrations produced when cuttings of these various minerals come into contact with a vibration producing element will differ.

In the simplest embodiment of the device one or more metal plates are suspended in a stream of the drilling mud so as to provide a tortuous path for the mud, thereby causing the cuttings to strike the plates and produce characteristic sounds for the particular type of mineral comprising the cuttings. It is preferred, however, that the vibrations be produced by continuously crushing the cuttings as they pass through the apparatus.

A technique developed by *H.-G. Doll; U.S. Patent 2,747,401; May 29, 1956; assigned to Schlumberger Well Surveying Corporation* is one for determining hydraulic characteristics of formations traversed by a borehole and more particularly relates to methods and apparatus for determining the fluid pressure, the permeability and the degree of hydraulic anisotropy of such formations.

In the production of natural fluids such as oil and gas, wells are drilled many thousands of feet into the earth. When a possible oil or gas bearing formation has been located in a well, as for example by electrical logging methods, it is highly desirable to determine the fluid pressure and the permeability of such a formation in order that the ability of the well to produce may be estimated prior to placing expensive production equipment in the well and at the surface. Further, most permeable formations are hydraulically anisotropic, that is, they have different permeabilities in vertical and horizontal directions respectively, and it is thus desirable that both the vertical and the horizontal permeability be determined for a formation in question.

The process is carried out by creating a pressure gradient in a zone within a selected formation and determining the fluid pressure at one or more points in the zone. In one embodiment, for example, the static pressure of a selected formation is determined at a given point within the formation. Preferably this measurement is taken by establishing as by the use of probe means, a fluid communication channel between the point in the formation and a suitable pressure responsive means in a borehole traversing the formation.

The pressure in the formation in the vicinity of the point is changed before, during or after the static pressure measurement to create the pressure gradient zone about the point, as by passing fluid into or extracting fluid from the formation.

Preferably this fluid flow creating the pressure gradient zone is established at a known rate and at a known distance from the measuring point. In certain cases this distance may be zero. The change in pressure gradient in the formation is measured, which datum is representative of both the actual and relative permeability of the formation.

If it is desired to learn the permeabilities of the formations in several different directions, thus revealing the degree of hydraulic anisotropy of the formation, these measurements may be made in several different directions.

In the constant search for petroleum bearing formations and optimum means for extracting the petroleum content of these formations, it is essential that as much information as possible concerning the nature of the strata lying beneath the surface of the earth be available. One of the most important problems confronting the subsurface engineer is the necessity of determining the same geological formation in each of several wells.

In order to thus correlate these geological formations from well to well, accurate data delineating one formation from another and recognizing the same formation in the wells, regardless of extraneous conditions, must be available. Of perhaps equal importance is the problem of determining whether the area in which an exploratory well is being drilled is a productive one and, if it is, whether the well is properly located. This problem can often be solved by a study of the characteristics of the formations traversed by a borehole, thus eliminating unnecessary drilling in areas devoid of petroliferous formations.

In addition, in areas where water-drive is contemplated as a secondary recovery method it is desirable to know the nature of the formation to be flooded in order to obtain optimum results from the operation. Knowledge of the nature of a particular formation is also highly valuable in cases in which the permeability of the formation is to be increased by chemical treatment, such as acidizing.

Much of the information necessary for a proper evaluation of the characteristics of the subsurface strata of the earth is now supplied by the well log, produced by observing certain properties, physical, chemical, or the like, of the various strata forming the wall of the well bore and plotting these properties on a graph showing the depth at which the formation having the particular property selected is located.

However, there is much valuable information regarding the nature of the subsurface formations which known methods of logging do not supply. The results obtained by these methods are often obscured by extraneous conditions brought into play by the methods of drilling utilized, and the data produced by conventional logging is often indistinct and incapable of accurate interpretation even by the most proficient observer.

Many of these difficulties can be overcome and various advantages realized by a technique described by *D.C. Bond; U.S. Patent 2,772,951; December 4, 1956; assigned to The Pure Oil Company* which involves producing a well log by plotting the cation exchange properties of the clay content of the subsurface strata against the depth at which the particular formation is found.

An apparatus developed by *M.C. Terry; U.S. Patent 3,028,542; April 3, 1962; assigned to Jersey Production Research Company* uses a technique for logging a borehole wherein earth formations are contacted with a reverse wetting surfactant consisting of a cationic salt. Measurements are made of the change in the resistivity of the earth formations brought about by the action of the surfactant. The surfactant alters the capillary forces within the earth formation so that the formation becomes preferentially oil wet. The resulting formation resistivity change in the vicinity of the borehole is very pronounced when the formation contains hydrocarbons such as petroleum and provides a positive indication of formations capable of producing hydrocarbons.

A technique developed by *R.W. Mangum; U.S. Patent 3,495,438; February 17, 1970; assigned to Hammit & Mangum Service Company, Inc.* is one for determining the porosity, permeability, and saturation of subsurface formations while drilling an oil well, wherein measured volumes of washed and screened cuttings are placed in a container. The container is filled with water, leaving a space at the top for an air cavity. A vacuum is placed on the container, drawing a vacuum on the contents therein. Gas is drawn from the sample and transferred to an accumulator chamber. After loading the accumulator chamber the collected gas is transferred into a gas chromatograph where it is analyzed to determine the amount and hydrocarbon content thereof. This initial gas withdrawal is plotted on a log form.

The vacuum is then released from the container and the sample of cuttings are pulverized by a blade therein to virtually powdered form. A vacuum is again drawn on the container and the gas is drawn therefrom into the accumulator from whence it is transferred to the gas chromatograph for analysis to determine the volume and composition thereof. This information is also plotted on a log form so that the gas given up by the cutting in its initial state and that given up when pulverized can be compared to thereby determine the porosity and permeability of the formation in its initial state. It may thus be determined that although hydrocarbons are present in productive quantities in the formation, the porosity and permeability thereof may be so tight as to prevent normal production.

The above indicated procedure is successively carried out as the well is drilled to continuously test the formations as they are encountered.

Circulating mud is also continuously tested by subjecting samples of gas taken therefrom to the chromatograph analysis, and such is plotted on a log form to compare the hydrocarbons picked up by the mud with those attained by the cutting analysis indicated above.

A system developed by *D.B. Murdock; U.S. Patent 4,157,659; June 12, 1979; assigned to Resource Control Corporation* is an oil well instrumentation system for measuring the pressure and temperature at various depths during drill stem tests. The measurements may be hydrodynamically analyzed to map the geological structure of an oil field in order to select drilling locations. The measurements may also be utilized to calculate properties of an existing oil well such as total production and optimum production rate and to determine the geologic properties of the structure in which the well is drilled such as porosity. The system includes a pressure sensor, a temperature sensor and a microprocessor based device receiving the outputs of the pressure and temperature sensors for displaying and recording periodic measurements.

Detecting Hydrocarbons

The ultimate result of prospecting for petroleum is the "show" of hydrocarbons at the well head.

A technique developed by *S.E. Buckley and W.D. Mounce; U.S. Patent 2,740,695; April 3, 1956; assigned to Esso Research and Engineering Company* involves the introduction into a well bore penetrating an earth formation of an oxidizing agent reactable with hydrocarbons or other carbonaceous matter in the formation and contacting the oxidizing agent with the formation containing the hydrocarbons or other carbonaceous matter. The reaction between the oxidizing agent and the petroliferous substance or hydrocarbon in the formation causes the generation of heat and the generation of an elastic wave, such as a sound wave or a pressure impulse.

The formation of heat or the generation of an elastic wave may be detected by thermometric methods or acoustical methods or by pressure indicating devices at the earth's surface which allows the presence of hydrocarbons to be indicated directly. Since the oxidizing reaction between the oxidizing agent and the hydrocarbon or petroliferous substances in the formation being tested may form oxidation products such as carbon dioxide, carbon monoxide and acids, the presence of such hydrocarbons may also be indicated chemically. In short, when carbon monoxide and carbon dioxide are formed, the presence of such gases may be determined chemically by means well known to the chemist and the presence of hydrocarbons in the formation thus be determined.

Likewise, the oxidation of hydrocarbons or petroliferous substances in or adjacent the formation by the oxidizing agent may cause the formation of acids which likewise may be determined chemically by titration or may be indicated by potentiometric methods. In any event it is contemplated that the reaction will cause one or more effects which may be detected and displayed whereby the presence of hydrocarbons in the formation is indicated.

The technique may be practiced by introducing an oxidizing agent into contact with an earth formation penetrated by a well bore. This may be done suitably by disturbing or disrupting the filter cake which ordinarily sheaths the permeable sections of the well bore. By disrupting the filter cake, the formation is exposed which may or may not contain hydrocarbons. The oxidizing agent is then introduced directly into contact with the exposed face of the formation and the reaction caused to proceed.

The oxidizing agent selected will determine whether or not it will be necessary to apply heat at the point where the oxidizing agent is contacted with the formation. By employing perchloric acid of a suitable strength of $HClO_4$ reaction may be obtained at temperatures as low as 40°C. At a temperature of 150°C. the reaction proceeds with moderate speed and at a temperature of 170°C the reaction between perchloric acid and hydrocarbons is violent. Actually, lower temperatures than 40°C may be encountered and induced.

The temperature of the earth formation may be substantially less than 30°C. Then, the temperature necessary to initiate the reaction may be reduced by employment of suitable catalysts. For example, by use of ceric ammonium nitrate dissolved in water and added to the perchloric acid it is possible to obtain an explosive reaction at temperatures as low as 40°C.

The quantity of gas evolved from a gas-containing increment of the drilling fluid will depend not only on the quantity of gas originally introduced into the increment by the drilling of a gas stratum but also on the gas-holding or gas-releasing properties of the fluid. The viscosity, gel-strength and thixotropic properties of the drilling fluid all are factors which determine the quantity of gas which will be released from the drilling fluid.

Thus, in the absence of some suitable standard of comparison, the mere measurement of the quantity of gas released in the trap by an increment of the mud will have no significant relation to the gas content of the stratum responsible for the gas in the increment. Nearly all conventional drilling fluids will retain the larger proportion of the gas originally introduced therein by the drilling of a gas stratum and will release only a relatively small proportion of the gas at the surface.

Thus, a technique developed by *R.W. Rochon; U.S. Patent 2,745,282; May 15, 1956; assigned to Monarch Logging Company, Inc.,* in order to provide the necessary standard of comparison which will render the gas detector readings meaningful of the relative richness of the gas strata penetrated by the drill, contemplates calibrating the gas trap in terms of the quantity of gas released from a given drilling fluid containing a known volume of gas.

In the drilling of oil and gas wells, particularly by the well-known rotary method, various means are employed to test the earth strata penetrated by the drill for their contents of oil and gas in order to determine commercial production possibilities of such strata. One commonly used testing device employed for this purpose is the "side-hole sampler" by which a small coring tool may be driven laterally into the earth formation forming the wall of the well bore for removing therefrom a small section of the formation. Such side-hole samples are generally in the form of a small solid core having a volume of a relatively few cubic centimeters.

Exploratory Well Drilling

The small cores are brought to the surface and examined to determine the properties of the strata from which they are taken. One of the tests is to determine the hydrocarbon gas content. Because the cores are small, the amount of gases present is usually small even in relatively rich strata, because most of the original gas is lost by expansion and leakage as pressure is reduced during passage from removal deep in the well to the surface. Generally, methods used to determine gas content are time consuming, laborious, and inaccurate, thus providing unreliable information.

A process by *R.W. Rochon; U.S. Patent 2,749,220; June 5, 1956; assigned to Monarch Logging Co., Inc.* is one where the hydrocarbon content of solid earth formation cores may be determined quickly and with a high degree of accuracy.

Generally stated, this method is one whereby a solid core is subjected to vacuum extraction in an evacuated chamber, air is then mixed with the extracted gas in the chamber to form therewith an air-gas mixture of known volume and the proportion of hydrocarbon gas in the mixture is then determined by suitable analytical means, as by transferring the air-gas mixture from the chamber directly to an instrument which is adapted to quantitatively determine the proportion of hydrocarbon gas in such air-gas mixtures. Such an instrument is preferably of the well-known "hot-filament" type conventionally employed for such determinations.

The most prevalent method of drilling oil wells is that known as rotary drilling wherein a column of weighted drilling fluid is circulated down through the drill stem, through ports in the drill bit at the bottom of the well, and back to the surface of the ground through the annular space between the drill stem and the wall of the borehole, carrying with it to the surface of the ground cuttings which are broken loose from the formations traversed by the drill bit.

One of the major problems associated with drilling by this method is that there is frequently no appreciable indication given at the top of the well when an oil-bearing formation is penetrated, due to the fact that the pressure exerted by the column of drilling fluid prevents the flow into the borehole of any substantial amount of fluids which might be contained in such formations. Such being the case, when this method of drilling is used there is considerable danger that a petroliferous zone will be unknowingly drilled through and completely passed up.

Numerous methods have been devised to meet this problem, and among those methods suggested is one which contemplates collecting successive increments of the cuttings, identifying each increment with the approximate depth from which it was obtained, and subjecting each increment to one or more tests for the purpose of determining whether or not hydrocarbons are present therein.

A method developed by *R.L. Slobod and S.H. Davis; U.S. Patent 2,749,748; June 12, 1956; assigned to The Atlantic Refining Co.* is one where cuttings brought to the top of the well in drilling fluid may be continuously analyzed in the order in which they arrive showing when an increment contains hydrocarbons. The apparatus comprises a shale-shaker where a portion of the cuttings may be passed in the order of arrival at the top. The cuttings are urged along the shaker and treated by streams of water, or other liquid, to remove drilling fluid. They are then passed on to a trough which conveys them to a chamber where they are comminuted to release any gases. The gases rise to the top and are withdrawn periodically or continuously, for hydrocarbon analysis.

To obtain a maximum recovery from the gas producing formations of a well, knowledge must be secured of the formation boundaries and preferably of the expectable rates of production, at least relatively among the formations of a given well. In the prior art, attempts have been made to secure information concerning the presence of gas-producing formations during the course of drilling by analyzing the drilling fluid which is circulated through the well as it passes from the well to a reservoir. It is evident, however, that detection of gas at a point so remote from the producing horizons cannot be expected to provide accurate and detailed indications of either formation boundaries or relative rates of expected productions.

An improved technique developed by *A. Blanchard; U.S. Patent 2,816,009; assigned to Schlumberger Well Surveying Corporation* is carried out by moving a bell-type chamber through drilling liquid within a well with the pressure in the chamber maintained to balance the hydrostatic pressure at successive levels in the well, allowing any hydrocarbons present in the drilling liquid to be evolved into such chamber, and recording their presence as the chamber moves past the producing formation.

More specifically, a bell-type housing including both a combustion chamber and a reference chamber are provided with a combustion-supporting gas introduced into the combustion chamber and an inert gas introduced into the reference chamber, each at a pressure sufficient to maintain a relatively constant liquid level in the chambers. Identical heating and temperature-sensing filaments are disposed in each chamber and connected together in arms of a Wheatstone bridge circuit.

To make this circuit accurately responsive to the presence of combustible hydrocarbons in the combustion chamber, the gases in the two chambers are substantially matched in thermal conductivity and have a low solubility to reduce their rate of flow through the chambers. Furthermore, to reduce the possibility of "zero drift" in the readings over a long period of time, provision is made for interchanging the functions of the chambers from run to run so that each may alternately be a reference chamber and a combustion chamber. During a given run, however, each chamber preferably serves only one function.

A method developed by *J.M. Hunt and R.N. Meinert; U.S. Patent 2,854,396; September 30, 1958; assigned to Jersey Production Research Company* is one in which various chemical and physical tests are utilized in order to evaluate a rock formation or sample in so far as its quality as a petroleum source rock is concerned. In connection with this evaluation technique, samples of rock that are suitable include the samples that are conventionally obtained during a conventional petroleum drilling operation. Samples that are particularly preferred in rotary drilling operations are those obtained by coring procedures, since core samples have been found to more truly represent a particular formation than do other samples, as for example, the cuttings that are continuously recovered by rotary drilling.

The cuttings from cable tool drilling are also suitable for use in this procedure. In accordance with this method, a portion of rock or a rock sample is first crushed and pulverized into extremely small particles, preferably less than about 40 microns in size. The pulverized rock is then refluxed for a period of several hours with an oil-miscible organic solvent in combination with a water-miscible organic solvent. This procedure has been found to extract the hydrocarbon components of a rock sample without chemically destroying or altering any of the hydrocarbons. The procedure, however, has also been found to extract compounds other than pure hydrocarbons, for example, carboids and various nitrogen, sulfur and oxygen compounds. These nonhydrocarbons are removed from the product of the refluxing step in a later step.

The oil-miscible organic solvent and the water-miscible organic solvent that are employed in the extraction step should preferably be of a type that boils below about 100°C and the two solvents should be substantially completely miscible with one another, especially at the refluxing temperature. It has further been found that the oil-miscible organic solvent should ordinarily constitute a major portion of the combined solvent and that this component should preferably be carbon disulfide or an aromatic hydrocarbon such as benzene or toluene.

The water-miscible organic solvent should be a low boiling, oxygenated aliphatic compound such as a ketone or alcohol. Preferred compounds are acetone, methyl ethyl ketone, diethyl ketone, methanol and ethanol. It will be particularly noted that it is generally preferred to employ one of the aforementioned ketones in combination with one of the alcohols, especially in about equal proportions.

Particularly effective combinations of the aforementioned solvents are solvent mixtures containing about 70 volume percent of benzene or carbon disulfide in combination with about 15 volume percent of acetone and 15 volume percent of methanol. Either one or both of these solvent combinations may be employed, and it has been found especially effective to employ one of the solvent combinations in a first extraction step and the other combination in a second such step. The combination of solvents used simultaneously as described above produces results not attainable by use of a single solvent alone or the sequential use of individual solvents.

The total extract from the extraction procedure is distilled to remove the solvent from the extract; and the last portions of solvent are preferably removed by evaporation at substantially atmospheric temperature.

Following this vaporization step, the rock extract is contacted with an organic solvent which is substantially completely miscible with petroleum at atmospheric temperatures. Such solvents include benzene, carbon disulfide, toluene, carbon tetrachloride, chloroform, ether and ethylene dichloride. This step has been found to dissolve all of the hydrocarbons that are contained in the original extract to the exclusion of materials that are commonly referred to as carboids. The insoluble carboids are separated from the resulting solution, and the solution is again vaporized in order to remove the solvent from the rock extract.

The residue from this vaporization step, which is now carboid-free, is next contacted at substantially atmospheric temperature with a paraffinic solvent containing from about 4 to 8 carbon atoms and preferably 7 carbon atoms. This procedure redissolves the hydrocarbons in the rock extract and ensures the precipitation of asphaltic compounds rich in nitrogen, sulfur and oxygen known as asphaltenes.

The resulting solution is then contacted with an adsorbent which is adapted to perform an adsorptive fractionation on the dissolved materials and to separate the hydrocarbons in the solution from the remaining sulfur, nitrogen, and oxygen-containing compounds that may be present in one or both of the solutions. Such a fractionation procedure is well known as such in the art and is popularly identified as a chromatographic procedure and employs adsorbents such as activated alumina, silica gel, magnesia, clays, and the like. An especially preferred adsorbent is alumina which has been activated by heating to about 450°C for a period of four hours. A desirable bed of activated alumina for the purpose has a length to diameter ratio of the order of about 10 to 1.

Having separated the hydrocarbons from the nonhydrocarbons in this chromatographic procedure, the hydrocarbons are isolated and weighed or measured in any conventional manner. One particularly effective way comprises successive elution with n-heptane and benzene. The hydrocarbons may be separated from the elutriants as by the vaporization procedure described earlier. The amount of hydrocarbons thus found in a rock is expressed hereinafter as barrels per acre foot (bbl HC/acre-ft).

An apparatus developed by *C.A. Youngman; U.S. Patent 2,883,856; April 28, 1959; assigned to The Atlantic Refining Company* is an apparatus for accurately detecting the presence of hydrocarbons in the return mud stream of a rotary drilling system.

It is very unusual to be able to determine the presence of such petroliferous material by ordinary visual or sensual means; therefore, several methods have been developed by which the presence of minute quantities of petroliferous material can be detected. One of these methods, known as chemical logging, has been employed for some time and involves collecting a sample of the fluent of the return stream after it has circulated in the well and before it has returned to the mud pits and thereafter analyzing the fluent for the presence of any hydrocarbon gas.

In another method the cuttings which are carried to the surface of the earth by the fluent and thereafter separated from the mud by means of a shale shaker at the mud pit, are collected and later analyzed for their hydrocarbon content.

Regardless of the success which has been experienced in the use of the above two methods, in many cases it has been found that neither of the two methods has proven satisfactory. In the first case, the mud frequently doesn't carry enough petroliferous material to evolve sufficient quantities of hydrocarbon gases for detection. In the second case, the cuttings in being separated from the mud often lose a great deal of their hydrocarbon content.

As a result, neither the mud or cuttings analyses have proven to be completely satisfactory for indicating "hydrocarbon shows," the term used when one of the methods indicates the presence of petroliferous material in a zone traversed by a borehole. Therefore, many oil-producing zones have been drilled through and completely undetected.

The improved process involves simultaneously and additively detecting hydrocarbon gases present in both the mud and cuttings of a rotary drilling system. This is accomplished by simultaneously catching both the mud and the cuttings in the return stream and thereafter pumping the mud and the cuttings to a grinder where the cuttings are comminuted in the presence of the mud with which they were collected and in the same order as they were collected. The gas evolved from both the mud and the cuttings is continuously analyzed to thereby determine whether a particular zone traversed by the borehole contains petroliferous material in sufficient quantities for producing hydrocarbons.

Such an apparatus is shown schematically in Figure 74. The central view shows the overall apparatus; the view at the upper left is a detail of the mud sampling point and the view at the lower right is a detail of the grinder used.

Figure 74: Atlantic Refining Co. Apparatus for Detecting Hydrocarbons in Drilling Mud and Cuttings

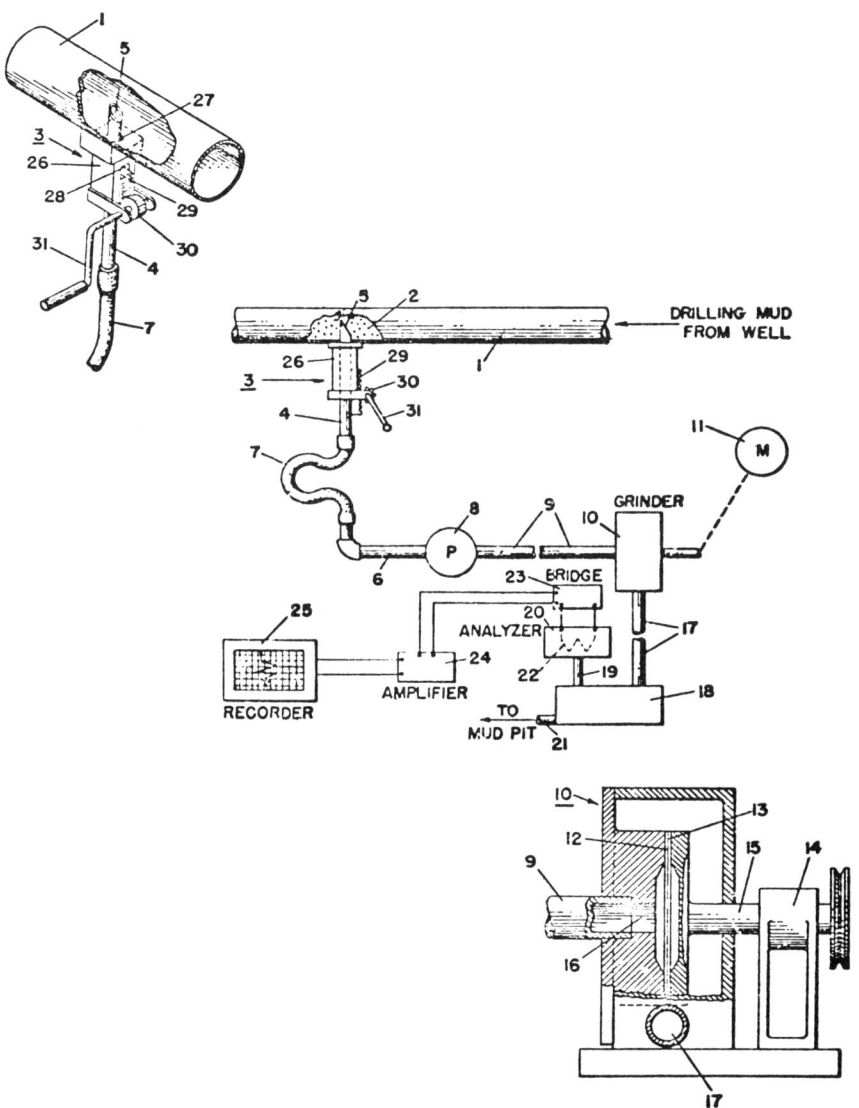

Source: U.S. Patent 2,883,856

Conduit 1 is shown representing the return line for the mud stream which comprises a fluent normally containing drilling mud and cuttings 2 dispersed in the mud. Sampling tube 4 which is a part of the preferred adjustable sampling tube assembly 3, is provided with an oblique opening 5 facing into the flow of the stream.

The position of opening 5 is adjustable in order to control the proportion of cuttings to mud caught by tube 4 and eventually passed to pump 8. Rubber coupling 7 provides a flexible connection between tube 4 and the conduit 6 leading into pump 8, to allow for such adjustment. Pump 8, which is any conventional type pump capable of handling both the mud and cuttings, is adapted for pumping the fluent through conduit 9 into grinder 10 which is the preferred grinder to be employed for the purpose of comminuting the cuttings in the presence of the mud with which they were collected.

The grinder comprises two grinding plates having serrated surfaces facing each other with one grinding plate 12 stationary and another grinding plate 13 rotatably mounted for rotation by motor 11. This rotatable mounting is accomplished by means of any conventional journal bearing box 14 in which shaft 15 is journaled and also axially adjustable for the purpose of adjusting the clearance between the grinding plates.

In the apparatus shown, an aperture 16 is provided through the center of stationary grinding plate 12 thus allowing the fluent containing both mud and cuttings to pass into the space between the serrations of the plates where comminution of the cuttings takes place while in the presence of the mud with which it was collected. After comminution both the mud and comminuted cuttings pass into conduit 17, in the same order in which the cuttings and mud were collected by tube 4.

The conduit is shown leading into trap 18 where the hydrocarbon gases present in the comminuted fluent are released and eventually passed through conduit 19 into analyzer chamber 20. The fluent is returned to the mud pits through line 21. Filament 22 located in the analyzer chamber is any conventional type of filament, preferably made of platinum, whose resistance varies in response to the concentration of combustible hydrocarbons in chamber 20. Thus, the filament connected in the electrical bridge 23, controls the reading of recorder 25 which is electrically connected to amplifier 24 and the bridge in any well-known manner. Accordingly the reading of the recorder indicates the amount of hydrocarbon gases present in the comminuted cuttings and mud.

Referring more specifically to the means employed for controlling the proportion of cuttings and mud pumped into grinder 10, there is shown the sampling tube assembly 3 for adjusting the height of opening 5 in the return stream.

The housing 26 is welded or fastened in any well-known manner to conduit 1. The interior of the housing communicates with the conduit through the aperture 27. Tube 4 is adjustably mounted in the housing and extends through the aperture. Any type of sealing means (not shown) is preferably located in the housing to prevent mud leaking out of the unit. The housing has a slot 28 on one side thereof in which rack 29, which is rigidly attached to tube 4, is adapted to slide. The position of the tube is thereby adjusted (either up or down) by movement of the rack in the slot, such movement being accomplished by means of the conventional rack and pinion arrangement shown, which comprises the pinion 30, rack 29 and crank 31.

It should be apparent from the drawing that tube 4 can be either raised or lowered by turning the crank, thereby providing means for adjusting the proportion of cuttings to mud entering opening 5 and eventually passing to grinder 10. As has been pointed out heretofore, opening 5 is preferably an oblique opening facing into the flow of the fluent stream, as shown. However, it may have other forms, e.g., the opening may be the end of an L-shaped member and form a 90° angle with the flow of the fluent stream or the end of a linear member with the opening parallel to the flow of the stream. Other geometric forms of the opening will be obvious. Also, the opening may be further modified to control the size of cuttings particles entering the sampling tube 4 by placing various mesh-size screens or the like over the opening.

In practice, the sampling tube is first adjusted so that the opening catches the correct proportion of cuttings and mud which grinder 10 is capable of handling. Thus the cuttings will always be comminuted in the presence of the mud with which they were collected. After entering the opening, the cuttings and mud flow through tube 4, coupling 7 and conduit 6 to pump 8. From the pump the fluent is pumped through conduit 9 into grinder 10 where the comminution of the cuttings takes place in the presence of the drilling mud and in the same order as they were collected by the tube.

Thereafter both the drilling mud and comminuted cuttings pass through conduit 17 into trap 18. At the trap hydrocarbon gases present in the comminuted cuttings and mud are evolved and pass through conduit 19 into analyzing chamber 20. As stated previously, the hot wire filament 22, whose resistance varies with the concentration of the hydrocarbon gases present in chamber 20, varies the balance of bridge 23 which in turn affects the reading of the recorder 25. Thus the recorder gives an indication of the combined hydrocarbon gas content of both the mud and the cuttings.

A technique developed by *C.J. Engle and S.J. Marwil; U.S. Patent 2,923,151; February 2, 1960; assigned to Phillips Petroleum Company* involves extracting gas from well drilling mud by passing the same tangentially into a hydraulic cyclone and removing the gas from the resulting vortex created therein. In one aspect it relates to analyzing at least a portion of the extracted gas for certain components, such as hydrocarbon gases, which would indicate the presence of oil and/or gas in the formations penetrated by the well. In another aspect it relates to a gas extraction process suitable for use in logging the hydrocarbons released from the cuttings into the drilling mud during drilling.

One aspect is to utilize the force of the conventional mud circulation pump to not only force the drilling mud into and out of the well, but to provide the force necessary to the operation of a hydraulic cyclone gas extraction unit.

A technique developed by *K.H. Schmidt; U.S. Patent 2,938,117; May 24, 1960; assigned to Petroleum Service and Research Corporation* is one for well logging wherein a quantitative analysis of hydrocarbons present is made to provide sufficient data to predict, independent of other methods such as core analysis or electrical logging, that a "sand" will produce gas, oil or water.

A technique developed by *J.W. Graham and J.S. Osoba; U.S. Patent 2,951,940; September 6, 1960; assigned to Jersey Production Research Company* is one in which the presence of crude oil in the earth's strata is detected or ascertained by operations on the very small rock chips carried to the earth's surface by drilling mud during drilling operations. The presence of hydrocarbons is detected by contacting the surface of a rock chip with a surface-active agent of the type that causes preferential wetting of the surface by hydrocarbons present therein (so-called "reverse-wetting" agents) and subjecting the chips to ultraviolet light.

By comparing the relative fluorescence of the surface of the chip before and after the contacting operation, a positive indication of the presence of hydrocarbons may be obtained.

It is preferable to obtain two rock chips from a small section of the earth's stratum being penetrated by the drilling bit. For this purpose, it is most desirable to obtain a chip large enough to yield two chips of approximately the same size; however, if convenient, it may be feasible to take two chips that are brought to the surface by the drilling mud at approximately the same time, this indicating they were severed from the earth by the drilling bit within the same stratum of the earth and within the same general region or section thereof. The surface of only one of the chips is contacted with the reverse-wetting agent, and direct comparison of relative fluorescence is made under ultraviolet light.

A method developed by *A. Blanchard; U.S. Patent 2,983,586; May 9, 1961; assigned to Schlumberger Well Surveying Corporation* involves the detection of oxidizable materials in earth formations traversed by a borehole and, more particularly, the detection of the presence of the oxidizable materials in the earth formations, on the sides of which a mudcake or other similar material has formed, by exposing portions of the earth formations to the direct influence of an oxidizing agent.

However, it has been found that the known systems have not proved particularly successful owing to the difficulty of injecting the oxidizing agent through the mudcake formed on the sides of the borehole into the formation under investigation.

This difficulty may be overcome in the method, by causing fluids present in the earth formations under investigation to flush out and away the mudcake formed on the borehole wall adjacent the area under investigation. This is accomplished by isolating an area of the borehole wall including the adjacent mudcake and providing a space contiguous the isolated area in which the pressure is lower than the pressure of fluids within the formation under investigation. This lower pressure is maintained for as long a period as is necessary to permit fluids from the formation under investigation to flush out the mudcake.

An oxidizing agent capable of reacting with petroliferous material, hydrocarbons, or other oxidizable substances is introduced into a space contiguous with the then exposed portions of the earth formations at a pressure at least equal, and preferably somewhat higher than, the pressure of the fluids present within the earth formations so as to enable a detectable reaction to occur between the oxidizing agent and any petroliferous material, hydrocarbon, or other oxidizable material that may exist in the formation. The occurrence of the reaction is then preferably detected by differential thermometric means responsive to a differential increase in temperature between a point within the influence of the reaction and a relatively remote point outside the influence of the reaction.

A technique developed by *E.J. Dower; U.S. Patent 3,049,409; August 14, 1962; assigned to Warren Automatic Tool Company* involves testing hydrocarbons contained in returning well drilling mud as an indication of whether earth formations being traversed have significant hydrocarbon content.

When hydrocarbons are returned with drilling mud, an immediate indication of the fact is desirable, as is also a continuing record throughout the drilling operation of whether and how much hydrocarbons are being returned together with an occasional test of samples from productive zones for determination and recording of the constituency and proportionality of different components. For such purpose, it is here proposed to employ a flowing column of carrier air for entraining hydrocarbons removed from returning mud and to mix the carrier air and any entrainments with a combustible inorganic gas for flame ionization. Inasmuch as electrical conductivity of a flame burning organic substance varies according to organic substance content of the mixture, a measurement of conductivity affords convenient indication of the characteristics of the mixture.

The ability to carry out analyses of hydrocarbon mixtures so as to determine with precision the separate constituents thereof has brought about a pressing need for a sampling technique which can provide a sample of the hydrocarbons present in a drilling mud in such a fashion that the relative proportion of the various single constituents are not varied from one sampling to another by random changes in the extraction process. Otherwise stated, with the analytical methods in use in the field in this art, many years ago, relatively "rough-and-ready" sampling devices and processes sufficed, because the analytical methods used were relatively unsophisticated.

These generally included merely a semiquantitative determination of methane on the one hand and of hydrocarbons heavier than methane on the other, in a gas sample obtained from a drilling mud. "Oily" hydrocarbons were determined by ultraviolet inspection, the oily droplets fluorescing readily. However, because of the advent of gas chromatography as well as of mass spectroscopy it is now possible in a field installation, and without an undue expenditure either for equipment or of skilled manpower, to carry out analyses of a scope and accuracy previously not thought possible, and in particular, it is possible to carry out highly quantitative analysis of the several individual species of hydrocarbons present in a mixture.

It has been found that when such analyses showing individual hydrocarbon concentrations are plotted as a function of depth, exceedingly useful information is obtained for logging purposes generally, and for correlating and productivity estimating purposes in particular. The situation is accordingly now reversed from what it was in the past, and the weak point in such mud logging is no longer the type and concentration of hydrocarbons present, but rather is now the sampling method employed, which must provide a sample which is truly representative of the hydrocarbons present in the drilling mud before the latter is tampered with.

Thus, an improved sampling method has been developed by *P.J. Moore; U.S. Patent 3,050,449; August 21, 1962; assigned to National Lead Company.*

A technique developed by *A.E. Worthington; U.S. Patent 3,118,299; January 21, 1964; assigned to California Research Corporation* involves examining samples obtained from subterranean formations to detect the presence therein of commercially valuable hydrocarbons containing molecules having at least five carbon atoms and having a boiling range of about 25°C (77°F) to 180°C (350°F) and known as gasoline fraction hydrocarbons.

Heretofore, in examining samples of earth formations contained in drilling fluid returns to find indications of penetration of oil bearing rocks, the practice was to examine such fluids and/or cuttings by one or both of two common methods. One method includes examination of drilling fluid returns for the presence of gases, e.g., methane, ethane, etc. In this system, drilling fluid or cuttings are subjected to agitation, evacuation and/or aeration, and vapors of the light hydrocarbons mixed with air are drawn off and sampled.

The presence of hydrocarbons in this air stream is indicated by measurement of the electrical conductivity of a hot wire, the wire serving as a catalyst for combustion of the gas-air mixture, thus causing the wire temperature to increase so that the presence or absence of hydrocarbon gases in the air is shown by an increase or decrease in electrical conductivity of the hot wire. By adjusting the temperature, either methane alone or methane plus ethane and higher MW hydrocarbons can be detected. Since methane is more abundant than ethane and other higher MW gases, even though the determination of methane by the hot-wire method is quite accurate, the determination of ethane and higher is inaccurate because of the small difference between two large numbers.

Unfortunately, also, high accuracy of measurement of methane is not warranted as a method of exploring for commercial petroleum accumulations, since "dry" gases consisting mostly of methane are frequently encountered in the earth in locations such as peat beds, coal seams, shales, dry-gas sands, and the like; thus in normally favorable sedimentary rocks, traces of such gases occur many times without any relation to accumulations of commercial oil.

For this reason, the accuracy of the measurement of methane does not indicate the presence or absence of commercial oil, and in part it can be misleading as to the need for performing formation fluid tests, either while drilling or after the well has been completed. Such formation tests are quite expensive in the total cost of drilling a well. Such high cost is due in part to the fact that it is usually necessary to set steel casing and achieve a suitable cement bond to seal the casing to the borehole.

The second method of drilling fluid logging common in field practice is to inspect the drilling fluid returns with ultraviolet light. The purpose of this inspection is to detect the presence of fluorescent materials that frequently are present in oil accumulations in the earth. Unfortunately, such fluorescent materials are also frequently encountered in other substances, not containing gasoline fraction hydrocarbons, such as those found in tar streaks and other petroleum deposits where all of the lighter components, specifically those that make an oil accumulation commercially interesting, have been lost or have migrated away from the deposit.

Exploratory Well Drilling

Various minerals, oils, gases, and such, used in the drilling operation, or even in the drilling fluid, may also be fluorescent. Accordingly, ultraviolet light inspection of the drilling fluid returns gives a far greater number of "oil indications" than the true number of potentially commercial petroleum deposits. Thus, ultraviolet methods for inspection of drilling fluid returns are as unsatisfactory for finding commercially valuable oil accumulations as the methane gas analysis system outlined above.

In contrast to the prior art methods of logging drilling fluid returns, the technique of determining the presence of significant quantities of gasoline fraction hydrocarbons in the drilling fluid returns provides a good indication and basis for distinguishing between commercial oil and dry gas or worthless remnant petroleum deposits.

A method developed by *G.W. Jamieson; U.S. Patent 3,118,738; January 21, 1964; assigned to Jersey Production Research Company* is an improved technique for the recovery and measurement of a gas in a drilling mud which, in essence, comprises out-gassing the mud in a degassing chamber utilizing carbon dioxide or an equivalent gas. The carbon dioxide is then removed and separated from the gas which is then accurately measured.

A system developed by *J.M. Horeth, R.H. Langenheim, and W.D. Howard; U.S. Patent 3,240,068; March 15, 1966; assigned to Esso Production Research Company* is designed to automatically analyze small drilling mud samples for their gas contents in terms of methane, ethane, propane, isobutane, normal butane and heavier hydrocarbons, such as heptanes and hexanes, and to record the analysis of these hydrocarbons by means of a recording gas chromatograph.

Various techniques and devices have been employed heretofore in the sampling and analysis of drilling muds. It is known in the art to volatize hydrocarbons from mud samples and to subject the hydrocarbons to analysis by gas chromatography.

A serious difficulty which has been encountered in the chromatographic analysis of hydrocarbon gases obtained from drilling mud in accordance with prior sampling techniques is baseline drift. The chromatograms obtained have been unreliable due to a persistent failure of the baseline to remain constant and stable throughout a repetitive analysis program. In an attempt to avoid this problem, it has been necessary to delay the analysis of each successive sample for a period of ten to twenty minutes. In actual field operations, such periods of delay are frequently intolerable.

This system is based in part upon the discovery that such baseline drift is caused by a contamination of the chromatographic column with small amounts of water and heavy hydrocarbons inadvertently introduced therein, along with the gaseous sample. Accordingly, there is provided a sampling apparatus which is capable of supplying a mud-gas sample free of contamination.

A scheme developed by *D.O. Seevers; U.S. Patent 3,287,088; November 22, 1966; assigned to Chevron Research Company* is one in which water suspected of containing liquid aromatic hydrocarbons is subjected to irradiation by ionizing radiation. Such ionizing radiation may be in the form of electromagnetic radiation, such as ultraviolet rays, x-rays, or gamma rays, or in the form of ionizing particles such as beta rays or alpha particles.

The irradiation is carried out for a predeterminable length of time sufficient to convert at least a portion of the hydrocarbons present in the water to a phenolic compound. The water samples are irradiated for a predeterminable length of time, based on the knowledge of the source strength and source geometry, to produce a predeterminable radiation dose. The irradiated water is then examined or analyzed to obtain a measure of the concentration therein of the phenolic compound.

The overall equipment arrangement for the conduct of such a scheme is shown in Figure 75. It portrays the analysis of water absorbed in a drilling fluid in the drilling of an exploratory well. Reference character **41** designates generally a derrick for supporting a drilling rig which includes a rotary table **42** driving a length of drill pipe **43**. A length of casing **44** surrounds drill pipe **43** between the ground and the floor of the drilling rig and is provided with a normally open blowout preventer **45** at approximately the surface of the ground. As is well known in the art, drilling fluid is circulated downwardly through the drill pipe, through openings in the drilling bit and back up the annulus between the drill pipe and the hole wall or casing to lubricate the bit, to maintain pressure on the drilled formations and to carry cuttings to the surface.

The drilling mud flows through the annulus between the drill pipe and casing and out through a conduit **47** to a suitable screening device **48**, such as a shale shaker, where the larger pieces are screened from the drilling fluid and the remaining fluid is returned to a mud sump **49**. A portion of this returning drilling fluid is withdrawn and analyzed to determine the concentration of aromatics.

Such withdrawal may be through a conduit **51** connected to conduit **47** and leading to a suitable separator **52**. The separator may be, as shown, a steam distillation device in which steam, entering through a conduit **53**, is bubbled through the drilling fluid in a series of baffle plates **54** to extract the more volatile constituents. These more volatile constituents are drawn off from the separator at the top thereof through a conduit **56** leading to a condenser **57**. The portion of the withdrawn drilling fluid which passes downwardly through the separator is returned to mud sump **49** through a conduit **58**.

As shown in the drawing, the condenser utilizes a cooling medium, such as water, which is circulated past the coils carrying the withdrawn fluid to condense the fluid to a liquid. The condensed liquid from the condenser is exposed to radiation of a predetermined dose in a suitable device.

In the illustrated embodiment, the condensed liquid passes through a coil **61** surrounded by a radioactive material **62** forming a source of gamma rays or x-rays. Source **62** is, in turn, surrounded by suitable shielding material **63**. The condensed liquid is thus exposed to gamma ray or x-ray radiation for a predetermined length of time in traversing coil **61**, to convert all or a portion of the aromatics therein to phenolic compounds. Since the strength of the source is known, the total dose may be closely regulated by regulating the rate of flow of the condensed liquid through the coil.

The irradiated liquid from the coil then has added thereto a suitable color reagent from a source **66** through a valve **67**. The mixture of irradiated liquid and color reagent then passes to colorimeter **27** where the colorimetric analysis is performed, in a manner similar to that described above, and the results thereof indicated by a pen **29** on a record strip **30** driven by a motor **32**.

Figure 75: Chevron Research Co. Process for Analyzing Drilling Fluid for Aromatic Hydrocarbons

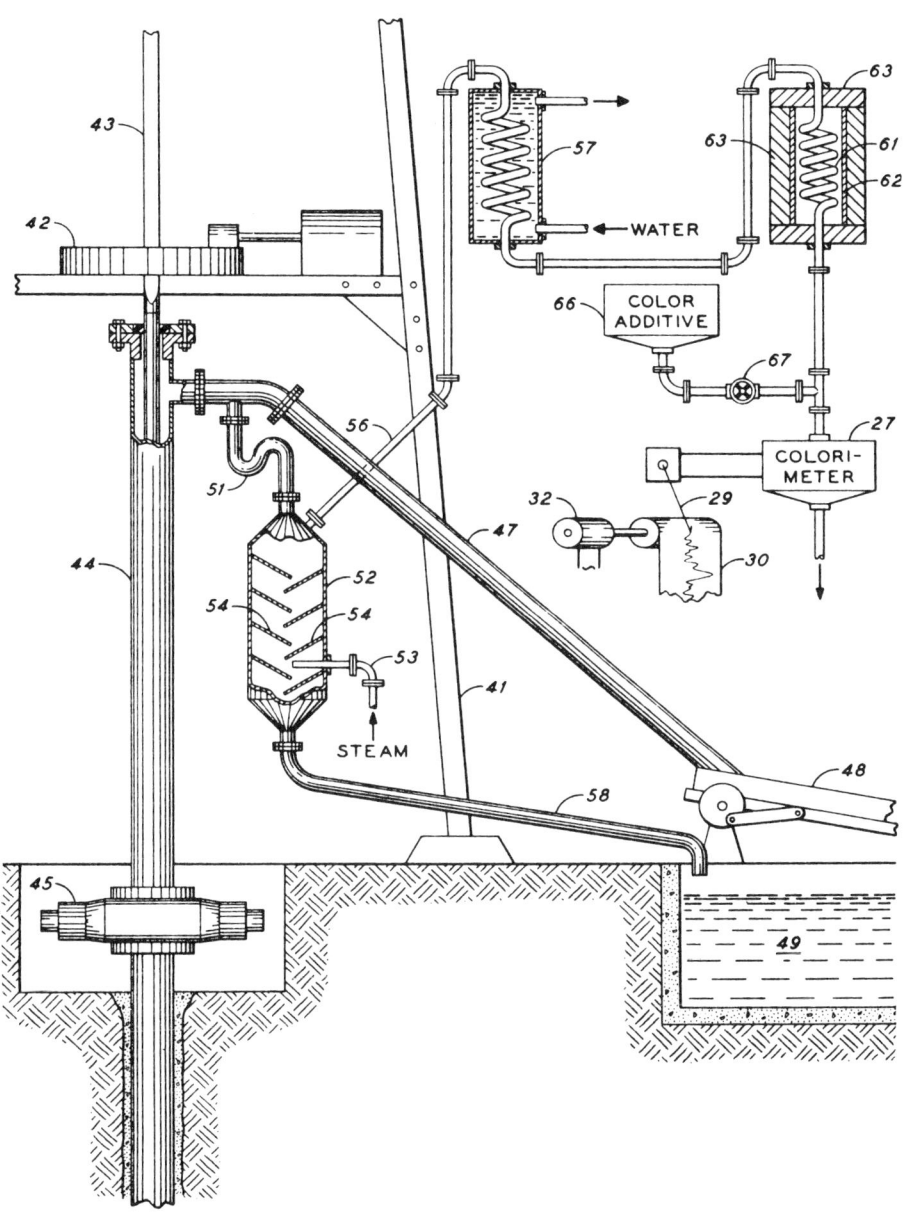

Source: U.S. Patent 3,287,088

The record on record strip 30 may be in terms of phenol concentration, from which the aromatic concentration may be determined by suitable correlation or alternatively, this correlation may be incorporated in the scale of record 30 and in the other elements of the recorder, so that the record strip reads directly in terms of aromatic concentration.

A technique developed by *O.T. Moore; U.S. Patent 3,386,286; June 4, 1968* is a hydrocarbon well logging method which includes, in addition to the usual depth measurements, mud flow measurements and detection of hydrocarbon gases in the diverted portion of the mud, the steps of measuring the total flow rate of mud, the flow rate of diverted mud and correlatively recording the total relative amount of gas for the mud flowing through increments of the well bore, in order to obtain the relative gas content of each volumetric increment of the formation through which the well is drilled. In addition the relative amount of total methane, ethane and propane, as well as the relative amount of normal butane, isobutane, normal pentane and isopentane may be recorded.

A method developed by *J.D. Schroeter; U.S. Patent 3,422,674; January 21, 1969* features an apparatus for continuously and simultaneously analyzing a sample of drilling fluid flowing through a mud ditch for the presence of oil content, gas content and electrical resistivity value. The detected amounts in which they are present are simultaneously logged on graphs for visual correlation which move at a rate corresponding with the depths from which the samples are taken.

A technique developed by *J.M. Horeth, W.D. Howard and R.H. Langenheim; U.S. Patent 3,462,761; August 19, 1969; assigned to Esso Production Research Company* involves recording the light hydrocarbon content of drilling fluids as a function of drilling depth. The chromatographic analyses of individual components of the hydrocarbons are transmitted to a recorder and individually recorded. The advance of the recorder is controlled by the drilling depth at the time of the analysis and in a more limited embodiment the advance of the recorder is controlled so that the analysis is reproduced on the depth scale at a point at which the hydrocarbons are introduced into the mud stream.

Previously proposed procedures for locating petroleum reservoirs measure the concentrations with location of the hydrocarbons that are contained in core samples, those contained in bit cuttings, those that become entrained in circulating liquid drilling fluids or muds, those that are dissolved or entrained in the fluids which flow into a borehole when the pressure within the borehole is reduced, etc. The concentrations at which the hydrocarbons are present in relatively shallow earth formations tend to be very small, generally less than one part per million.

Because of this, such previously proposed exploration procedures have tended to require costly operations, such as the careful collecting and preserving and analyzing of whole samples of the earth formations; or have tended to provide measurements that are difficult to correlate with respect to each other or with respect to other information, such as seismic information, relating to the region being explored.

A technique developed by *J.B. Turner and H.L. Wise; U.S. Patent 3,645,131; February 29, 1972* is one in which earth formations are sampled by gas drilling boreholes into them while circulating a gaseous drilling fluid that is initially substantially free of a selected mobile reservoir fluid. The concentration of the mobile reservoir fluid in the earth formations around the borehole is measured by measuring the difference in its concentration in equivalent portions of the drilling fluid as the drilling fluid enters and leaves the borehole. Such measurements indicate the concentration of the mobile reservoir fluid with areal location and/or depth within the earth formations.

Figure 76 shows a suitable form of apparatus for the conduct of such a technique. It shows a gas drilling system comprising a rig structure **1**, a drilling assembly **2**, a gaseous drilling fluid input line **3** (which is connected to a compressor or blower, not shown) a bit-cuttings separator **4**, and a gas discharge sample line **5**.

Figure 76: Schematic of Soil Sampling for Hydrocarbons in Connection with Gas Drilling Operation

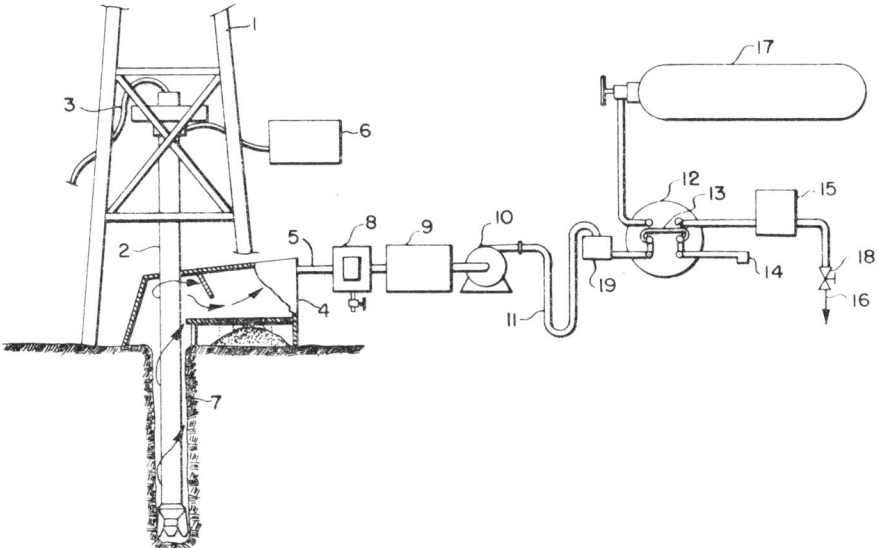

Source: U.S. Patent 3,645,131

A chromatograph **6** is arranged to measure the concentration of a selected mobile reservoir fluid in the inflowing gaseous drilling fluid before it has been circulated into borehole **7**. A liquids and solids separator **8** is connected to receive and treat a portion of the gas being discharged from the bit-cuttings separator. The gas treated by separator **8** is supplied to a manifold **9**. The manifold preferably contains a filter means, such as a micropore dust filter (not shown), for further cleaning the gas. A transfer pump **10** is connected to displace a sample stream of gas from the manifold through a drying means **11** and into a multiport valve **12**. This valve, in position shown, allows a sample stream of gas to flow through a sample trap loop **13** and exit through port **14**.

The position of the ports in the valve can be arranged so that the fluid in the trap loop is displaced through chromatograph **15** and discharged through port

16. The displacement of the fluid from trap loop **13** is preferably effected by allowing an inflow of clean gas from a source **17** and outflow, ultimately through port **16**, as indicated by the arrows. The pressure of the fluid displaced into chromatograph **15** is preferably regulated by a throttle valve **18** arranged upstream of discharge port **16** to maintain a pressure between that of a relatively high pressure at clean air source **17** and atmospheric pressure.

In a preferred arrangement, check valve **19** is placed upstream from multiport valve **12** to restrict backflow without significantly impeding forward flow. Pump **10** is preferably operated at substantially atmospheric pressure. If the sample loop is filled with clean gas at a higher pressure when it is connected to the outflow from the check valve, the higher pressure is dissipated by a purge through discharge port **14** and the loop is refilled with undiluted sample gas at substantially atmospheric pressure.

In this process the measurement of the concentration of the selected fluid (such as a hydrocarbon) in the circulating drilling fluid, can be made by means of substantially any measuring device which is relatively sensitive and accurate with respect to concentrations of less than about one part per million. A chromatograph comprising a combination of a chromatographic column, a flame ionization detector, and an electrometer is a preferred type of such a measuring device.

The sampling and analyzing of the drilling fluid before and after its circulation into the borehole can be performed by means of one or more such measuring devices. It is desirable that, at substantially each increment of depth of the borehole, at least one measurement before and after such circulation is made on the same portion (or on substantially equivalent portions) of the circulating gaseous drilling fluid.

In the system of Figure 76, for a given rate of drilling fluid circulation, a portion of an inflowing gas on which a measurement is made as the gas enters the drill string assembly **2** (by means of chromatograph **6**) will, after a determinable length of time, become the gas that is filling the trap loop **13** of valve **12**. The gas in the trap loop is isolated for measurement (by means of chromatograph **15**) by changing the post arrangement of valve **12**.

In field operations utilizing chromatographic measurements, it has been found that samples of about 10 cc in volume are adequate, and such measurements can readily be repeated as often as every 15 seconds. At a typical drilling rate of about 6 feet per minute such a sampling rate provides a measurement that corresponds to each yard of advance of the borehole.

The measurements corresponding to each increment of depth, that is, the measurements of the difference between the hydrocarbon concentrations of equivalent portions of the circulating drilling fluid before and after they have flowed through the borehole, tend to be selectively responsive to the newly exposed earth formations near the bottom of the borehole and relatively unaffected by high or low concentrations that were present in formations encountered at shallower depths. The consistency of such a selectivity of response is enhanced by controlling the circulation rate of the drilling fluid so that the rate is substantially the same while each portion of the drilling fluid on which a measurement is to be made is being circulated through the borehole.

As a borehole is drilled, at each depth to which the hole is extended, some of the rock solids are released, as bit cuttings, and some remain, as rocks exposed along the wall of the borehole. The surfaces of both the released and exposed rock solids are contacted by a relatively large volume of circulating drilling fluid. Where the drilling fluid is a gas that was substantially free of hydrocarbons at the time it entered the borehole, hydrocarbons tend to be rapidly desorbed from the contacted rock surfaces and entrained in the circulating gas.

The rate of desorption is initially very high and, within a few seconds, tends to decline to a rate that is much lower. Because of this, the magnitude of the hydrocarbon content of the outflowing portions of such a drilling fluid tends to be selectively responsive to the amount of hydrocarbon desorbed, or otherwise released, from the newly exposed earth formations near the bottom of the borehole.

Thus, the difference in the hydrocarbon content of a portion of the circulating gaseous drilling fluid before it enters and after it leaves the borehole tends to be the hydrocarbon content of the newly released and exposed rocks that are located at a depth near that to which the borehole is then extended. When such measurements are timed so that they sample the amounts of each of a plurality of hydrocarbons that are being desorbed at each of a series of increasing depths, the results show both the variation with location of the concentration of each hydrocarbon and the tendency for some of the upward migrating hydrocarbons to be adsorbed ahead of others, and thus to be retained at a lower depth, while the hydrocarbons were migrating up through the earth formations into which the borehole is extended.

BIBLIOGRAPHY

(1) U.S. Department of the Interior, "Suggestions for Prospecting," Washington, D.C., U.S. Geological Survey (1978).
(2) U.S. Department of the Interior, *United States Mineral Resources*, Geological Survey Professional Paper No. 820, Washington, D.C., U.S. Geological Survey (1973).
(3) U.S. Department of Commerce, *Technology Assessment and Forecast: Eighth Report*, Washington, D.C., Patent and Trademark Office (December 1977).
(4) Moody, G.B., Editor, *Petroleum Exploration Handbook*, New York, N.Y., McGraw-Hill Book Co. (1961).
(5) Muir, W.L.G., *Coal Exploration: Proceedings of the First International Coal Exploration Symposium, London, England, May 18-21, 1976*, San Francisco, Miller Freeman Publications, Inc. (1976).
(6) Kuzvart, M. and Bohmer, M., *Prospecting and Exploration of Mineral Deposits*, New York, Elsevier Scientific Publishing Co. (1978).
(7) Peters, W.C., *Exploration and Mining Geology*, New York, John Wiley and Sons (1978).
(8) Levinson, A.A., *Introduction to Exploration Geochemistry*, Calgary, Applied Publishing Ltd. (1974).
(9) Campbell, M.D., Editor, *Geology of Alternate Energy Resources in the South-Central United States*, Houston, Texas, Houston Geological Society (1977).
(10) Brooks, R.R., *Geobotany and Biogeochemistry in Mineral Explorations*, New York, Harper and Row (1972).
(11) Smith, W.L., Editor, *Remote Sensing Applications for Mineral Exploration*, Stroudsburg, Pa., Dowden, Hutchinson and Ross, Inc. (1977).
(12) Nettleton, L.L., *Gravity and Magnetics in Oil Prospecting*, New York, McGraw-Hill (1976).
(13) Allaud, L.A. and Martin, M.H., *Schlumberger: The History of a Technique*, New York, John Wiley and Sons (1977).
(14) Morse, J.G., Editor, *Nuclear Methods in Mineral Exploration and Production*, New York, Elsevier Scientific Publishing Co. (1977).

(15) Condon, J., *Proceedings of the Second International Coal Exploration Symposium, Denver, Colorado, October 1-4, 1978*, San Francisco, Miller Freeman Publications, Inc. (1979).

COMPANY INDEX

The company names listed below are given exactly as they appear in the patents, despite name changes, mergers and acquisitions which have, at times, resulted in the revision of a company name.

Addeco, Inc. - 266
Aircraft Engineering and Maintenance Co. - 87
Amoco Production Co. - 55, 201, 209, 246, 247, 248
Applied Invention Corp. - 156
Arthur D. Little, Inc. - 101
Atlantic Refining Co. - 63, 68, 88, 271, 274
Atlantic Richfield Co. - 64, 143, 194, 195, 201, 246, 252, 255
Barringer Research Ltd. - 16, 19, 23, 56, 70, 72, 90, 132
Bendix Corp. - 226
British Petroleum Co., Ltd. - 120
California Research Corp. - 77, 113, 179, 253, 280
Canadian Airborne Geophysics Ltd. - 129
Canadian Minister of Mines and Technical Surveys - 97
Centre National de la Recherche Scientifique - 115
Charles Stark Draper Laboratory, Inc. - 102
Chevron Research Co. - 48, 98, 137, 165, 247, 254, 281
Cities Service Research and Development Co. - 12, 45
Columbia Scientific Industries, Inc. - 19
Compagnie Generale de Geophysique - 112, 113, 258, 261
Consolidated Engineering Corp. - 33
Continental Oil Co. - 30, 211, 230, 243, 249, 250
Cornell University - 142
Crossland Licensing Corp. Ltd. - 99
Dow Chemical Co. - 23
Dresser Industries, Inc. - 142, 165, 167, 171, 179
Electricite de France - 156
AB Elektrisk Malmletning - 129
Ensign-Bickford Co. - 189
Entreprise de Recherches et d'Activites Petrolieres Elf - 194
Entreprise de Recherches et d'Activites Petrolieres (ERAP) - 260
Esso Production Research Co. - 250, 281, 284
Esso Research and Engineering Co. - 266, 269
Exxon Production Research Co. - 139, 203, 211
General Electric Co. - 9, 178
Geomet Mining and Exploration Co. - 28, 69, 142
Geo-Nav, Inc. - 135
Geophysical Exploration Co. - 125
Geophysical Research Corp. - 243
Geophysical Survey Systems Inc. - 135
Geosource, Inc. - 234, 251
Geo Space Corp. - 234
Gesellschaft fur Kernforschung mbH - 13
GTE Sylvania Inc. - 259

Company Index

Gulf Research & Development Co. - 62, 96, 98
Halliburton Co. - 134
Hamilton Brothers Oil Co. - 211
Hammit & Mangum Service Co., Inc. - 268
Hercules Inc. - 188, 193, 194
Huntec (70) Ltd. - 226
Hydroacoustics Inc. - 220, 227
Imperial Chemical Industries, Ltd. - 191
Institut Francais du Petrole - 39
Institut Francais du Petrole, des Carburants et Lubrifiants - 37, 194, 227, 255
Jersey Production Research Co. - 230, 246, 268, 272, 278, 281
Labofina SA - 38
Lost River Mining Corp. Ltd. - 24
McCollum Geological Explorations, Inc. - 186
Mobil Oil Corp. - 2, 34, 46, 52, 63, 67, 68, 86, 93, 168, 178, 239, 241, 248, 254, 255, 261
Monarch Logging Co., Inc. - 270, 271
National Lead Co. - 33, 280
Nordel Corp. - 117
Nucom Ltd. - 132
Occidental Oil Shale, Inc. - 10
Ormad Systems - 89
Pan American Petroleum Corp. - 35, 46, 54, 63, 252
Petroleum Service and Research Corp. - 277
Petty Geophysical Engineering Co. - 250
Petty-Ray Geophysical, Inc. - 203, 240, 253
PGAC Development Co. - 120, 156, 172
Phillips Petroleum Co. - 31, 37, 54, 55, 77, 83, 120, 248, 264, 277
Pure Oil Co. - 42, 43, 97, 268
Rayflex Exploration Co. - 34
Resource Control Corp. - 269
Ruhrkohle AG - 7
Schlumberger, Ltd. - 172
Schlumberger Technology Corp. - 171, 176, 180, 237, 255
Schlumberger Well Surveying Corp. - 134, 179, 266, 272, 278

Scintrex Ltd. - 125, 127
Seiscom Delta Inc. - 202, 220, 238, 239
Seismic Computing Corp. - 251
Selco Exploration Company Ltd. - 135
Senturion Sciences, Inc. - 202
Shell Development Co. - 156, 179
Shell Oil Co. - 35, 53, 57, 66, 215, 233, 256
Sinclair Research, Inc. - 56
Societe Nationale des Petroles d'Aquitaine - 201, 215, 251
Societe Nationale Elf Aquitaine (Production) - 253
Socony Mobil Oil Co., Inc. - 39, 41, 74, 75, 135, 162, 174
Socony-Vacuum Oil Co. - 116, 248
Standard Oil Co. (Indiana) - 85, 212, 258
Standard Oil Development Co. - 246
Stanolind Oil & Gas Co. - 179
Sun Oil Co. - 56, 78, 81, 83, 246
Symynex - 166
Terradex Corp. - 38, 58, 147
Texaco, Inc. - 53, 134, 150, 157, 172, 173, 179, 180, 188, 192, 202, 218, 234, 253, 254, 258, 260
Texas Co. - 77, 142, 171, 179, 253, 254
Texas Instruments Inc. - 153, 207, 209, 211, 240, 246
Texas Pacific Oil Co., Inc. - 104
Union Oil Company of California - 212
U.S. Atomic Energy Commission - 165
U.S. Energy Research and Development Administration - 92
U.S. Secretary of the Interior - 25, 26, 28, 156
U.S. Secretary of the Navy - 99, 101, 226, 227, 253
Varian Associates - 97, 131
Vedeckovyzkumny uhelny ustav - 6
Warren Automatic Tool Co. - 279
Wells Surveys, Inc. - 153, 155, 162, 165, 166, 167, 172, 176
Western Company of North America - 178
Western Geophysical Company of America - 202, 205, 215, 230, 235, 245
Westinghouse Electric Corp. - 148
Wyler AG - 103

INVENTOR INDEX

Abbott, F.R. - 226
Aiken, C.B. - 134
Aine, H.E. - 37
Airhart, T.P. - 201
Aitken, A.R. - 246
Albertson, M.M. - 179
Allen, L.S. - 168
Allison, J.B. - 53
Alter, H.W. - 38, 58
Anstey, N.A. - 220, 238, 239
Apenberg, W. - 13
Arnold, D.M. - 157, 172, 180
Atwood, G.R. - 134
Badger, A.S. - 234
Bailey, J.R. - 202, 255
Baird, C. - 217
Barbier, M. - 215, 251
Barkalow, D.E. - 142
Barringer, A.R. - 16, 19, 23, 70, 72, 90, 132, 135
Bartz, G.L. - 264
Bayhi, J.F. - 203
Bays, M.G. - 202
Bedenbender, J.W. - 207, 211
Belcher, D.J. - 142
Berry, J.E. - 255
Biederman, E.W., Jr. - 45
Billings, G.K. - 56
Bissada, K.K. - 53
Black, H.D. - 101
Blake, F.G. - 247, 248
Blanchard, A. - 272, 278
Blau, L.W. - 246

Böhme, G. - 13
Bond, D.C. - 42, 268
Bonner, T.W. - 162, 174
Born, W.T. - 243
Borst, R.L. - 54
Bouyoucos, J.V. - 220, 227
Boyd, J.F. - 218
Bradley, J.S. - 63, 64, 68, 88
Braham, R.M. - 253
Brant, A.A. - 125
Bray, E.E. - 34, 39, 41, 52
Brillaud, A.R. - 83
Briner, S. - 97
Brittian, R.W. - 240
Broding, R.A. - 209, 212
Brown, J.B. - 266
Brown, L.R. - 81
Brown, R.J.S. - 247, 248
Buck, S.W. - 102
Buckley, S.E. - 269
Buckner, G.O., Jr. - 134
Bucy, J.F., Jr. - 253
Cagniard, L. - 115
Caldwell, R.L. - 162, 174, 178
Cardwell, W.T., Jr. - 179
Carman, W.H., Jr. - 201
Carter, R.F. - 254
Cartier, W.O. - 99, 132
Cassand, J. - 227
Chase, H.H. - 75
Chew, R.T., III - 10
Cholet, J. - 194, 260
Clews, D.R. - 56

Coggeshall, N.D. - 62
Cooper, J.E. - 67
Cowles, C.S. - 215, 233
Craig, C.B. - 64
Crawford, J.M. - 250
Cuykendall, T.R. - 142
Dahm, C.G. - 248
Daigle, E.E. - 57
Daniel, D.B. - 135
Davis, J.B. - 46, 63, 74
Davis, S.H. - 271
De Falco, R.J. - 83
Dehmelt, H.G. - 97
Del Grande, N.K. - 92
Delignieres, R. - 255
Dennis, C.L. - 86, 241
Denny, C.G. - 142
Desbrandes, R. - 255
Dick, C.W. - 202, 215
Dillingham, M.E. - 178
Dlouhy, L. - 6
Doll, H.-G. - 134, 266
Donaldson, C.A. - 135
Doty, W.E.N. - 250
Douros, J.D., Jr. - 81
Dower, E.J. - 279
Dowling, D.J. - 134, 218
Dransfield, C.D. - 201
Duffey, D. - 165
Dunlap, H.F. - 63
Durand, B. - 37
Duschatko, R.W. - 55
Egan, E.F. - 179
Egleson, G.C. - 23
Elliott, J.W. - 237
Ellis, L.G. - 246
Engle, C.J. - 277
Erdman, J.G. - 55
Erich, O.G., Jr. - 212
Escaron, P.C. - 255
Espitalie, J. - 37, 39
Evjen, H.M. - 117
Fail, J.-P. - 227
Fair, D.W. - 211
Fallgatter, W.S. - 45
Fanger, H.-U. - 13
Farr, J.B. - 247, 258
Fearon, R.E. - 155, 162, 165, 172, 176
Felice, P.E. - 148
Fiat, G. - 89
Field, H.S. - 198
Fitch, H.L. - 194

Foote, R.S. - 153, 156, 253
Forgotson, J.M. - 124
Foster, M.R. - 248
Franklin, A.A. - 194
Frederickson, A.F. - 54
Friedman, G.M. - 54
Gabillard, R. - 255
Gant, O.J., Jr. - 143
Gavrilov, J.G. - 15
Geiger, A.R. - 109
Gerrard, J.A.F. - 246
Gilbert, E.A. - 125
Givens, W.W. - 2, 178
Glaser, B. - 13
Godfrey, D.E. - 9
Goodman, C. - 176, 179
Goupillaud, P.L. - 249, 250
Gournay, L.S. - 86, 93
Graham, J.W. - 278
Gray, P.R. - 31
Grayson, J.F. - 85
Green, E.L. - 253
Greene, E.F., Jr. - 254
Groth, P.K.H. - 85
Gunter, J.R. - 68
Haas, H. - 7
Hain, K. - 13
Hall, C.S., Jr. - 34
Hall, E.M., Jr. - 230, 235
Hall, H.E., Jr. - 172, 173
Hamilton, F.F. - 211
Hammond, J.H., Jr. - 243
Hanson, W.E. - 62
Hardison, J.E. - 201
Harrell, J.W. - 86
Harris, R.A. - 253
Harvey, H.A. - 99, 132
Hassler, G.L. - 156
Hawes, W.S. - 234
Heacock, R.L. - 53
Heinze, B. - 45
Helling, P. - 7
Herzog, G. - 142, 179
Hitzman, D.O. - 77, 83
Hochheimer, H.-J. - 7
Hoehn, G.L., Jr. - 162
Hollingsworth, T.J. - 250
Holser, W.T. - 137
Holub, F.F. - 38
Hood, A. - 53
Hoover, H., Jr. - 33
Hopkinson, E.C. - 171
Horeth, J.M. - 281, 284

Horvitz, L. - 35, 58
Hoss, R.C. - 179
Howard, W.D. - 281, 284
Howell, E.P. - 143, 255
Hower, J., Jr. - 54
Hoyte, A.F. - 156
Hubener, J. - 13
Huckabay, W.B. - 34
Hufstedler, A.G. - 209
Hunt, H.B. - 104
Hunt, J.M. - 272
Hutchins, R. - 226
Itria, O.A. - 188, 192, 202
Jacob, H.-L. - 7
Jakosky, J.J. - 124
Jamieson, G.W. - 281
Janssen, H. - 120
Johnson, C.H., Jr. - 153
Johnson, J.S. - 211
Johnson, V.R. - 246
Johnston, O.A. - 202
Johnstone, C.W. - 180
Jones, H.J. - 246
Jones, S.B. - 137, 179
Judson, R.D. - 247
Kahma, A.A. - 129
Kalb, B.J. - 160
Kaufman, A.A. - 127
Kehler, P. - 156
Kelly, G.H. - 207
Kelly, S.R. - 189
Kelsey, M.C. - 34
Keowski, J.W. - 253
Kermabon, A.J. - 166
Kirby, R.A. - 203
Kunetz, G. - 113
Kvenvolden, K.A. - 67, 68
Lakin, H.W. - 25, 28
Landrum, R.A., Jr. - 209
Langenheim, R.H. - 281, 284
Lankford, F.L., Jr. - 234
Laporte, J.-L. - 39
Larson, R.R. - 188
Larson, R.S. - 227
Lee, B.D. - 188
LeMoal, R. - 260
Leplat-Gryspeerdt, P.A. - 38
Lindblom, E.D. - 129
Lindsey, J.P. - 248
Link, A.J. - 157
Loofbourrow, R.J. - 253
Louage, F. - 255
Luehrmann, W.H. - 64, 68

Luke, R.R. - 57
Maddox, J., Jr. - 77
Madec, M. - 39
Madsen, A. - 87
Malarcher, F.L. - 240
Malarky, I.R. - 247, 248
Mangum, R.W. - 268
Manin, M. - 261
Marquis, F. - 39
Martin, J.E. - 226
Martinez, P., Jr. - 156
Marwil, S.J. - 277
Mason, C.M. - 56
Mayne, W.H. - 186, 234, 251
McAuliffe, C.D. - 48
McCollum, B. - 185, 186, 250
McDermott, E. - 34
McKay, A.S. - 171, 173
McLaughlin, G.H. - 99, 132
McNeel, W.O. - 234
Meinert, R.N. - 272
Merritt, J.W. - 156
Michon, D. - 258
Mifsud, J.F. - 211, 250
Miller, D.E. - 230
Miller, G.K. - 259
Milly, G.H. - 28, 69, 142
Milochik, S.W. - 85
Mintrop, L. - 185
Mitchell, D.K. - 198
Mollere, J.C. - 193, 194
Monaghan, R. - 142
Moore, O.T. - 284
Moore, P.J. - 33, 280
Moore, T.F. - 63, 88
Morey, R.M. - 135
Morgan, J.D. - 250
Morrow, W.C. - 189
Mott-Smith, L.M. - 203
Mounce, W.D. - 269
Muffly, G. - 96
Muir, T.G. - 227
Mulligan, R.P. - 150
Murdock, D.B. - 269
Murphy, J.A. - 157
Musgrave, A.W. - 254
Nakagawa, H.M. - 25, 28
Nash, A.J. - 55
Neeley, W.P. - 239
Neufeld, J. - 179
Neuman, C.H. - 165
Nicol, J. - 99
Norelius, R.G. - 167

Inventor Index

Oberdorfer, P.E., Jr. - 78
O'Brien, J.P. - 189
Orton, T.W. - 226
Osoba, J.S. - 278
Ovchinnikova, T.M. - 15
Overton, H.L. - 54
Paap, H.J. - 180
Paitson, J.L. - 202, 205
Panteleimonov, V.M. - 15
Parker, M.L. - 205
Parrack, A.L. - 260
Pauletich, J. - 195
Peraldi, R. - 253
Peterson, C.M. - 53
Phelan, S.R. - 124
Piety, R.G. - 120
Pirson, J.E. - 113
Pirson, S.J. - 113, 179
Plum, W.B. - 217
Pogorski, L.A. - 36
Pollock, H.C. - 178
Pontecorvo, B. - 166
Prech, V. - 13
Prescott, B.O. - 66
Prescott, H.R. - 243
Preston, J.N. - 103
Puranen, M. - 129
Quay, R.G. - 240
Ray, C.H. - 240
Raymond, R.L. - 81
Reed, D.H. - 194, 195, 201
Reid, W.P. - 24
Reynolds, R.C., Jr. - 54
Rhodes, J.R. - 19
Riggs, E.D. - 246, 252
Rittenhouse, G. - 66
Robinson, W.A. - 99, 132
Rochon, R.W. - 270, 271
Roe, G.D. - 64
Roetter, M.F. - 99
Rogers, J.R. - 258
Ronka, V. - 129, 130
Rosenfeld, W.D. - 77
Ruddock, K.A. - 97, 131
Ruehle, W.H. - 261
Rugen, D.F. - 78
Ryss, J.S. - 15
Sack, H.S. - 142
Salvatori, H. - 245
Sarrafian, G.P. - 246, 253
Saussier, D. - 260
Savit, C.H. - 202
Sayous, L. - 201

Scherbatskoy, S.A. - 153, 156, 172
Schlumberger, C. - 111, 116, 117
Schlumberger, M. - 112
Schmidt, G.W. - 63
Schmidt, K.H. - 277
Schmuck, R.H. - 230
Schneider, W.A. - 240, 254
Schorno, K.S. - 37
Schroeter, J.D. - 284
Schultz, W.E. - 172
Schuster, N.A. - 167
Schwartz, R.J. - 180
Searcy, F.L. - 243
Seevers, D.O. - 281
Seigel, H.O. - 125
Senftle, F.E. - 156, 165
Sengbush, R.L. - 248
Serson, P.H. - 97
Sewell, B.W. - 266
Shapiro, S. - 99
Sieberg, R.D. - 19
Silverman, D. - 202, 209, 212, 246, 248, 252, 255
Siska, L. - 6
Skelton, J.D. - 246
Skrabis, A. - 6
Slack, H.A. - 43, 97
Slater, B.R. - 209
Slaven, T.L. - 193
Slobod, R.L. - 63, 271
Smith, H.D., Jr. - 172
Smith, M.E. - 37
Smith, M.P. - 180
Staron, P. - 258
Stas, B. - 6
Stauber, S. - 103
Steele, D. - 191
Stegmaier, W. - 13
Stone, J.E. - 211
Stratford, W.M. - 142
Straus, A.J.D. - 255
Summers, G.C. - 116
Sundberg, K. - 128, 129
Supernaw, I.R. - 157
Swift, G. - 142, 165
Taner, M.T. - 251
Taylor, M.C. - 19
Teichmann, C.F. - 142
ten Brink, K.C. - 179
Terry, M.C. - 268
Thai, N. - 226
Thayer, J.M. - 165, 176
Thomas, G.S. - 120

Thompson, R.R. - 35, 46, 55
Threadgold, P. - 120
Tittle, C.W. - 172
Tittman, J. - 171
Treitel, S. - 248
Turner, J.B. - 57, 284
Unterberger, R.R. - 137, 253
Updegraff, D.M. - 75
Vacquier, V.V. - 96
Vagner, J. - 13
Varian, R.H. - 97
Vogel, C.B. - 256
Von Thuna, P.C. - 101
Voronin, D.V. - 15
Wack, B.F. - 156
Walling, D. - 245
Walters, A. - 66
Ward, F.N. - 26
Ward, W.J., III - 147
Warren, R.K. - 139
Waters, K.H. - 249
Watson, R.J. - 248
Weaver, R.R. - 209
Weber, R.M. - 211

Weisz, P.B. - 41
Westervelt, P.J. - 227
Westkaemper, J.C. - 19
Widmyer, R.H. - 179
Wiggins, P.F. - 165
Wilson, O.L. - 12
Wimberley, J.W. - 30
Wise, D.H. - 160
Wise, H.L. - 35, 57, 66, 284
Worthington, A.E. - 280
Wülk, B. - 7
Wyckoff, R.D. - 96
Yarbrough, H.F. - 63
Yost, W.J. - 135
Youmans, A.H. - 165, 167, 171
Young, R.C. - 19
Youngman, C.A. - 68, 274
Yungul, S.H. - 98, 113
Zemanek, J., Jr. - 241
Zilinskas, G. - 226
Zimmerman, J.R. - 34
Zonge, K.L. - 127
Zurflueh, E.G. - 98

U.S. PATENT NUMBER INDEX

1,163,469 - 111	2,390,433 - 155	2,777,799 - 74
1,599,538 - 185	2,406,870 - 96	2,781,453 - 142
1,672,495 - 185	2,473,469 - 248	2,816,009 - 272
1,675,121 - 186	2,508,772 - 166	2,854,396 - 272
1,678,489 - 129	2,515,500 - 165	2,856,536 - 179
1,719,786 - 116, 117	2,518,513 - 96	2,861,921 - 75
1,724,495 - 186	2,555,209 - 96	2,862,106 - 172
1,724,720 - 186	2,561,490 - 97	2,867,728 - 178
1,748,659 - 129	2,562,961 - 142	2,871,105 - 33
1,820,953 - 129	2,582,314 - 134	2,875,135 - 77
1,913,293 - 111	2,586,667 - 113	2,880,142 - 77
2,003,780 - 243	2,611,004 - 125	2,883,856 - 274
2,008,698 - 243	2,644,130 - 116	2,884,534 - 176
2,034,447 - 112	2,657,380 - 135	2,918,579 - 63
2,046,843 - 243	2,677,801 - 115	2,921,003 - 77
2,087,120 - 245	2,685,038 - 179	2,923,151 - 277
2,094,691 - 150	2,688,124 - 250	2,929,984 - 129
2,148,422 - 246	2,700,734 - 179	2,931,974 - 99
2,192,404 - 124	2,732,906 - 186	2,933,609 - 167
2,197,453 - 156	2,733,353 - 179	2,933,923 - 85
2,219,273 - 153	2,740,291 - 266	2,934,652 - 162
2,220,070 - 134	2,740,292 - 266	2,938,117 - 277
2,243,729 - 246	2,740,695 - 269	2,947,870 - 156
2,275,748 - 176	2,742,575 - 39	2,948,811 - 174
2,284,990 - 112	2,745,282 - 270	2,951,940 - 278
2,305,384 - 33	2,747,401 - 266	2,957,083 - 156
2,314,597 - 124	2,749,220 - 271	2,959,240 - 230
2,320,643 - 179	2,749,748 - 271	2,968,022 - 253
2,342,626 - 117	2,755,388 - 41	2,972,733 - 253
2,347,794 - 120	2,761,977 - 171	2,983,586 - 278
2,349,712 - 162	2,767,320 - 62	2,991,364 - 179
2,352,993 - 179	2,772,951 - 268	3,015,060 - 132
2,365,763 - 160	2,773,991 - 41	3,016,961 - 173

3,022,140 - 54	3,307,912 - 46	3,685,345 - 35
3,028,313 - 78	3,329,580 - 83	3,690,164 - 255
3,028,542 - 268	3,341,706 - 142	3,691,378 - 171
3,033,287 - 42	3,343,917 - 54	3,691,517 - 246
3,033,654 - 43	3,345,137 - 48	3,697,938 - 251
3,033,761 - 81	3,379,884 - 179	3,697,939 - 254
3,042,857 - 130	3,386,286 - 284	3,700,407 - 35
3,043,908 - 87	3,389,257 - 178	3,711,765 - 54
3,047,836 - 246	3,395,987 - 28	3,714,811 - 57
3,048,235 - 188	3,397,040 - 25	3,719,453 - 55
3,049,409 - 279	3,422,674 - 284	3,722,271 - 58
3,050,148 - 188	3,428,431 - 56	3,729,704 - 247
3,050,449 - 280	3,428,806 - 156	3,730,683 - 28
3,065,149 - 81	3,432,807 - 248	3,734,489 - 69
3,068,401 - 120	3,434,800 - 26	3,745,445 - 125
3,070,745 - 97	3,446,597 - 34	3,747,055 - 254
3,075,607 - 246	3,452,210 - 172	3,758,846 - 15
3,096,254 - 83	3,455,144 - 63	3,759,617 - 70
3,100,258 - 179	3,457,044 - 63	3,768,302 - 16
3,105,934 - 135	3,461,291 - 176	3,774,146 - 250
3,108,220 - 131	3,462,761 - 284	3,780.301 - 172
3,112,443 - 134	3,463,922 - 156	3,801,281 - 55
3,118,299 - 280	3,480,396 - 68	3,806,795 - 135
3,118,738 - 281	3,490,032 - 98	3,817,328 - 165
3,120,428 - 34	3,495,438 - 268	3,820,390 - 124
3,134,957 - 253	3,496,350 - 52	3,823,319 - 171
3,143,648 - 68	3,508,877 - 53	3,825,751 - 153
3,149,068 - 45	3,514,693 - 115	3,829,768 - 99
3,153,147 - 88	3,524,195 - 252	3,833,087 - 194
3,174,910 - 83	3,524,346 - 63	3,834,122 - 53
3,180,983 - 34	3,529,282 - 248	3,835,710 - 36
3,182,743 - 250	3,539,299 - 35	3,846,631 - 156
3,184,598 - 172	3,546,454 - 167	3,847,549 - 37
3,188,558 - 113	3,550,073 - 248	3,852,659 - 132
3,188,559 - 113	3,555,409 - 134	3,858,167 - 6
3,200,251 - 167	3,561,546 - 64	3,858,171 - 195
3,210,652 - 125	3,562,504 - 253	3,862,576 - 36
3,240,068 - 281	3,571,591 - 64	3,865,467 - 101
3,241,100 - 253	3,584,292 - 97	3,866,111 - 139
3,254,959 - 45	3,596,089 - 54	3,866,174 - 251
3,259,878 - 250	3,597,727 - 247	3,866,709 - 211
3,262,050 - 120	3,599,175 - 250	3,868,222 - 23
3,263,161 - 97	3,609,363 - 142	3,872,293 - 253
3,273,967 - 12	3,609,366 - 180	3,882,446 - 240
3,278,746 - 89	3,620,591 - 252	3,884,324 - 211
3,281,333 - 77	3,622,968 - 246	3,885,160 - 178
3,285,698 - 56	3,629,800 - 254	3,888,122 - 101
3,286,163 - 137	3,638,020 - 165	3,890,501 - 180
3,287,088 - 281	3,645,131 - 284	3,893,539 - 203
3,294,972 - 165	3,652,980 - 249	3,896,413 - 218
3,300,641 - 45	3,653,837 - 23	3,896,898 - 203
3,302,706 - 46	3,680,040 - 248	3,899,768 - 240
3,305,317 - 67	3,681,028 - 56	3,908,789 - 192

U.S. Patent Number Index

3,909,776 - 209	3,979,300 - 180	4,071,755 - 157
3,912,041 - 193	3,979,715 - 209	4,081,675 - 264
3,912,042 - 195	3,979,724 - 255	4,093,420 - 85
3,913,060 - 227	3,981,379 - 201	4,100,481 - 93
3,916,370 - 239	3,982,606 - 255	4,100,991 - 201
3,919,547 - 156	3,983,957 - 209	4,102,428 - 189
3,919,684 - 195	3,984,805 - 209	4,102,429 - 201
3,921,126 - 249	3,986,163 - 237	4,103,158 - 166
3,924,260 - 253	3,987,677 - 58	4,106,908 - 38
3,926,054 - 102	3,990,034 - 226	4,114,086 - 127
3,931,609 - 239	3,991,625 - 103	4,114,722 - 211
3,938,072 - 217	3,991,850 - 255	4,114,723 - 205
3,939,941 - 191	3,992,693 - 226	4,116,300 - 211
3,942,003 - 13	3,993,973 - 226	4,121,464 - 109
3,942,606 - 215	3,997,022 - 194	4,122,431 - 253
3,943,436 - 113	4,004,267 - 251	4,125,823 - 230
3,944,019 - 195	4,005,289 - 92	4,126,203 - 230
3,946,831 - 227	4,005,290 - 168	4,132,943 - 86
3,948,177 - 194	4,006,794 - 202	4,132,974 - 194
3,949,352 - 256	4,006,795 - 220	4,134,097 - 233
3,949,831 - 227	4,007,803 - 201	4,134,098 - 261
3,950,695 - 132	4,008,784 - 202	4,134,099 - 234
3,952,281 - 260	4,010,413 - 135	4,136,754 - 261
3,952,833 - 194	4,011,539 - 258	4,137,751 - 19
3,953,171 - 37	4,011,540 - 258	4,143,552 - 9
3,953,827 - 260	4,011,924 - 215	4,143,736 - 211
3,957,438 - 24	4,016,951 - 202	4,143,737 - 212
3,957,439 - 66	4,016,952 - 201	4,144,520 - 234
3,961,187 - 90	4,017,731 - 143	4,146,870 - 261
3,961,306 - 238	4,020,447 - 258	4,146,871 - 261
3,961,307 - 7	4,020,919 - 212	4,146,872 - 254
3,961,683 - 255	4,022,064 - 103	4,147,228 - 220
3,962,674 - 255	4,023,413 - 103	4,149,805 - 10
3,965,412 - 98	4,026,382 - 198	4,152,504 - 234
3,967,190 - 127	4,026,383 - 207	4,153,415 - 39
3,967,235 - 241	4,051,372 - 37	4,156,138 - 148
3,968,855 - 193	4,056,969 - 19	4,157,659 - 269
3,970,428 - 72	4,059,760 - 157	4,159,463 - 212
3,971,319 - 188	4,064,436 - 147	4,159,464 - 235
3,974,476 - 215	4,065,972 - 38	Reissue 17,242 - 186
3,975,157 - 37	4,066,891 - 31	Reissue 24,226 - 172
3,976,161 - 201	4,066,892 - 2	Reissue 24,464 - 135
3,978,446 - 259	4,067,693 - 30	Reissue 24,797 - 172
3,979,140 - 202	4,068,160 - 104	

NOTICE

Nothing contained in this Review shall be construed to constitute a permission or recommendation to practice any invention covered by any patent without a license from the patent owners. Further, neither the author nor the publisher assumes any liability with respect to the use of, or for damages resulting from the use of, any information, apparatus, method or process described in this Review.

OFFSHORE OIL TECHNOLOGY

Recent Developments 1979

by M. William Ranney

Energy Technology Review No. 38
Ocean Technology Review No. 8

From the time the first submersible drilling structure was used in the Gulf of Mexico in the late 1940s, technology for offshore drilling developed rapidly to service the needs originated with this new era in oil and gas exploitation.

Today engineers are continuing to design new apparatus and systems to improve the world's supply of these resources. This text keeps the reader abreast of these latest developments.

The first chapter reviews the activity in offshore drilling to date, discussing the mobile and fixed structures and equipment employed. Subsequent chapters detail approximately 170 recent processes, many of which are even now being implemented throughout the industry. These processes include the key areas of platform erection, drilling operations, wellhead construction, and storage facilities. Over 80 figures and illustrations accompany the process descriptions.

The partial, condensed table of contents presented below with chapter headings and some subtitles demonstrates the comprehensive range of this text. The numbers in parentheses signify the total number of processes covered in each chapter.

1. **OFFSHORE DRILLING OVERVIEW**
   Background—History and Geology
   Survey Vessels
   Exploratory Drilling Rigs
   Development and Production Systems
   Transshipment and Storage
   Onshore Support Needs

2. **PLATFORM CONSTRUCTION (56)**
   Telescopically Mated Stanchions
   Hollow-Based Five-Column Tower
   Vertically Movable Legs
   Cellular Foam for Jacket Legs
   Structure for Earthquake-Prone Regions
   Anchoring System for Soft Substrates
   Self-Elevating Platforms
   Jacking Units for Template Legs
   Platform Self-Mobile on Land
   Vertically Moored Platform
   Grouting Systems
   Reinforced Elastomeric Seal
   Sealing Concentric Tubular Inserts
   Hollow Underwater Anchoring Apparatus
   Multiple Wells Through Single Caisson
   Parallelepipedic Box Deck Structure
   Multiple Platform Crane
   Interconnected Prefabricated Units

3. **MARINE RISERS AND WELL CONDUCTORS (24)**
   Hydraulic Heave Compensator
   Pressure-Compensated Dual Riser
   Risers for Platform Structures
   Well Conductors
   Dog-Leg Segment for Directional Drilling
   Well Conductor Pipe in Soft Floor
   Cutting and Recovery of Surface Casing

4. **DRILLING PROCESSES (16)**
   Fluid-Actuated Counterbalancing Means
   Drilling Fluid Diverter System
   Drilling Tool to Curb Debris Buildup
   Running and Retrieving Tools

5. **WELLHEAD APPARATUS (34)**
   Flow Control System
   Atmospheric Pressure Operation
   Reentry Guidance Apparatus
   Portable Atmospheric Cellar
   Christmas Tree Junction Housing
   Replacing Broken Guidelines
   Flow Lines and Pipe Connectors
   Using Two Flow Line Mating Vessels
   Tension Control in Flow Line Connection
   Dry Isolated Chambers in Base
   Guiding Technique for Diving Bell
   Self-Propelled Trenching Apparatus
   Multipressure, Single-Line Supply Means

6. **ARCTIC REGION OPERATIONS (18)**
   Movably Mounted Ice Cutter
   Fill Material on Natural Ice
   Ice Platform
   Ice Anchor
   Bottom-Supported Vessel
   Monopod Platform on Arctic Floor
   Casting Concrete Tanks Under Ice

7. **STORAGE AND PROCESSING (20)**
   Marine Pier with Submerged Storage
   Storage Tank Submarine
   Caisson-Defined Storage Area
   Mobile Processing Plant
   Underwater Sphere for Processing
   Oil Recovery from Underwater Fissures
   Automated Subsea Control System
   Work Arm System in Submerged Chamber
   Platform to Lower Blow-Out Preventer

ISBN 0-8155-0741-0

399 pages

CRUDE OIL DRILLING FLUIDS 1979

by Maurice William Ranney

Chemical Technology Review No. 121
Energy Technology Review No. 35

The drilling of oil wells into subterranean reservoirs presents many problems concerning the choice of a drilling fluid. Rotary drilling rigs are used nowadays almost universally throughout the world's oil regions. Perhaps the greatest advantages of rotary drilling over other methods is that the well bore is kept full of liquid during drilling. A weighted fluid (also called drilling mud) in the bore hole serves two important purposes: by its hydrostatic pressure it blocks the entry of formation fluids into the well, thus preventing blowouts and gushers. In addition, the drilling mud carries the crushed rock to the surface, so that the drilling is continuous until the bit wears out.

This apparent simplicity of the process is complicated by many factors, such as sealing the productive formation with a casing, which must have openings to allow the oil to enter the well under controlled conditions.

Many new, potentially oil-producing areas are being explored under oceans, and in arctic and desert regions. Increasing emphasis is also placed on secondary recovery techniques. Research and development activity therefore centers on improvements in drilling fluids, stimulation methods and secondary recovery processes involving water and polymer flooding techniques.

This book, focusing on the U.S. patent literature since 1974, encompasses over 250 processes. A partial and condensed table of contents follows here. Chapter headings and important subtitles are given. Numbers in parentheses indicate the number of processes per topic.

1. **DRILLING FLUIDS PART I (72)**
   Fluid Loss Control
   Lost Circulation
   Viscosity Control
   Polysaccharides
   Surfactants
   Salt Tolerance
   Foam Fluids
   Permafrost Formulations
   Clay Treatment

2. **DRILLING FLUIDS PART II (26)**
   Lubricity
   Corrosion & Scale Inhibitors
   Neutralization
   Gas Scavenging
   Weighting Agents

3. **CONSOLIDATION & PLUGGING (42)**
   Furfuryl Alcohol Composition
   Sealing Techniques
   Polyvinylpyrrolidone
     for Thickening Preflush
   Plugging Compositions
   Cationic Wax Emulsions
   Acrylic-Epoxy Emulsions
   Polyethylene & Polyamides
   Mg-Abietate + Lignosulfonates

4. **FRACTURING & STIMULATION (21)**
   Acidizing Compositions
   Fracture Fluids
   Thickened Alcohols
   Self-Braking Viscous Solutions
   Polymer Dispersions
   Friction-Reducing Agents
   Crosslinked Gel to Carry
     Propping Agent
   Detonation Processes

5. **SECONDARY OIL RECOVERY & FLOODING PROCESSES (46)**
   Surfactant Systems
   Cationic Compositions
   Polymeric Compositions
   Use of Biopolymers
   Cellulose Ethers + Polyvalent Metals
   Chelating Agents
   Sodium Silicate Solutions

6. **TAR & HEAVY CRUDES (12)**
   Thermal Recoveries
   Superheated Water
   Steam Soak Process
   Combustion & Retorting

7. **WELL CEMENTING (35)**
   Fly Ash + Lime
   Magnesium Oxide + Slag
   Cement Mixture + NH_4Cl
   Encapsulated Accelerator
   Resulfonated Lignosulfonates
   Drilling Fluid plus
     Cement Compositions
   Aqueous Sodium Silicate
     for Squeeze Cementing
   Cementing in
     Permafrost Formations

Note: These recent processes provide hundreds of formulations, techniques and new developments which will be used to efficiently recover crude oil from subterranean oil reservoirs in the coming years.

ISBN 0-8155-0732-1

348 pages

REPROCESSING AND DISPOSAL OF WASTE PETROLEUM OILS 1979

by L.Y. Hess

Pollution Technology Review No. 64
Chemical Technology Review No. 140

Recovery and judicious disposal of waste oil have become financially rewarding practices by reason of resource conservation and environmental protection.

Dirty and contaminated lubricating oil still has high energy values. It can be re-refined or used as a feedstock for making other petroleum products. Industrially, it can be reclaimed to nearly original quality by simple equipment.

Sufficiently worthwhile disposal practices include road oiling, combining with fuel oil, use as auxiliary fuel in municipal incinerators, and landspreading with decomposition by suitable microorganisms.

This book is based on the latest government research reports and U.S. patents which contain practical technological process information. A condensed table of contents follows here.

1. **SOURCES AND CHARACTERISTICS**
   Generation and Disposal Figures
   Automotive Service Centers
   Commercial Truck Fleets
   Railroad Service Centers
   Aviation Service Centers

2. **RECOVERY PROCESSES**
   Acid/Clay Process
   Extraction Processes
   Distillation

3. **PROPRIETARY PROCESSES (118)**
   Treatment with Chemicals
   Solvent Extraction
   Adsorption
   Distillation & Hydrogenation
   Filtration
   Other Separation Techniques
   Application of Electric Fields

4. **RESEARCH PROJECTS**
   Innovative Techniques
   Ion Exchange Percolation
   Chelation
   Comprehensive Waste Oil Facility

5. **DISPOSAL PRACTICES**
   Incineration and Use as Fuel
   Environmental Aspects

6. **DISPOSAL OF RECOVERY RESIDUES**
   Waste Products Generated
   Acid Sludge
   Caustic Sludge
   Spent Clay
   Distillation Bottoms
   Characterization of Wastewaters
   Marine Waste Oil Processing Facility
   The Re-Refining Facility
   Environmental Assessment

7. **DISPOSAL AND RECYCLING OF AIR FORCE WASTE OILS**
   Generation of Waste POLs
   Kelly AFB Survey
   Andrews AFB Survey
   Available Disposal/Recycle Techniques

8. **COMBUSTION STUDIES OF WASTE POLs**
   Air Force Experimental Studies
   Combustion Test Equipment
   Energy Recovery
   Waste Oil Burn-Off in Coast Guard Power Plants
   Waste Lube Oil Added to Diesel Fuel

9. **MUNICIPAL INCINERATOR FUEL**
   Physical and Combustion Properties of Waste Oil
   Heating Requirements
   Phenomena Occurring in a Refuse Bed
   Transferring Heat Flux
   Mixing Waste Oil Directly into Refuse
   Utilizing Auxiliary Burners
   Impact on Air Quality

10. **DISPOSAL BY LANDSPREADING**
    Land Disposal Practices
    Microbiological Studies
    Shell Project Description and Objectives
    Oil Decomposition Rate and Effect on Fertilizer
    Metals Contents of Soils
    Microbial Action
    Rainfall Runoff
    Union Carbide Study
    Combined Oil/Nitrate Application
    Summary

ISBN 0-8155-0775-5

322 pages

OIL SHALE AND TAR SANDS TECHNOLOGY 1979

RECENT DEVELOPMENTS

by M.W. Ranney

Energy Technology Review No. 49
Chemical Technology Review No. 137

Oil Shale and Tar Sands are a potential source for huge supplies of crude petroleum that for many years has attracted government agencies and oil companies. The current energy crisis is providing added incentive for the development of an industry that can release this oil to produce significant quantities of hydrocarbons.

One indication of the new phase, perhaps, is that the Government of Alberta has decreed a name change for the tar sands. It has decided that all bituminous deposits are *Oil Sands*.

Recovery and conversion processes are relatively well developed. Both above ground processing and in situ processes are important, depending on the location of the sands. Many of these processes, as described in this book, have reached or passed the pilot plant stage.

The solutions to the technical difficulties associated with converting the crude material to a finished product is covered in this book. Also, environmentally there is a major concern with the impact of large-scale oil shale processing. But the technology is there and is getting better.

The table of contents below gives chapter headings and **examples of some** subtitles. The number of processes per topic is in parentheses.

1. **OVERVIEW**
   Oil Shale Described—Location, Extraction
   Assessment of Available Oil
   Canadian Tar Sands Described—Deposits
   In situ Processing; The Future
   U.S. Tar Sands—Major Findings

2. **OIL SHALE RETORTING (42)**
   Gas Combustion
     Retorting-Gasification with Recycle Gas
     Monitoring Density and Gravity
   External Heat Pyrolysis
     Oil Adday and Rock Pressure Relation
     Misting of Reactor Recycle Gas
     Reducing Gas Stream Particulate Content
     Hydrostatic Sealing Technique
     Cyclone Reactor-Separator
     Quenching Reactor Effluent Streams
   Solid Heat Transfer Processes
     Segregating Heat Carrier Flows
   Miscellaneous Recoveries
     Microwave Energy
     Preretorting Arsenic Content Reduction
     Bacterial Treatment of Retort Water

3. **SHALE PROCESSING IN SITU (51)**
   Formation of Retort Cavities
     Columnar Voids
     Multiple Gallery-Type Retort Zones
     Successive Fragmentizing
   Retorting Processes
     Ignition Techniques
     Laser Retorting
     Emulsion Breaking Technique
   Treatment of Off-Gases from Retort
     Oxidation of Hydrogen Sulfide
   Miscellaneous in situ Recovery Processes
     Molecular Sulfur and Benzene Recovery

4. **SHALE OIL REFINING AND PURIFICATION (14)**
   Retorting & Hydrocracking of Spent Shale
   Fluidized Bed Hydroretort
   Using Sand in Oil Extraction & Cracking
   In situ Hydrogen Generation
   Sulfur Removal with Polysulfides

5. **IN SITU TAR SAND TREATMENT (39)**
   Hot Aqueous Fluid Drive
     Cone-Shaped Gravel Pack
     Vibrating Probe to Fluidize Sand
     Hydraulic Mining Using Steam Jets
   Solvent Leaching and Fluid Flow
     Aromatic Solvent, CO_2 plus Solubilizer
     Solvent-Saturated Gas & Steam Injection
     Chlorofluorocarbon Solvents
   Oxidation and Combustion
     Superheated Steam and Combustion Drive
     N-Methyl-2-Pyrrolidone/Furfural Use
   Electrode Wells

6. **TAR SANDS SEPARATION (49)**
   Hot Water Processes
     Extraction Cell with Deflection Baffles
     Ketones to Demulsify
   Cold Water Processes
   Solvent Extraction
     Countercurrent Flow Using Hydrocarbon
     Ultrasonic Energy and Stirring
     Recovery of Zircon and Rutile
   Coking Using Hot Solids Recycle
   Sludge Accumulation and Compaction
   Clarification of Clay-Containing Water

ISBN 0-8155-0769-0

430 pages

SOLAR CELLS FOR PHOTOVOLTAIC GENERATION OF ELECTRICITY 1979
MATERIALS, DEVICES AND APPLICATIONS
by Marshall Sittig

Energy Technology Review No.48

The direct conversion of sunlight into electricity by the use of semiconductor devices has an intriguing potential. Quiet and reliable, they can convert sunlight directly into electricity without the use of moving parts. Considerable emphasis in this book is on obtaining high current densities and the resultant economics.

Originally the high cost of producing photovoltaic cells did confine them to relatively exotic uses—powering space craft, for example. But silicon, the prime raw material, is cheap and abundant, and the technology for mass production, especially of the so-called ribbon cells, is available, as described in this book.

More research is needed to improve the conversion efficiency from sunlight to electrical current, but the way is also shown clearly in this book, e.g., by integrating finished solar cells into whole systems suitable for large scale applications.

Once the critical cost figures are reached, the market for solar cells should grow dramatically. Installation of whole banks of these cells on individual homes and office buildings, together with large arrays as regional electrical generating stations, should provide electricity at rates competitive with those now provided by utilities to many commercial and residential customers in the future.

The scope of this book is shown by the following table of contents, which presents chapter headings and subtitles. The numbers in parentheses indicate the number of processes per topic.

1. **INTRODUCTION**
   Fundamentals of Solar Cells
   Development Prospects

2. **APPLICATION AREAS (30)**
   Spacecraft
   Household Heating and Cooling
   Community Energy Systems
   Automobile Propulsion and Cooling
   Wrist Watches
   Beacons at Remote Locations
   Miscellaneous Applications

3. **MATERIALS FOR SOLAR CELLS (40)**
   Silicon
     Slicing or Sawing Crystals
     Ribbons
     Filament Form
     Tubular Form
     Microspheres
   Polycrystalline Silicon
   Silicon-Hydrogen "Alloys"
   Germanium
   Cu, Cd, Pb Sulfides and Tellurides
   Gallium, Indium Arsenides and
     Phosphides
   Mixed Semiconductor Compounds
   Oxide Ceramics
   Organic Media

4. **MATERIALS AVAILABILITY**

5. **TOXIC HAZARDS**

6. **SOLAR CELL FABRICATION (134)**
   From Silicon Wafers
   Thin Film Devices
   Vertical Multijunction Devices
   Horizontal Multijunction Cells
   Configurations to Improve Efficiency
   Attachment of Electrical Leads
   Antireflection Coatings
   Environmental Protection from Heat,
     Radiation, Pollutants and Corrosives
   Mass Production Techniques

7. **TESTING (2)**

8. **ENHANCEMENT (31)**
   Luminescent and Fluorescent Means
   Optical Concentrators
   Solar Tracking Devices

9. **MOUNTINGS (22)**
   Flat Plate Arrays (Panels)
   Shingle Arrangement
   Tubular Surface Mountings

10. **PHOTOELECTRIC COGENERATORS—HYBRID SYSTEMS (6)**

11. **PHOTOEMISSIVE CELL DEVICES (2)**

12. **SOLAR CELL ENERGY STORAGE (8)**

13. **PHOTOELECTROCHEMICAL ENERGY CONVERSION (11)**

14. **ECONOMICS OF SOLAR CELLS**
    Materials and Production Costs
    Market Forecasts

15. **BIBLIOGRAPHY**

ISBN 0-8155-0765-8   350 pages

METHANOL TECHNOLOGY & APPLICATION IN MOTOR FUELS 1978

Edited by J. K. Paul

Chemical Technology Review No. 114
Energy Technology Review No. 31

This book contains detailed descriptive information relating to methanol production technology from unusual sources, the utilization of methanol as an automotive fuel, and the conversion of methanol into gasoline. The first chapter is an overview and serves as an introduction to the subject.

The next three chapter are feasibility studies and discuss the production of methanol from coal, solid waste, and natural gas. This book does not discuss in any detail the production of methanol from liquid hydrocarbons, since one of the reasons for the future utilization of methanol in automobiles is to lessen the reliance upon petroleum hydrocarbons. Regarding natural gas, methanol can be produced from natural gas at a distant location, and shipped as a liquid at potentially less cost and less danger than the shipping of liquefied natural gas (LNG). Regarding coal, the cost of methanol appears to be comparable on an energy-equivalent basis to the production of synthetic gasoline and substitute natural gas from coal.

The next two chapters relate to the use of methanol/gasoline blends in automobiles and the use of 100% methanol as a vehicle fuel. Small amounts of methanol can be added to gasoline for use in current engines, however, the use of 100% methanol as a vehicle fuel does require engine changes, including modifications to the carburetor, and the necessity for a greater heat supply to the intake manifold.

The last two chapters relate to the production of gasoline from methanol, and these two chapters are based primarily on the efforts being conducted by the Mobil Corporation in this direction.

This book is based on federally funded studies, U.S. patents and other sources. It does convey the impression of considerable progress in methanol technology.

1. OVERVIEW
Methanol Production
Extraction
Fermentation
Syntheses
From Synthesis Gas
Oxidation of Hydrocarbons
By Irradiation of CO_x
Organic Feedstocks
Inorganic Origins
Raw Materials
  Nonrenewable Sources
  Renewable Sources
Methanol as a Fuel
Combustion Emissions
Suitable Engines
Electrochemical Oxidation of Methanol
Human Toxicity:
  Visual Impairment & Blindness from Methanol

2. METHANOL FROM COAL
Coal Gasification and Methanol Synthesis
DuPont Feasibility Study
Sasol Type Process Study
Badger Conceptual Design

3. METHANOL FROM SOLID WASTE
Conversion of Municipal Solid Waste
Pyrolysis Procedures
Suitability Specifications for Municipal Wastes
Conversion of Wood Wastes

4. ALASKAN METHANOL
Synthesis & Conversion Plants
Pipeline (Alyeska)
Pumping Power Requirements
Other Transportation
Other Transportation
Methanol Crude Separations at Los Angeles
Cost of Supply Gas
Slug Flow Interface
Overall Cost Estimates

5. METHANOL/GASOLINE BLENDS
5%, 10%, 15%, 20% Methanol
Missouri U. Study
Simulation Procedures
Engine Adjustments
Bartlesville Energy Research Center Study
Vehicle Optimizations Needed
Performance Mapping
U. of California Study
Engine Configurations and Operations
Optimum Amounts in Blends

6. 100% METHANOL MOTOR FUEL
Engine Modifications
Fuel-Air Injection Systems
Cold Start Difficulties
Thermochemical Engines

7. CHEMICAL CONVERSION: METHANOL TO GASOLINE
Mobil Oil Corp. Process
Fixed Bed Pilot Plant
Vehicle Studies with Synthetic Gasoline from CH_3OH

8. METHANOL TO GASOLINE
Proprietary Processes
Mobil Oil Corp. Patents
Ethyl Corp. Patents
Others

ISBN 0-8155-0719-4

470 pages

ETHYL ALCOHOL PRODUCTION AND USE AS A MOTOR FUEL 1979

Edited by J.K. Paul

Energy Technology Review No. 51
Chemical Technology Review No. 144

This book is a companion volume to our *Methanol Technology and Application in Motor Fuels*. Ethanol is available from so-called "renewable raw materials," i.e., agricultural crops such as corn, grain, potatoes, sugar beets and sugar cane, or other biomass and suitable garbage.

In this book an attempt has been made to present an economic assessment of possible modes of preparation of ethanol from various forms of biomass, natural resources and their waste materials or by-products. A chapter on current technology is also included. The present and potential availability of biomass from sugar crops, grains and grasses and silvicultural forms is considered.

Current crop production, proposed crops grown specifically for energy production and crop wastes and residues are discussed. Finally, to determine the actual practicality of fueling motor vehicles with ethanol, either 100% or in blends, several sets of engine test data are reviewed. The results seem favorable, 10-20% ethanol blends performing very similarly to straight gasoline with slight gains in octane rating and mileage.

1. **NEAR-TERM POTENTIALS**
   Alcohol Fuel in Automobiles
   Costs of Biomass-Based Fuels
   Resources Required
   Government Role

2. **ECONOMIC ASSESSMENT**
   Mitre Study
   Comparison of Process Economics
   Raphael Katzen Study
   California Energy Commission Study

3. **ETHANOL FROM MUNICIPAL WASTES**
   Enzymatic Hydrolysis of Cellulosic Wastes
   Plant Conversion Requirements

4. **AVAILABILITY OF SUGAR CROPS**
   Sugarcane
   Sweet Sorghum
   Sugar Beets
   General Route from Sugar Crop to Fuel

5. **AVAILABILITY OF GRAINS AND GRASSES**
   Criteria for Crop Selection
   Grain and Grass Crops

   Plant Species Relative to Production for Energy
   Grasses and Grains Presently Grown in the U.S.
   Documented Plant Candidates for Expanded Utilization

6. **AVAILABILITY OF WOOD BIOMASS**
   Silvicultural Energy Farms
   Utilization of Forest Residues
   Mill Residues

7. **ETHANOL PRODUCTION TECHNOLOGY**
   Fermentation
   Distillation
   Cellulose Conversion
   Dartmouth College and Bureau of Solid Waste Management
   U.S. Army—Natick Laboratories
   University of California at Berkeley (Wilke)
   University of Pennsylvania (Humphrey) and General Electric Company
   Purdue (Tsao)
   Gulf Oil Chemicals Company
   Various Methods

8. **NEBRASKA GASOHOL PROGRAM**
   Background
   Formation of Agricultural Products Industrial Utilization Committee
   Two Million Mile Road Test
   Consumer Acceptance
   Food and Fuel for the Future
   Feasibility Studies
   Formation of National Gasohol Commission

9. **ETHANOL USE IN BRAZIL**
   Alcohols as Fuels in Brazil
   Prospects of Alcohols as Fuels
   Otto Engines with Ethanol Blends
   Otto Engines with Straight Ethanol
   Energetics and Economics of Cassava and Sugarcane Fuel Ethanol

10. **MOTOR VEHICLE OPERATION WITH ETHANOL AND ETHANOL BLENDS**
    Changes in Fuel/Air Ratio
    Emission and Fuel Economy
    Performance Test Data
    Dual Fuel Diesel Application
    EPA Gasohol Test Program
    BERC Testing Program
    Brazilian Engine Calibration Test Program

ISBN 0-8155-0780-1

350 pages

COAL LIQUEFACTION PROCESSES
1979
by Perry Nowacki

Chemical Technology Review No. 131
Energy Technology Review No. 45

With the precarious position of petroleum supplies in the world market and the unfavorable balance of payments caused by overdependence on foreign sources, it may be expedient to exploit coal further as a source of liquid fuels. In the United States coal is the most abundant natural energy resource. Through its conversion to liquid fuels, it could supply a portion of the energy required for transportation, heating, etc.

Coal can furnish a complete range of liquid fuels, as well as chemical feedstocks. To achieve maximum yields, however, judicious adjustment of temperature, pressure, time and catalyst, as elaborated in this book, is essential. The text reviews the histories of important domestic and foreign coal liquefaction processes—pyrolysis, solvent extraction, and catalytic and indirect liquefaction. It details the chemistry of the processes and their inherent problems, also discussing economics, developmental status, and environmental impacts.

Chapter headings and **examples of some** important subtitles follow below:

1. **INTRODUCTION**
2. **COAL RESOURCES**
3. **COAL LIQUEFACTION CHEMISTRY**
4. **OVERVIEW OF PROCESSES**
5. **PYROLYTIC PROCESSES**
   Char Oil Energy Development (COED)
   Clean Coke
   TOSCOAL
   Occidental Flash Pyrolysis
   Coalcon Hydrocarbonization
   Oak Ridge Hydrocarbonization
   Flash Hydropyrolysis of Lignite
   Flash Hydrogenation
   Lurgi-Ruhrgas Flash Carbonization
6. **SOLVENT EXTRACTION**
   Solvent-Refined Coal (SRC) Process
     Pilot Plants—Ft. Lewis & Wilsonville
   Consol Synthetic Fuel Process
     Cresap Pilot Plant
   Exxon Donor Solvent Process
   CO-Steam Process
   Supercritical Gas Solvent Extraction
   UOP Solvent Extraction
7. **CATALYTIC LIQUEFACTION**
   H-Coal Process
     Trenton, N.J. Pilot Plant
   Synthoil Process
   Clean Fuel from Coal Process
   Zinc Halide Coal Liquefaction
   Dow Coal Liquefaction
8. **INDIRECT LIQUEFACTION**
   Fischer-Tropsch Process
     SASOL Commercial Plant, South Africa
   Mobil M-Gasoline Process
9. **COAL LIQUIDS FROM H-COAL**
   Analysis and Properties of Products
   Distillate Separation & Characterization
10. **SOLVENT-REFINED EXTRACT—COMMERCIAL, EXPANDED-BED HYDROPROCESS**
11. **ECONOMICS OF SRC PROCESS**
    Reclaiming, Preparing, Liquefying Coal
    Hydrogenation of Fuel Oil and Naphtha
    Stack Gas Desulfurization
    Hydrogen Manufacture
    Capital Investment and Costs
12. **ENGINEERING ASSESSMENT OF SYNTHOIL PROCESS**
    Feed Slurry Pumping and Heating
    Reactor Design
    Removal of Spent Catalyst
    Low-Temperature Carbonization
13. **FISCHER-TROPSCH DESIGN & COST**
    Recovery of Heat, Liquids, Sulfur
    Gasification and Methanation
    Water Systems
14. **EQUIPMENT AND MATERIALS**
    Pumps, Heat Exchangers, Fired Heaters
    Vessels and Pipe Systems
    Instrumentation and Metallurgy
15. **ENVIRONMENTAL, HEALTH AND SAFETY ISSUES**
    Solid Waste Management
    Hazards in Product Transfer and Storage
    Federal Legislation
16. **COMPARISON OF PROCESSES**
    Technical and Economic Aspects
    Probability of Commercial Success

ISBN 0-8155-0756-9

339 pages

THE PETROLEUM REFINING INDUSTRY
ENERGY SAVING AND ENVIRONMENTAL CONTROL
1978

by Marshall Sittig

Energy Technology Review No. 24
Pollution Technology Review No. 39

With hardly any new oil refining facilities being built in the United States, this book is especially timely, because existing refineries must necessarily undertake much restructuring and retrofitting to conform to changing technology and product demand.

Some process changes, particularly upgrading of gasoline fractions, due to the elimination of tetraethyl lead, demand more energy; yet savings are possible by switching large process heaters and boilers from gas to liquid fuels. Raising the octane numbers by stricter catalytic reforming produces less gasoline and more sulfur-containing waste oil.

New techniques call for hydrosulfurization units using hydrogen. Hydrogen is dissolved in the liquid product and mixed with purge gas. This is applicable to crudes and wastes. Up to 85% sulfur in the catalytic cracker feed can be removed, and 90% removal of the sulfur in the residuum is realized in an environmentally acceptable manner.

This book demonstrates several processing strategies for wastewater and sulfur emissions. The high amounts of H_2S and NH_3 present in wastewater from Mid-East crudes often makes recovery of these chemicals economically attractive and avoids the effluence of pollutants.

Many other approaches to saving energy and avoiding pollution at profit level in existing and projected refining units are presented in this volume, which is based on various technological studies and U.S. patents. A partial, condensed table of contents follows here.

1. BASIC REFINERY MODULES

2. AIR POLLUTION CONTROL
Regulatory Constraints
Handling of Crude
Thermal Cracking
Coking
Deasphalting
Catalytic Cracking
Reforming
Hydrogen Production
Sulfur Recovery
Sludge Incineration
Flares & Blowdowns

3. WATER POLLUTION CONTROL
Desalting of Crudes
Fractionation of Crudes
Cracking Operations
Hydrotreating
Lubricating Fractions
Hydrogen Manufacture
Regulatory Constraints
In-Plant Control
Pretreatments
End-of-Pipe Control

4. SOLID WASTE DISPOSAL
Wastewater Residuals
Crude Tank Bottoms
Separator Sludges
Filter Cakes
Coke Fines
Factors Affecting
   Solid Waste Generation

5. REFINERY NOISE CONTROL

6. ENERGY CONSERVATION
Fired Heater Improvements
Improving Boiler Efficiency
Heat Exchanger Improvements
Air Cooler Improvements
Piping & Pump Improvements
Fractionating Tower Improvements
Tank Mixing
Insulation
Instrumentation
Vacuum Systems
Steam Systems
Cracking Unit Improvements
Electric Load Leveling
Power Factor Improvement
Energy Recovery
   from Process Streams
Loss Control
Catalyst Developments
Better Processing Techniques
1980 Energy Use Target
Some Projections
   Beyond 1980 to 1990

7. REFINERY PROCESS OPTIONS
ENERGY vs. ENVIRONMENTAL
Systems Analysis: Procedures
Systems Analysis: Interpretations
Direct Combustion of Asphalt
Hydrocracking of
   Heavy Bottoms
Flexicoking
Electric Power from
   Asphalt Combustion
Hydrocarbon Production by
   Partial Oxidation of Asphalt

ISBN 0-8155-0694-5

374 pages